STEM Education - Recent Developments and Emerging Trends

Edited by Muhammad Azeem Ashraf and Samson Maekele Tsegay

Published in London, United Kingdom

STEM Education - Recent Developments and Emerging Trends
http://dx.doi.org/10.5772/intechopen.107815
Edited by Muhammad Azeem Ashraf and Samson Maekele Tsegay

Contributors
Agnes A. Ewuru, Agnes L. Okute, Alba Guzmán-Duque, Ann Dijeh, Calvin Okello, Chairil Faif Pasani, Charles Tushabomwe-Kazooba, Christine Challen, Danyal Farsani, Edet E. Okon, Emilia A. Belciak, Emilia M. Iacobelli, Francesca Nneoyi-Egbe, Georgina Mokgaetji Mokganya, Geri Salinitri, Imelda Kemeza, Innocent Zitha, Irfan Ahmed Khan, Ita E. Okon, Jean-Luc Dorier, Jennifer Miller, Jr, Kay Sidebottom, Kinga Kovácsné Pusztai, Leander Kwabena Brown, Luisa Chalarca-Guzmán, Majid Nawaz, Maria B. Onabe, Marie Bassford, Mary A. Iyam, Maud Chanudet, Mfon E. Usip, Michael R. L. Odell, Muhammad Azeem Ashraf, Muhammad Imran Ahmad, Nada Abdelhai, Orifha Sinthumule, Rizky Amelia, Rubie Ibuola, Sabine Apfler, Samson Maekele Tsegay, Solomon A. Tesfamicael, Stephen Bepeh Undie, Stéphane Favier, Teresa J. J. Kennedy, Umar Daraz, William C. Kyle, Younas Khan

First published in London, United Kingdom, 2024 by IntechOpen
IntechOpen is the global imprint of INTECHOPEN LIMITED, registered in England and Wales, registration number: 11086078, 167-169 Great Portland Street, London, W1W 5PF, United Kingdom

British Library Cataloguing-in-Publication Data
A catalogue record for this book is available from the British Library

Additional hard and PDF copies can be obtained from orders@intechopen.com

STEM Education - Recent Developments and Emerging Trends
Edited by Muhammad Azeem Ashraf and Samson Maekele Tsegay
p. cm.

This title is part of the Education and Human Development Book Series, Volume 14
Topic: Education
Series Editor: Katherine K.M. Stavropoulos
Topic Editor: Delfín Ortega-Sánchez

Print ISBN 978-1-83769-255-2
Online ISBN 978-1-83769-256-9
eBook (PDF) ISBN 978-1-83769-257-6
ISSN 2755-9513

For EU product safety concerns:
IN TECH d.o.o., Prolaz Marije Krucifikse Kozulić 3, 51000 Rijeka, Croatia,
info@intechopen.com or visit our website at intechopen.com.

IntechOpen Book Series

Education and Human Development

Volume 14

Aims and Scope of the Series

Education and Human Development is an interdisciplinary research area that aims to shed light on topics related to both learning and development. This Series is intended for researchers, practitioners, and students who are interested in understanding more about these fields and their applications.

Meet the Series Editor

Katherine Stavropoulos received her BA in Psychology from Trinity College, in Connecticut, USA and her Ph.D. in Experimental Psychology from the University of California, San Diego. She completed her postdoctoral work at the Yale Child Study Center with Dr. James McPartland. Dr. Stavropoulos' doctoral dissertation explored neural correlates of reward anticipation to social versus nonsocial stimuli in children with and without autism spectrum disorders (ASD). She has been a faculty member at the University of California, Riverside in the School of Education since 2016. Her research focuses on translational studies to explore the reward system in ASD, as well as how anxiety contributes to social challenges in ASD. She also investigates how behavioral interventions affect neural activity, behavior, and school performance in children with ASD. She is also involved in the diagnosis of children with ASD and is a licensed clinical psychologist in California. She is the Assistant Director of the SEARCH Center at UCR and is a faculty member in the Graduate Program in Neuroscience.

Meet the Volume Editors

Dr. Muhammad Azeem Ashraf is an associate professor at the Institute of Education, Hunan University, China. He obtained a bachelor's degree in clinical medicine from Shihezi University, China, in 2008 and a doctoral degree in educational leadership and policy (comparative education) from Beijing Normal University, China, in 2018. He has served as visiting faculty at Henan Institute of Science and Technology, China, from 2016 to 2019 and a lecturer at Inner Mongolia University for Nationalities, China, from 2008 to 2011. He is the English editor of *University Education Science*. Dr. Ashraf has been a principal investigator of several projects funded by the National Natural Science Foundation of China and the Ministry of Science and Technology, China. He has published more than forty papers in international peer-reviewed journals and books.

Dr. Samson Maekele Tsegay is a research fellow at the School of Education, Anglia Ruskin University, UK. He obtained an MA and Ph.D. in Educational Leadership and Policy (Comparative Education) on a Chinese Government Scholarship at Beijing Normal University, China. In 2020, he also completed another Ph.D. in Education and Social Justice at the University of Roehampton, UK. He was a recipient of the distinguished Sacred Heart Scholarship from Roehampton University, UK. Before joining Beijing Normal University, Dr. Tsegay worked for about 10 years at the University of Asmara, Eritrea, and the National Board for Higher Education. He carried out various research projects and published many book chapters and articles in peer-reviewed journals. Moreover, Dr. Tsegay serves as an editorial board member for different academic journals, such as *Teaching in Higher Education* and *Access: Critical explorations of equity in higher education*.

Contents

Preface XV

Chapter 1 1
The Goals of Science Education Should Be Linked to the Central Tenets of Sustainable Development, Equity, and Social Justice
by William C. Kyle, Jr. and Emilia A. Belciak

Chapter 2 23
Introduction to the STEM Student Success Model
by Leander Kwabena Brown

Chapter 3 41
Integrating STEM: An Interdisciplinary Approach to PreK-12 Education
by Teresa J. Kennedy and Michael R.L. Odell

Chapter 4 59
Addressing the Gender Gap in the STEM Field
by Rubie Ibuola

Chapter 5 69
Bridging the Gap: Progressive Teaching Strategies for Gender Equity in STEM Education
by Umar Daraz, Younas Khan, Muhammad Azeem Ashraf and Samson Maekele Tsegay

Chapter 6 89
Thinking, Alternatively: The Application of Art, Philosophy and Holistic Practice in the Teaching of Science, Technology, Engineering and Mathematics (STEM)
by Christine Challen and Kay Sidebottom

Chapter 7 103
Evaluating STEM Lecturers' Experiences with Digital Assessments and Continuity of Digital Transformation in Higher Education
by Innocent Zitha, Georgina Mokgaetji Mokganya and Orifha Sinthumule

Chapter 8 117
STEM Integration in Resource Constrained Environments
by Majid Nawaz, Irfan Ahmed Khan and Muhammad Imran Ahmad

Chapter 9 131
Developing Collaborative Skills through STEM Approach
by Chairil Faif Pasani and Rizky Amelia

Chapter 10 147
Improving the Teaching of Mathematical Problem Solving – A Collaborative Research Based on Theoretical Findings from Two PhD Dissertations
by Maud Chanudet, Jean-Luc Dorier and Stéphane Favier

Chapter 11 165
Collaboration in Science, Technology, Engineering, Mathematics (Stem) Education and Technical, Vocational Education and Training (TVET) for Graduates' Competence in Nigeria
by Edet Okon, Maria B. Onabe, Mary A. Iyam, Ann Dijeh, Francesca Nneoyi-Egbe, Stephen Bepeh Undie, Ita E. Okon, Agnes L. Okute, Mfon E. Usip and Agnes A. Ewuru

Chapter 12 181
Cross-Cultural Experiences of Canadian Science Educators Visiting the Sister Schools in Chongqing, China: A Cross-Cultural Experience
by Emilia M. Iacobelli and Geri Salinitri

Chapter 13 193
Enhancing Educators' Cultural and Digital Literacies through Makerspace Development Activities
by Jennifer Miller

Chapter 14 205
Creating a Culturally Responsive Mathematics Education: The Case of Gebeta Game in Ethiopia
by Solomon A. Tesfamicael and Danyal Farsani

Chapter 15 223
The Importance of Activity-Based Learning for Mastering Mathematical Tasks during Distance Learning in Austria
by Sabine Apfler

Chapter 16 237
Innovative Methods in Computer Science Education
by Kinga Pusztai

Chapter 17 255
Integration BIM and Emerging Technologies in Architectural Academic Programs
by Nada Abdelhai

Chapter 18 279
Current and Emerging Trends within Higher Education That Enhance Diversity and Inclusivity in Undergraduate STEM Provision
by Marie Bassford

Chapter 19 295
Educational Innovation in Higher Education with the Use of Management Simulators for the Development of Competencies
by Alba Guzmán-Duque and Luisa Chalarca-Guzmán

Chapter 20 313
STEM Education and the Role of the HEAC in Uganda
by Charles Tushabomwe-Kazooba, Imelda Kemeza and Calvin Okello

Preface

The book arises from the global attention that Science, Technology, Engineering, and Mathematics (STEM) education has received in the past few decades. Governments, policymakers, researchers, and educators have continuously called for improvements in STEM teaching and learning so that students can learn the most advanced knowledge in STEM disciplines. As a result, educational institutions and stakeholders worldwide have been attempting to integrate STEM subjects into their curriculums and in-learning services so that learners can meet the challenges of increasing economic competition and advancing technology and innovation. In light of these developments, this book explores the recent developments and emerging trends in STEM education. This book is designed for instructors and students involved in the most recent and progressive practices for supporting STEM teaching and learning at different levels of education. It is important to note that the STEM disciplines are not separate from each other but are linked to each other. Thus, a cohesive interdisciplinary approach is required to understand STEM disciplines in educational institutions. The chapters in the book are written by authors from various countries and diverse educational settings, making this volume one of the best collections of its kind with chapters that will resonate with international audiences.

Some of the book's chapters focus on particular subjects within STEM disciplines, while others discuss STEM education in general. Furthermore, some chapters analyze various STEM disciplines within each other or combine STEM education in relation to other disciplines such as culture, arts, and vocational education. Chapter 1 explains why science educators must become active agents in facilitating young learners' engagement in acquiring the knowledge and skills to address global challenges. Chapter 2 provides critical components of the student success model for underrepresented minority undergraduate students in STEM. Chapter 3 examines STEM as a meta-discipline by explaining its origin in the United States, definitions of STEM literacy, and implementation models for STEM integration in PreK-12 Education. Chapter 4 discusses the gender gap in STEM with a focus on the United Kingdom. It explores the significant factors causing the gender gap from primary school to career and provides possible solutions to bridge the gap. Chapter 5 empirically investigates the impact of progressive teaching methods on gender equity in STEM education in Pakistan.

Chapter 6 investigates the challenges and opportunities for working in transdisciplinary ways across STEM disciplines, focusing mainly on the role of art and philosophy. Chapter 7 is an empirical study that examines South African rural university lecturers' experiences with digital assessments and their perspectives on the continuity of digital transformation in higher education in the post–COVID-19 era. Chapter 8 focuses on the challenges and opportunities of STEM integration in resource-constrained environments through the authors' personal experiences implementing STEM in Pakistan. Chapter 9 reviews the literature on developing technology-based collaborative skills through the STEM approach, focusing on integrating technology into education. Chapter 10 includes two studies on mathematical problem-solving at different school levels in Geneva, wherein the first study examines teachers' assessment practices in mathematics problem-solving courses, and the second study examines the students' experience in problem-solving situations. Chapter 11 discusses the collaboration between STEM education and technical and vocational education in Nigeria to equip students with the most advanced theoretical knowledge and practical skills.

Chapter 12 is a case study of two in-service science teachers' experiences teaching science in Canada and sister schools in China. Chapter 13 examines the impact of the mobile makerspace program on educators' self-efficacy toward integrating technology and culturally responsive teaching using pop-up activities. Chapter 14 describes an example of creating a culturally responsive mathematics education in Ethiopia using a locally available game called Gebeta. Chapter 15 is based on findings from a research project in Austria that investigated the features of activity-based, tools-supported mathematics instruction that helped students master mathematical tasks during distance learning. Chapter 16 documents the author's experience of trying innovative methods through digital storytelling and gamification in teaching undergraduate students. Chapter 17 examines the importance of adopting building information modelling (BIM) with emerging technologies in architectural academic programs. Chapter 18 argues the opportunities to create a more diverse and inclusive STEM higher education community by providing different types of support to minority students. Chapter 19 is an empirical study examining university students' perceptions of developing individual and group-level managerial competencies using simulators. Chapter 20 uses the Higher Education Access Certificate (HEAC) program in Uganda as a case study to examine its effects on increased enrollment in STEM disciplines and how it can be implemented for better outcomes.

In summary, the book provides diverse perspectives and recent developments in STEM education across the globe. This book is an ideal reference for students and teachers seeking guidance in improving STEM teaching and learning practices. Moreover, the book is essential for university teachers to use as a textbook for training students with the most recent practices in STEM education. Readers who want to learn more about STEM education and its integration with other disciplines will also benefit from this text.

Muhammad Azeem Ashraf
Institute of Educational Science,
Hunan University,
Changsha, China

Samson Maekele Tsegay
School of Education,
Anglia Ruskin University,
Cambridge, United Kingdom

Chapter 1

The Goals of Science Education Should Be Linked to the Central Tenets of Sustainable Development, Equity, and Social Justice

William C. Kyle, Jr. and Emilia A. Belciak

Abstract

In this chapter, we offer a rationale for why science educators ought to become active agents in facilitating the engagement of youth as they acquire the knowledge and skills to address global challenges facing humanity. The youth of today are demanding action. Science educators have a responsibility to facilitate learners - and their communities - as they transform and reinvent the world they are inheriting. Science education is a primary means of investing in human resources. The youth of today must be able to address complex everyday issues that are yet unforeseen. This is not a modest goal, but it must be accomplished. All learners must have access to an equitable education. We must bridge the divide and facilitate dialog between formal and informal/free-choice educators, as well as disciplinary and interdisciplinary science educators. Today's youth recognize the consequence of failing to transform towards a more sustainable, equitable, and socially just future is profound.

Keywords: sustainable development, environmental education, global challenges, equity, social justice

1. Introduction

Herein we will highlight the 2030 Agenda for Sustainable Development [1] in the context of current and emerging global challenges [2] and the need to transform education. It is important to note that nearly a quarter of the global population is composed of youth ages 10–24. This amounts to 1.8 billion people [3, 4]. Additionally, many underserved people in the global population are youth [4], with 90% of the global youth population living in less developed countries (LDCs).

By 2050, action is needed on environmental issues, biodiversity, and climate change (this is particularly true since July 2023 was the hottest month on record; see, [5]) and, significant investment is needed in education/literacy, health care and nutrition, addressing poverty alleviation and hunger, decreasing under- and unemployment, and enhancing food production/productivity. The World Youth Report on Youth and the 2030 Agenda for Sustainable Development examines the mutually supportive roles of the new agenda and current youth development efforts [6]. Youth are not mere beneficiaries of the 2030 Agenda, rather they have a critical role in the implementation of the sustainable development goals (SDGs). The active engagement of youth in sustainable development efforts will be imperative to achieving the goals of the 2030 Agenda.

We wish to further develop the notions presented by Kyle [7] when he asserted:

> *Science educators ought to be at the forefront of ensuring the education discourse is oriented towards the goals, aspirations, desires, and needs of all 1.8 billion youth versus adhering to a bureaucratic characterization of science in which universal standards, goals, objectives, and accountability measures are imposed upon learners, teachers, and administrators by policymakers and politicians. All too often, as a result of universalism and standardization, learners experience an education in science disassociated from the contextual realities of life (p. 2).*

Moreover, we must ensure the goals of science education address sustainable development, empowerment, equity, social justice, and social transformation, thereby ensuring an informed, ecologically/environmentally literate, thoughtful, and empathetic citizenry. Empowerment – particularly in LDCs - is correlated with poverty alleviation and economic growth. An education in science must be contextualized and connected to the lives and world experiences of learners while taking into consideration issues of place-based locality, as well as social, civic, and cultural values [7].

2. The origin of the environmental education movement

The practices of sustainability education and legislation have traceable roots as ancient as the Western Han Dynasty (145–86 BCE) when regulations surrounding sustainable management of trees, swamps, and mountains were recorded [8]. Works attributed to Plato and Aristotle in the fifth century BCE show scholarly debate over foundational environmental issues, such as the impact of the introduction of invasive species (rats), the existence of ecological niches and keystone species, and the interdependence of species. "The environmental movement had no clear beginning. The movement did not begin in one country and then spread to another; it emerged in different places at different times, and usually for different reasons" ([9], p. 1). Following the advent of colonialism, the prevailing thought of colonial settlers was the relative abundance of natural resources, which gave them the right to use and kill without limit and regard for any notion of conservation. Key moments of environmentalism include South Africa, 1658: proclamations to curb reckless forest clearing and hunting practices; Australia,1803: warnings about deforestation leading to soil erosion; South Africa, 1846: environmental legislation, and several international efforts to protect and preserve nature began in the 1900s [9]. This period marked the beginning of international collaboration to curb ecological devastation with environmental efforts in Britain, New Zealand, and Australia. In 1909, the World's first international environmental agreement saw Britain, France, Germany, Belgian Congo, Italy, and Portugal sign the Convention for the Preservation of Animals [9]. During the eighteenth century, western philosophers of the industrial period, such as Pinchot, Thoreau, and Muir, were credited with developing sustainability and conservation concepts mirroring those in ancient Confucian writings [8]. Such endeavors typically had little success [9]. The formation of the International Congress for the Protection of Nature in 1948 (now the IUCN) was interrupted by the outbreak of world war. It was unable to do much other than collect data on human impacts, save for some work in ornithological conservation. By 1943–1948, alarm was raised regarding the perils of self-indulgence, ignorance of natural laws, and the need to bring human demands upon nature into conformity with natural limitations [10]. In 1947, The International Office for Protection of Nature (now IUCN) hosted a conference with 70 delegates from 24 countries - the largest environmental conference yet, which was still ultimately fruitless.

The late 1940s and early 1950s saw the beginning of environmental consciousness in the US as issues of public health and environmental concerns increasingly captured the public's attention. For example, in 1952, a report by the President's International Materials Policy Commission (see Ref. [11]) urged the US to look to renewable energy resources (solar and wind power) and urged ecological prudence to protect natural resources into the twenty-first century (see, Ref. [10, 12]). Increased environmental literacy meant legislative efforts to curb air pollution also began to take shape [13]. A flurry of books offered urgency to environmental issues of the times (see, for example Ref. [10, 14–16]), yet approaching 75 years later youth of today must confront similar catastrophic issues. In the fall of 1953, courses on the science of ecology were starting to be offered in US colleges and universities. Advanced degrees in the emerging field were growing in number. The catalyst was a groundbreaking textbook, *Fundamentals of Ecology*, by Eugene Odum [17], who noted:

> *The roots of the science of ecology are old; they go deeply into the substrate of modern science and support a large and important superstructure of applied sciences. Partly because of its wide scope, few attempts have been made to present ecology as a whole and to place it in a logical perspective in relation to other divisions of biology and to other fields of learning ([17], p. v).*

The origin of environmental science, and subsequently environmental education, can be traced to the 1960s. At this time, a growing awareness of the need for interdisciplinary studies analyzing environmental issues and concerns was taking shape. Simultaneously, primarily in Western developed nations, the passage of environmental laws and protections fed an increasing public awareness of such issues. These laws focused on a wide range of issues encompassing air and water quality, waste management and cleanup, natural resource management, and protection of biodiversity [7]. Today, environmental laws are part of the fabric of most nations, as well as the basis of international law and treaties. The environmental movement and environmental education (EE) arose as a result of increased public awareness. In the US, and many developed nations, Rachel Carson's [18] *Silent Spring* inspired the public's interest and engagement with environmental issues.

3. Preparing for the future

Even with all this attention being drawn to environmental issues, today's youth are poised to inherit a world on the precipice of dramatic ecological upheaval. The Intergovernmental Panel on Climate Change (IPCC) [19] paints a dire picture of an intensifying climate crisis with complex impacts on human well-being. It underscores the urgency of the situation, highlighting how greenhouse gas emissions must be capped by 2030, and nearly halved by 2050. The report sets a goal of averting a global average temperature increase above the 1.5-degree tipping point that could result in irreversible damage to natural systems and catastrophic loss of human life [19]. Achieving such a dramatic change in greenhouse gas production will require rapid, substantial, and sustained changes in societies across the globe. The development of requisite technology, policies, knowledge, and systems will require tremendous investment and cooperation from all fields, including education [19]. Science educators thereby have crucial roles to play in equipping youth for participation in this rapid evolution of society.

On a global scale, the geographic regions and sociopolitical groups of the world most vulnerable to the impacts of climate change and ecological devastation are also the ones with the largest per-capita population growth [20, 21]. These regions and populations are disproportionately facing the losses and damages associated with climate change while bearing the least of the responsibility for causing the calamity, an issue exposed front and center at the COP27 conference in Egypt [22]. The International Energy Association's data reveal the handful of nations responsible for generating the most carbon emissions outpace the ones with the least production of CO_2 by over 1000 percent [23]. Further, the Global North is responsible for 92% of carbon emissions [23, 24]. Taken together, this means most of the Earth's population is about to inherit a planet reeling from the consequences of the choices of a handful of wealthy nations. It also means wealthy nations are going to need to enact dramatic changes to address climate change. Another consequence of this pattern is the LDNs will be pressured to catapult into a high-tech future while leap-frogging the scaffolds of fossil fuels if they are to obtain equitable access to economic prosperity [20]. Educators, both formal and informal, have a powerful role to play in preparing youth for changing this reality. They have the power to foster creativity and encourage curiosity in students. Educators must choose to confront the realities of our changing world head-on while helping students understand its relevance to their communities [25, 26]. If they do not, then students will simply replicate existing conditions or be forced to develop these skills and acquire this knowledge without their support: a daunting and potentially insurmountable hurdle.

The educational community has a responsibility to prepare youth to fully understand issues of sustainability. In fact, in a recent UNESCO study, 90% of youth agreed schools should be sites where people get prepared for issues of climate change [27], with 70% reporting they cannot explain climate change and can only explain its broad principles, or possess limited knowledge about climate change. This discrepancy highlights the need for educators to fulfill the moral imperative of preparing young people for the future. Moreover, youth activism reveals a distinct interest of youth in protecting the future of our planet. The school strike for climate change, a youth-led movement to draw attention to climate change by walking out of class in protest of the inaction of leaders towards addressing such issues, drew over a million students from hundreds of nations [28]. Subsequent movements such as '#Fridays for the future' continue to branch off the success of these endeavors in mobilizing continued youth resistance. Despite the urgency of the situation and the fervor of young people's resistance, educators, politicians, and administrators expressed concern over whether the protests were an infringement upon one's education (see Ref. [29]). In response to this concern, Kyle [29] expressed:

> *I wonder why raising consciousness regarding the global crisis facing humanity is an infringement upon one's education. I hope educators will have the wisdom and*

> *courage to respond to the demands of youth, transform educational practice, and engage with a research agenda focused on the global challenges facing humanity (p. 150).*

Concerns around sustainability and social justice are a high priority for this generation of youth. By focusing on the interests of youth, educators can strengthen their engagement with sustainability, conservation, equity, and social justice, and facilitate a deeper understanding of the topics they are exploring in their academic pursuit.

Yet, in many nations, the moral imperative of adequately preparing youth to address climate change is regarded as secondary compared to the immediacy of school rankings and educator evaluations as measured by standardized test scores. This view of schooling does not need to be the case. The integration of sustainable development goals into science education standards and curricula can be a mutually beneficial enterprise, supporting the attainment of standards, while engaging students in meaningful learning focused upon real-world challenges, thereby preparing students for the world they will inherit. For the youth of today, the future is now [7].

4. Emphasizing standardized testing and standardizing inequality

Standardized test scores remain the penultimate measure by which politicians and administrators evaluate the relative successes of instructors and institutions [30]. The misuse, overuse, and overemphasis of standardized testing had multiple deleterious consequences for science and sustainability education. It functionally reinforces an outdated model of learning, emphasizing performance on singular, summative exams as indicators of educational attainment. These assessments typically fail to capture a glimpse of the true breadth of skills and knowledge students must acquire to participate fully in scientific discourse and affect substantive change [30–32]. For example, even though well-designed summative and end-of-course exams may be used to measure students ability to apply labels to elements of ecosystems and explain the ways in which they interact with one another [33], they may not measure whether students can apply this knowledge to make decisions about their own consumption of resources and engagement in practices that can help or harm the ecosystems they have influence over. This structure is designed to assess and incentivize students' ability to function within the existing systems, which has led humanity to the very brink of disaster. It is not intended to assess their ability to improve on these systems, a skill vital to the future of our planet and humanity. Additionally, many schools and educators opt to reduce the quality and quantity of science, history, and arts instruction to boost scores for the higher stakes math and reading exams [31–34]. This is especially true in the post-pandemic era, as schools scramble to make up for education gaps due to disrupted learning during the COVID-19 pandemic. Students deserve better than this.

An overemphasis on standardized tests also contributes to a reduction in diversity in the STEM workforce as white, male students are more likely to score well on such tests and thus self-select into STEM fields, even though the exams have been demonstrated to be a poor predictor of success in STEM majors [30]. Such a trend operates in contrast to the goals of equity in education presented in the SDGs. Unequal STEM field participation and education are evident in research from around the world along lines of race, ethnicity, gender, and class [35–38]. The trend of disproportionate underrepresentation leaves humanity at the disservice of missing the valuable contributions in diversity of thought, skills, and perspectives offered by the disenfranchised populations. Diverse teams regularly outperform heterogeneous teams in problem solving [39]. It also leaves these vulnerable groups at the disadvantage of having less of a voice at the table, and therefore fewer opportunities for crucial self-advocacy [24, 40]. Interestingly, researchers examining the practices of 1,951 companies across 24 industrialized economies linked increased gender diversity to a 5% decrease in a firm's carbon emissions [41]. This suggests organizations can mitigate their climate change impacts as a side effect of focusing on boosting equity in the composition of their workforce, demonstrating another way in which sustainability goals are linked. Still, women and other oppressed populations across the world continue to face exclusion and under-investment in science education. Educators have a responsibility to be aware of and take effective steps to disrupt this harmful trend. Indeed, research has identified various modalities by which educators can achieve more equitable outcomes for their students [24, 42]. Culturally responsive, authentic learning experiences emerge as a trend in much of this research.

We raise the issues of emphasizing standardized testing and standardizing inequality herein as we wish to focus upon the contradictory roles schools play in society. Schools generally reproduce the inequities and inequalities we see in society and the economy [43]. And, while schools play a key role in reproducing social inequality, they simultaneously play a contradictory role with respect to society and social equality by allowing resistance to this reproduction. Students - and educators - do and must resist the social programming of schooling; and schools and individual classrooms must exist as contested sites of production of justice seeking to care for students, school staff, and the community in ways that build critical consciousness and critical social action [44].

The COVID-19 pandemic, which significantly disrupted education for students worldwide, has deepened many educational disparities [45, 46], exacerbating the disadvantages faced by vulnerable students. Furthermore, the pandemic has laid bare the perils of misinformation and mistrust in science [47, 48]. Intensifying climate change and ecological disruptions erode the natural safeguards to mitigate spillover events at the same time humanity is increasing travel and urbanization, amidst warnings of pandemics becoming a more frequent feature of the human experience [49]. Yet, in the scramble to make up for education gaps, many schools around the world have decreased investment in science education, opting to align content towards exams and backwards planning around standardized assessments [50]. We must choose to reimagine education to better cultivate the skill sets and content knowledge that can be harnessed to create a more sustainable future, instead of falling into the trap of replicating extant systems. COVID-19 has provided the silver lining of forcing rapid and sometimes revolutionary evolution of educational practices. Educators around the world have radically adapted their pedagogy [46, 48, 51]. While in-person field trips were canceled in many countries, educators explored alternative ways of connecting students to outside resources through video chats with local museums or using students' own backyards and homes as learning tools. Now is the time to harness the innovative spirit and move towards decolonized, equitable, and experiential learning practices to usher in a new era of sustainability [50, 52].

5. Transforming the world toward a more sustainable future

Students need to develop a variety of complex skills and values predicated on critical thinking and the synthesis of interdisciplinary content to be successful in transforming the world towards a more sustainable future [24, 52, 53]. They need to be able to evaluate the veracity of claims presented to them by candidates for elected office, weighing their personal values against those of the candidates. They need to be able to analyze the complex interrelationships of factors involved in the production and consumption of goods against their values so they may make responsible decisions around their participation in commerce, unphased by greenwashing practices. They also need the skills to problem solve, design, and evaluate solutions, work as members of a team, and communicate their ideas so they may shape the future of science and technology [54]. These skills can either be nurtured by or suppressed by educators.

Paulo Freire's emancipatory, citizenship-focused framework for education emphasized the power of dialogics, as both a means of acquiring knowledge and an epistemological relationship, a way of knowing in and of itself [55]. When education is exercised as a mere conveyance or passage of information from educators to students, rather than an exercise towards preparation for the social functions of citizenship, he argued, it is being wielded as a tool of oppression. In *Teaching to Transgress*, bell hooks emphasize the liberatory power of education. She describes the contrast between how the anti-hegemonic politics of her early all-black education contrasted with the stifling white-centric atmospheres of the classrooms she encountered throughout her collegiate and graduate education experience [56]. Many modern well-meaning science educators and institutions engage in practices resembling the latter, often through the lens of positive intent, believing preparing students to score high on standardized exams will launch them into positions of power and thereby liberate them. This oversimplified misunderstanding of the dynamics of oppression fails to nurture the critical thinking, self-study skills, intrinsic motivation to continue learning, and contextual understanding needed to foster active participation in shaping the world. What use will the possession of large sums of 'banking knowledge' (see Ref. [55]) afford students when they are demoralized and dehumanized by the learning experience? Students deserve the power to confront sustainability, equity, and social justice issues from an interdisciplinary framework. They need educational opportunities to empower them to be socially responsible agents of change through critical, action-oriented, and reflective pedagogy [57]. The pedagogical practices of Socio Scientific Issues education (SSI) and Science, Technology, Society and Environment (STSE) are approaches seeking to provide these opportunities by exploring explicitly the interconnectedness of science and society while helping students develop their own morals and ethical frameworks for decision making [58]. Students harness the engaging power of controversy to stimulate curiosity and rich educational discussions in the classroom. Moreover, students marvel at the novel opportunity to develop and share their own informed opinions.

6. The crossroads of science, politics, and culture

Debate exists regarding whether incorporating the sustainable development goals in science education - regarded by some as representing an imposition of Western neoliberalism and by others as anti-imperialist and decolonializing [59] - needlessly politicizes the discipline. After all, science is inherently political [60]. From Copernicus to the Manhattan Project, to Wangaari Mathi, and Anthony Fauci, scientists have always contributed to and engaged in political discourse. As Backhaus [60] notes "Science produces knowledge, knowledge provides options for politics, and politics then enables and finances (or impedes and starves) science" (p. 310). Tiptoeing around the eggshells of public opinion is antithetical to scientific praxis and effective liberatory teaching. Science Education cannot be apolitical, and when it pretends to be, it is most likely acting to reinforce hegemonies [56, 60]. The avoidance of politically sensitive topics in science education is futile and hinders advances in the field, tacitly reinforcing harmful power structures.

Science is inseparable from culture, whose influences shape the types of questions researchers ask, the framework for thinking about and investigating them, the lenses through which data are collected and interpreted, the perspective through which it is disseminated and understood [25, 61]. Educators must enable students to navigate the boundaries and borders between science and their lived experiences [62, 63]. The epistemology, beliefs, and practices underpinning Western science constitute a culture unto itself. Teachers can act as guides - helping students explore, comparing this to a border crossing, or an admission ticket, which demands socialization into unfamiliar ways of knowing, doing, and being. For much of its history, mainstream western science education was heavily influenced by didactic teaching methods and strict logical positivism, divorcing it from the dynamics of socio-political context, and meaningful relevance to the realities of students [64–66]. It also artificially separated the various science disciplines from one another [61]. This approach created a homogenized and sanitized representation of science as a field, ignored the diversity of learners in schools, removed the contextual factors that ground and incentivize learning [61, 66, 67] and operated counter to student's effective meaning-making and development of critical thinking skills [25]. Critics of this approach, such as Dewey, argued science as a practice is an interdisciplinary, iterative, inquiry-based experience intertwined with culture and values, thus science education ought to emphasize preparing students to engage in this process [61]. For example, students who are taught how to dissect the impacts of excess fertilizer runoff on invertebrate health from this perspective will likely emerge capable of successfully filling out a diagram connecting the cause and effects of this problem. However, to fix this problem, they need to be able to communicate effectively about this problem to people who may have differing opinions and perspectives, dissecting the interplay of a variety of factors. Simply developing an understanding that X harms Y and affects Z in multiple ways does not consider the fact that people who cause X are complex individuals acting within even more complicated social contexts. Students must be challenged to imagine how to effectively reshape these social contexts. However, the positivist, reductionist framework continues to plague modern science education, accentuated and exacerbated by the imposition of standardization and aggressive testing measures. This is incompatible with the demands of the modern world, which require students to adapt to rapidly developing situations affected by politics, knowledge, and technology [60, 64, 65, 67].

Today's youth are inheriting the inequities and injustices that previous generations failed to rectify, many of which have complex intersectional interactions with one another. Imperialism and colonialism precipitated the emergence of the Anthropocene [68] as the wanton exploitation of indigenous peoples and resources simultaneously ushered in devastation of societies and the ecosystems upon which they depended.

In December of 2022, the UNFPA published a brief [69] documenting the intersections of race, gender, environmental and climate justice. The brief summarizes research and expert testimonies from Asia, Africa, North and South America, as well as Australia. It underscores the importance of addressing intersectionality in sustainability issues. Systemic discrimination due to race, disability, and ethnicity is intricately linked with issues of health, well-being, safety, environmental degradation, and climate change. These connections extend to displacement, vector-borne diseases, adverse birth outcomes, deforestation, and numerous other interconnected issues. The consequences of emissions from wealthier nations lead to melting ice, rising sea levels, and increased flooding in poorer nations. In the aftermath of these calamities, women and children are at higher risk of health issues, as well as violence, including sexual assault. On a global scale, women and gender minorities of the African diaspora are significantly more likely to experience adverse health outcomes from environmental pollution when compared to peers from other backgrounds [69]. Even in the wealthiest nations, such as the United States, Black women, and children are significantly more likely to be exposed to pollution and persistent environmental pollutants, while being less likely to have access to green spaces and decent healthcare, leading to shorter lifespans. Similar patterns are evident for indigenous populations around the world affecting every walk of life [70].

An international study of 10,000 youth in 42 countries found youth around the world are experiencing significant levels of anxiety and despair, often feeling betrayed by their leaders [71]. Youth in nations most impacted by climate change, especially poorer nations, were most susceptible. Thanks to the efforts of activists, the tides may be beginning to turn on these trends. Ineza Umuhoza Grace is an ecofeminist who inspires young people to take action. She is passionate about empowering women to lead change towards sustainability. Her youth-and-woman-led Rwandan nonprofit Green Protector fights for environmental improvement. In 2022, Umuhoza played a pivotal role in advocating alongside numerous youth activists at COP27, the United Nations' worldwide conference on climate change, to demand a fund dedicated to addressing losses and damages stemming from natural disasters linked to climate change. Her effort paid off. World leaders agreed to make contributions to begin offsetting climate change's effects on the most vulnerable nations. Recognizing the potential of youth and the power of education, she also started an organization called The Green Protector that works to inspire other Rwandan youth to protect the environment. She is a recipient of the 2023 Global Citizen Award [72]. Similar efforts are being undertaken by youth activists worldwide, providing an invaluable opportunity for educators to connect demoralized youth to effective models and role models of change.

7. Acting locally to educate globally

Educators do not need to reinvent the wheel to accomplish the lofty goal of fostering active engagement of youth in sustainable development. Youth already possess the interest; educators just need to work within their local contexts to overcome defeatism and bridge the gap between prescribed curricular content and community/societal needs. Models and examples for how to effectively employ the SDGs towards connecting the realities of human lived experiences to environmental science abound [25, 54, 62, 73].

In light of the 17 Sustainable Development Goals (SDGs) [74] and the European Parliament resolution on the climate and environment emergency [75], Rodrigues [51] emphasizes how fostering emotional connections with nature is imperative to addressing the environmental challenges facing humanity and the climate change emergency. She states:

> *Exploring local environmental problems, using nature as a classroom and inspiration across different disciplines, promoting field trips to nearby natural areas or recreating ecosystems in the school playground, and involving students and their families in projects with visibility, usefulness, and value to the community can foster this emotional connection. (Educational Practice Agenda Proposal and Implications section)*

She presents a compelling narrative demonstrating the power of community-centered, intergenerational learning and illustrating how it can have a profound impact on all involved. This approach fosters a sense of mutual respect and communal understanding, strengthening the ability of stakeholders to address important issues [51]. She emphasizes empowering education as dialogic co-construction of meaning. The circle of this dialog extends to parents/grandparents, and the community. Through such engagement, students, family, and community members develop complex conceptual relationships and construct notions of what knowledge is relevant to address identified issues, how to take action to bring about change, and how to mitigate the global problem of climate change within their local community [51]. In the absence of these vital factors, students often emerge feeling powerless, anxious, paralyzed and defeated [69]. Current patterns of inequity are unacceptable and unsustainable. Students today exist within these contexts and will receive the responsibility for subverting or recreating them. We can aid students by introducing them to the data, tools, and skills of sustainability activism. It is not enough for science educators to prepare students to understand ecological issues, sustainability, and conservation. They must also help students understand how these issues fit within the social contexts of their daily lived experiences while teaching them how to be active agents of social transformation.

Climate change education can have significant impacts on individual carbon emissions by influencing behavior [76]. Students should learn how to mobilize and engage with their communities to produce lasting, meaningful change. Climate change education is effective when the program conscientiously connects climate change to the daily lives of students, engages them in cultivating stewardship, and engages students in creating change in the community [25, 76]. Each of these components can be implemented in various educational settings: e.g., schools, homes, libraries, museums, zoos, conservation organizations, and/or community centers. Collaboration between diverse community organizations and stakeholders can have a powerful impression on youth, strengthening their capacity to make deep connections and build a sense of belonging. When the members of the community come together to show youth they care about them and are willing to invest in them, young people can be inspired to reciprocate such altruistic investments.

A particularly stirring example of an effective approach can be found in a school in rural Malawi. The school took part in a participatory action research project under the auspices of Project SUSTAIN [77, 78]. Students engaged in transformative reforestation, creating sustained community development, while revitalizing an ecologically devastated area. They accomplished this by using Participatory Rural Appraisal [79] tools, including asset mapping. The project resulted in a revitalized community, improving both economic and ecological conditions for all of its inhabitants. This astonishing transformation was elicited by the facilitation of dialogic, place, problem, and phenomenon-centered pedagogical practices as championed by Freire [55] and other critical pedagogs. This technique delivered the benefits of recontextualizing environmental science content knowledge within the community's needs of reducing poverty and increasing equitable access to resources for all the villagers [57].

The teacher could have stuck to the status quo but instead empowered students to collaboratively improve the existing conditions within their community.

Another inspiring example of effective sustainability education emanates from Project SUSTAIN, which worked to develop a program to utilize science education to empower female learners in rural South Africa to reach their full potential [80]. By working in collaboration with a variety of community stakeholders, including students, teachers, local government, and village elders, the researcher was able to co-produce an educational program to revitalize the science education practices of the local school and cement the value of such education in a generation of teachers, students, and community members. Together, the community was able to discover the power of scientific knowledge to produce tangible improvements in their lives, bolstering support for enabling female students to continue in the program long-term, and improving the educational experiences of all involved. Sustainable development goals such as promotion of gender equality, and inclusion of all in quality equitable education guided the development of the program. This produced a demonstrably more valuable learning experience for the learners.

The theme of this section - ACTING LOCALLY TO EDUCATE GLOBALLY - need not only be inspired by school-based initiatives or by educators. Rather, individuals and communities can be the inspirational trailblazers. A book that will surely inspire youth of today and hope for the future, while facilitating a sense of empowerment and inspiration amongst girls with the notion of 'thinking outside of the box' and 'dreaming big' is *Girls Who Green the World: Thirty-four Rebel Women Out to Save Our Planet* [81]. The book features such leaders/trailblazers/activists as Catherine Coleman Flowers (an Environmental Health Activist and Founder of the Center for Rural Enterprise and Environmental Justice, p. 29); Rhea Mazumdar Singhal (Founder and CEO of Ecoware, p. 59); Komal Ahmed (Founder of Copia; solving world hunger and food waste, p. 67); Caroline Danehy (Cofounder and CEO of Fair Harbor; recycling plastic bottles into bathing suits, p. 97); Varshini Prakash (Cofounder and Executive Director of Sunrise Movement; the organization's movement in 2020 made a difference in electing Green New Deal champions, p. 141); Rhiana Gunn-Wright (Director of Climate Policy at Roosevelt Institute and Architect of the Green New Deal, p. 163); and Leah Thomas (Founder of Intersectional Environmentalism; bringing justice to the environmental movement, p. 179).

Xiuhtezcatl Martinez is a 23-year-old indigenous climate activist and hip-hop artist noted for his leadership in the global youth environmental movement. His upbringing in Aztec tradition taught him to value nature and humanity's role in it, inspiring him to begin activism at the age of 6, when he delivered his first speech on environmental concerns. Martinez also formerly served as the Youth Director of Earth Guardians until 2019, an organization dedicated to teaching youth across the world how to use civic engagement and the arts to help solve environmental issues [82]. He explains his passion as follows:

> *I want to inspire people to bring activism into what they love to do. Whether you are an entrepreneur, poet, scientist, athlete, or artist, you can find ways to use what you are passionate about to have an incredible impact. Our movement desperately needs a diversity of new ideas and ways of thinking to bring forth real change. That's why we need more than traditional activist tactics. Some of the most impactful work being done in my generation is from people spreading awareness through art, music, and other creative means ([82], p. 212).*

In this chapter, we also highlight that social movements often capture the humanizing, lived experiences of individuals within local communities. Méndez [83] notes "environmental justice groups center their advocacy on community-specific public health campaigns" (p. xii). He offers an urgent and timely analysis of the contentious politics of incorporating environmental justice into global climate change policy. He tells the compelling story of people, place, and power in the context of climate change and inequality. He explores the California (USA) environmental justice movement and the advocacy work on climate change and articulates the perspectives and influence low-income people of color bring to their local communities while highlighting the importance of incorporating local knowledge, culture, and history into policymaking to address the global complexities of climate change and the threats facing local communities.

8. Places of empowerment and agency

A place-based perspective challenges the facilitating pedagogue to connect learning to the places where students are engaging with learning. This is especially vital for oppressed and displaced people, who face tumultuous relationships with the lands upon which they live [84]. It can be enhanced, where available, by technology. Facilitators can start by engaging students in open-source citizen science projects. For example, they can utilize freely available citizen science technology, such as the iNaturalist app [85] to identify local flora and fauna, or the marine debris tracking app [86] to collect data on types of trash pollution present in their communities. Now, students are authentically participating in larger-scale studies by contributing to the expansion of evolving data sets. To take things to the next level, educators can engage students in dialogs, shifting the cognitive load to students. Instead of the teacher delivering the answers, they can charge students with making meaning of the data sets and noticing patterns. Then, teachers can help students contextualize those patterns by connecting students to other data sets or research, such as localized or international reports on environmental racism or eco-justice. Finally, students can begin designing community-based solutions to local issues revealed by the data with which they interact. Students can practice communication and outreach skills by contacting organizations and community stakeholders, thereby gaining assistance in their efforts. The teacher's role now shifts to guiding the learning and research process, providing resources, asking questions, and helping to edit student's communication for clarity. The best part is students can sharpen a whole host of content knowledge and skills while meaningfully contributing to their communities.

Educators should identify ways in which this placed-based notion of contributing to climate change and sustainability can become school-community-based initiatives. Schools and communities ought to partner with local businesses/industries/NGOs, social service agencies, and/or partner schools internationally to identify ways in which the local communities can work towards climate change solutions. Transformative action research projects should be developed so that schools and communities share their initiatives and accomplishments. Such school-private climate governance initiatives would ensure students and communities become active agents in the process of transformation [29]. Ideally, such an educational orientation would facilitate the emergence of youth-led community organizing, a movement to empower youth while enabling them to make substantive contributions to their communities [87–89]. Educational experiences ought to be intergenerational, community-based, oriented towards self- and social empowerment, and transformative [29, 78]. Youth have a significant role in contributing to the social and environmental changes that must transpire. The overarching goal of environmental education is to educate citizens who can address environmental issues facing humanity that are yet unforeseen and bring about social transformation. Such an orientation to teaching and learning should enhance the scientific and citizenship skills amongst learners, as well as critical political literacy with respect to local and global inequalities [29].

There is no shortage of issues students may choose to address. Students are constantly surrounded by sustainability-related phenomena, although they sometimes need to be primed to notice how they are related to science and sustainability. Sustainability issues in their community can include low biodiversity around their homes, food insecurity, high prevalence of a debilitating health condition, or plastic waste in the vicinity of their local community. If you simply ask students how their community could be improved, then they will have a plethora of ideas; the educator's role is to facilitate fitting their ideas into the grand scheme of community development/enhancement. Finally, and most powerfully, educators can work to support students in designing and implementing a solution to the issues identified. This crucial step makes the difference between fostering awareness and preparing students to reimagine and reconstruct the world around them. Solutions can start small; nevertheless, they may require the participation of various stakeholders and community associations. Bringing these individuals and institutions together results in strengthened community partnerships, emboldened students, and long-lasting relationships toward community revitalization. While students may not remember every lesson throughout their educational journey, the emotional impressions and skills gained through their place-based action-oriented projects will leave a lasting legacy [73].

9. Conclusion

Colonial legacies permeate science education around the world [70], including in sub-Saharan Africa [87], North and South America [89], the Caribbean [40], and Australia. Science educators can tap into the suppressed cultures, identities, knowledge, and perspectives. Indeed, colonial legacies insidiously shape all communities, with implications for indicators of well-being, such as health, happiness, and education [68]. Where educators serve homogeneous and/or privileged communities, they still have the responsibility to help students deconstruct the messages they have received about sustainability. For example, students can explore the origin of the resources they use daily, such as textiles or food. Often, they will find developed nations continuing colonial practices of exploitation, to the detriment of poorer nations, in the same destructive and self-indulgent fashion criticized by activists in the 1940's [40, 68]. It is up to educators to assist students in reinventing the world, to trust them as they meaningfully contribute to a greener planet, and to follow their lead in ending these self-perpetuating cycles [88].

Youth today are being affected by the choices of politicians, policymakers, and corporate leaders. Youth are calling on each of those entities and educators to make substantive changes towards sustainable practices. We must act, and we must equip youth so they can act responsibly as well. The issues framing the United Nations Climate Change Conference (COP 28) in Dubai, United Arab Emirates, from 30 November to 12 December 2023, are the very issues youth of today will need to act upon in the future, such as global climate action, industry decarbonization and net-zero emissions, energy transition, food, nature, and innovative finance (see Ref. [90]).

Though the challenges of the future loom large and difficult to predict, the youth of today will be responsible for shaping the future. Educators must have the courage and humility to embrace this challenge in collaboration with our communities. Though it exacerbated educational inequities, COVID-19 offered us a glimpse into the future with reduced fossil fuel use and a greener future (see Refs. [91–93]). With the reduction in the use of fossil fuels, *vis-a-vis* reduced industry and vehicles on the road; clear, smog-free skies appeared over much of the planet. The global shutdown revealed a greener future is possible! As McDonald notes [94],

> *Green technology such as wind and solar are already less costly than fossil fuels with other forms following suit. It is also one of the fastest developing sectors of the economy, with tremendous potential for growth as current products improve their performance and new products come online. That means jobs, and more jobs, and a stronger economy...A new green age is upon us ([94], p. 13).*

The future truly is now as the youth of today have the opportunity to co-construct a new Green Age.

Conflict of interest

The authors declare no conflict of interest.

Author details

William C. Kyle, Jr.* and Emilia A. Belciak
College of Education, University of Missouri - St. Louis, St. Louis, MO, USA

*Address all correspondence to: bill_kyle@umsl.edu

References

[1] United Nations. Transforming our world: The 2030 agenda for sustainable development [internet]. Transforming our World: The 2030 Agenda for Sustainable Development. United Nations General Assembly. 2015. Available from: https://documents-dds-ny.un.org/doc/UNDOC/GEN/N15/291/89/PDF/N1529189.pdf?OpenElement [Accessed: July 20, 2023]

[2] Glenn J, Gordon T, Florescu E. 2009 State of the Future. Washington, DC: The Millennium Project; 2009

[3] Commonwealth Secretariat. Global Youth Development Index and Report 2016: Commonwealth Secretariat. London, UK: The Commonwealth; 2016

[4] UNFPA. State of World Population - the Power of 1.8 Billion: Adolescents, Youth and the Transformation of the Future. New York: United Nations Population Fund; 2014

[5] World Meteorological Organization. July 2023 is Set to be the Hottest Month on Record [Internet]. public.wmo.int. 2023. Available from: https://public.wmo.int/en/media/press-release/july-2023-set-be-hottest-month-record

[6] United Nations. The World Youth Report on Youth and the 2030 Agenda for Sustainable Development. New York: United Nations; 2018

[7] Kyle WC Jr. Expanding our views of science education to address sustainable development, empowerment, and social transformation. Disciplinary and Interdisciplinary Science Education Research. 2020;**2**:2. DOI: 10.1186/s43031-019-0018-5

[8] Fang WT, Hassan A, LePage BA. Philosophy and history of environmental education. Sustainable Development Goals Series. 2022:25-48

[9] McCormick J. Reclaiming Paradise. Bloomington, IN: Indiana University Press; 1991

[10] Vogt W. Road to Survival. New York: William Sloan & Associate; 1948

[11] Office of the Historian, Foreign Service Institute United States Department of State. Foreign Relations of the United States, 1952-1954, General: Economic and Political Matters, Volume I, Part 2 - Office of the Historian [Internet]. history.state.gov. Available from: https://history.state.gov/historicaldocuments/frus1952-54v01p2/d16

[12] Brinkley D. Silent Spring Revolution: John F. Kennedy, Rachel Carson, Lyndon Johnson, Richard Nixon and the Great Environmental Awakening. New York: Harper Collins; 2022

[13] Jacobson MZ. Air Pollution and Global Warming: History, Science and Solutions. Cambridge, UK: Cambridge University Press; 2012

[14] Osborn F. Our Plundered Planet. Boston, MA: Little, Brown and Company; 1968

[15] Leopold A. A Sand County Almanac. New York: Oxford University Press US; 2020

[16] Douglas W. Of Men and Mountains. New York: Harper; 1950

[17] Odum E. Fundamentals of Ecology. Philadelphia: W. B. Saunders; 1953

[18] Carson R. Silent Spring. New York: Houghton Mifflin; 1962

[19] Langsdorf S, Löschke S, Möller V, Okem A, Rama B, Belling D, et al. Climate Change 2022: Impacts, Adaptation and Vulnerability Working Group II Contribution to the Sixth Assessment Report of the Intergovernmental Panel on Climate Change. Geneva, Switzerland: IPCC [Internet]; 2022. Available from: https://www.ipcc.ch/report/ar6/ wg2/downloads/report/IPCC_AR6_ WGII_FullReport.pdf

[20] Saha CK, Carter AV. Phase-out or lock-in fossil fuels? Least developed countries' burning dilemma. The Extractive Industries and Society. 2022;**11**:101140

[21] Environment UN. Emissions Gap Report 2022 [Internet]. UNEP - UN Environment Programme. 2022. Available from: https://www.unep.org/emissions-gap-report-2022

[22] United Nations. Decision/CP.27 Sharm el-Sheikh Implementation Plan [Internet]. 2022. Available from: https://unfccc.int/documents/624444 [Accessed: August 5, 2023]

[23] Cozzi L., Chen O., Kim H. The World's Top 1% of Emitters Produce over 1000 times more CO2 than the Bottom 1% – Analysis [Internet]. International Energy Agency 2023. Available from: https://www.iea.org/commentaries/the-world-s-top-1-of-emitters-produce-over-1000-times-more-co2-than-the-bottom-1

[24] Hickel J. Quantifying national responsibility for climate breakdown: An equality-based attribution approach for carbon dioxide emissions in excess of the planetary boundary. The Lancet Planetary Health [Internet]. 2020;**4**(9):399-404. Available from: https://www.thelancet.com/journals/lanplh/article/PIIS2542-5196(20)30196-0/fulltext

[25] Svarstad H. Critical climate education: Studying climate justice in time and space. International Studies in Sociology of Education. 2021;**30**(1-2):214-232

[26] Wals A, Corcoran P. Re-orienting, re-connecting, and re-imagining: Learning based responses to the challenge of (un)sustainability. In: Learning for Sustainability in Times of Accelerating Change. The Netherlands: Wageningen Academic Pub; 2012. pp. 19-32

[27] UNESCO. Youth Demands for Quality Climate Change Education [Internet]. 2023. Available from: https://unesdoc.unesco.org/ark:/48223/pf0000383615

[28] BBC. School strike for climate: Protests staged around the world. BBC News [Internet]. 2019. Available from: https://www.bbc.com/news/world-48392551

[29] Kyle WC Jr. Youth are demanding action regarding climate Change: Will educators have the wisdom and courage to respond? APEduc Journal [Internet]. 2020;**1**(1):150-160. Available from: https://apeducrevista.utad.pt/index.php/apeduc/article/view/49/11

[30] Rachel R. Standardized Tests: Effects on Science Education and Diversity in Science [Internet] [Honors Thesis]. [University of South Dakota]. 2021. Available from: https://red.library.usd.edu/honors-thesis/153

[31] Anderson KJB. Science education and test-based accountability: Reviewing their relationship and exploring implications for future policy. Science Education. 2011;**96**(1):104-129

[32] Fontaine C. The myth of accountability: How data (mis)

use is reinforcing the problems of public education. SSRN Electronic Journal. 2016:13. Available from: https://papers.ssrn.com/sol3/papers.cfm?abstract_id=3126460

[33] National Research Council. A Framework for K-12 Science Education: Practices, Crosscutting Concepts, and Core Ideas [Internet]. Washington, D.C.: National Academies Press; 2012. Available from: https://www.nap.edu/read/13165/chapter/12

[34] Diamond JB, Spillane JP. High-stakes accountability in urban elementary schools: Challenging or reproducing inequality? Teachers College Record. 2004;**106**(6):1145-1176

[35] Fry R, Kennedy B, Funk C. STEM Jobs See Uneven Progress in Increasing Gender, Racial and Ethnic Diversity Higher Education Pipeline Suggests Long Path Ahead for Increasing Diversity, Especially in Fields like Computing and Engineering [Internet]. Vol. 1. Unite States: Pew Research Center; 2021. p. 2021. Available from: https://www.pewresearch.org/science/wp-content/uploads/sites/16/2021/03/PS_2021.04.01_diversity-in-STEM_REPORT.pdf

[36] Majee U, Ress S. Colonial legacies in internationalisation of higher education: Racial justice and geopolitical redress in South Africa and Brazil. Compare: A Journal of Comparative and International Education. 2018;**50**(4):463-481

[37] UNESCO International Centre for Technical and Vocational Education and Training. Boosting Gender Equality in Science and Technology: A Challenge for TVET Programmes and Careers [Internet]. Unesco.org. 2023. Available from: https://unesdoc.unesco.org/ark:/48223/pf0000374888 [Accessed: August 5, 2023]

[38] Zeichner N. Racial Equity in Education: Reflections on South Africa. Education Week [Internet]. 2018. Available from: https://www.edweek.org/teaching-learning/opinion-racial-equity-in-education-reflections-on-south-africa/2018/12

[39] Hong L, Page SE. Groups of diverse problem solvers can outperform groups of high-ability problem solvers. Proceedings of the National Academy of Sciences [Internet]. 2004;**101**(46):16385-16389. Available from: https://sites.lsa.umich.edu/scottepage/wp-content/uploads/sites/344/2015/11/pnas.pdf

[40] Ferdinand M. Decolonial Ecology: Thinking from the Caribbean World. Cambridge: Polity Press; 2022

[41] Altunbas Y, Gambacorta L, Reghezza A, Velliscig G. Does gender diversity in the workplace mitigate climate Change? SSRN Electronic Journal. 2022:45. Available from: https://papers.ssrn.com/sol3/papers.cfm?abstract_id=4046329

[42] Tanner KD. Structure matters: Twenty-one teaching strategies to promote student engagement and cultivate classroom equity. CBE—Life Sciences Education [Internet]. 2013;**12**(3):322-331. Available from: https://www.lifescied.org/doi/full/10.1187/cbe.13-06-0115

[43] Apple MW. Can Education Change Society? New York: Routledge; 2012

[44] Au W. Unequal by Design: High-Stakes Testing and the Standardization of Inequality. New York: Routledge; 2009

[45] Pietro D., Biagi G. The Likely Impact of COVID-19 on Education: Reflections based on the Existing Literature and Recent International Datasets. JRC Technical Report [Internet]. 2020. Available from: https://publications.

jrc.ec.europa.eu/repository/bitstream/JRC121071/jrc121071.pdf

[46] Stracke CM, Burgos D, Santos-Hermosa G, Bozkurt A, Sharma RC, Swiatek CC, et al. Responding to the initial challenge of the COVID-19 pandemic: Analysis of international responses and impact in school and higher education. Sustainability [Internet]. 2022;**14**(3):1876. Available from: https://www.mdpi.com/1487372

[47] Siani A, Green I. Scientific misinformation and mistrust of COVID-19 preventive measures among the UK population: A pilot study. Vaccine. 2023;**11**(2):301

[48] UNESCO. Education in a Post-COVID World: Nine Ideas for Public Action International Commission on the Futures of Education [Internet]. Paris, France: UNESCO; 2020. Available from: https://en.unesco.org/sites/default/files/education_in_a_post-covid_world-nine_ideas_for_public_action.pdf

[49] Haileamlak A. Pandemics will be more frequent. Ethiopian Journal of Health Sciences [Internet]. 2022;**32**(2):228. Available from: https://www.ncbi.nlm.nih.gov/pmc/articles/PMC9175207/ [Accessed: August 5, 2023]

[50] Cairns R. Exams tested by Covid-19: An opportunity to rethink standardized senior secondary examinations. Prospects. 2020;**51**:331-345

[51] Rodrigues M. Learn with meaning(s), act locally, educate for global emergency. APEduC Revista/APEduC Journal. 2020;**1**(1):168-174

[52] Huish R. Global citizenship amid COVID-19: Why climate change and a pandemic spell the end of international experiential learning. Canadian Journal of Development Studies/Revue Canadienne d'études du Développement. 2021;**42**(4):1-18

[53] Misiaszek G. Educating the Global Environmental Citizen: Understanding Ecopedagogy in Local and Global Contexts. Routledge: Taylor and Francis; 2012

[54] Stevenson R, Brody M, Dillon J, Wals A. International Handbook of Research on Environmental Education. New York: Routledge; 2013

[55] Freire P. Pedagogy of the Oppressed [Internet]. New York, NY: The Continuum International Publishing Group Inc.; 2005. Available from: https://envs.ucsc.edu/internships/internship-readings/freire-pedagogy-of-the-oppressed.pdf

[56] hooks, b. Teaching to Transgress [Internet]. NewYork, NY: Routledge Taylor & Francis Group; 1994. Available from: https://academictrap.files.wordpress.com/2015/03/bell-hooks-teaching-to-transgress.pdf

[57] Saiti A, Kyle WC Jr, Sinnes A, Nampota D, Kazima M. Developing Relevant Environmental Education in a Rural Community in Malawi. Revista Brasileira de Pesquisa em Educação em Ciências [Internet]. 2014;**14**(2):185-198. Available from: https://periodicos.ufmg.br/index.php/rbpec/article/view/4360 [Accessed: August 5, 2023]

[58] Zandvliet D. Chapter 20 enhancing environmental learning through controversy. In: Learning for Sustainability in Times of Accelerating Change. The Netherlands: Wageningen Academic Publishers; 2012. pp. 317-326

[59] Arora-Jonsson S. The sustainable development goals: A universalist promise for the future. Futures. 2023;**146**:1-12. Available from:

https://www.sciencedirect.com/science/article/pii/S0016328722001872#bibliog0005

[60] Backhaus T. Acknowledging that science is political is a prerequisite for science-based policy. Integrated Environmental Assessment and Management. 2019;**15**(3):310-311

[61] Sun YH. Why teach science with an interdisciplinary approach: History, trends, and conceptual frameworks. Journal of Education and Learning. 2017;**6**(4):66

[62] Tippins DJ, Mueller MP, Eijck MW, van, Adams J. D. Cultural Studies and Environmentalism: The Confluence of EcoJustice, Place-Based (Science) Education, and Indigenous Knowledge Systems. Dordrecht: Oslo, Norway: Springer; 2010

[63] Odegaard M. The Drama of Science Education [Dissertation]. Oslo, Norway: University of Oslo; 2001

[64] Kyle WC Jr. Science education in developing countries: Access, equity, and ethical responsibility. Journal of the Southern African Association for Research in Mathematics and Science Education. 1999;**3**(1):1-13

[65] Kyle WC Jr. The road from Rio to Johannesburg: Where are the footpaths to/from science education? International Journal of Science and Mathematics Education. 2006;**4**(1):1-18

[66] Jakins P, Sobel J. Preface. In: Cultural Studies and Environmentalism: The Confluence of EcoJustice, Place-Based (Science) Education, and Indigenous Knowledge Systems. Dordrecht: Springer; 2010

[67] Bleier M. Beyond neat classifications: A case for the In-Betweens. Cultural Studies of Science Education. 2023;**18**(2):483-497

[68] Bonneuil C, Fressoz J-B. The Shock of the Anthropocene. Brooklyn, NY: Verso Books; 2016

[69] UNFPA. In our words: Voices of Women of African Descent for Reproductive and Climate Justice [Internet]. United Nations Population Fund. 2022. Available from: https://www.unfpa.org/publications/our-words-voices-women-african-descent-reproductive-and-climate-justice [Accessed: August 5, 2023]

[70] UN. State of the World's Indigenous Peoples (SOWIP). United Nations for Indigenous Peoples [Internet]. www.un.org. 2021. Available from: https://www.un.org/development/desa/indigenouspeoples/publications/state-of-the-worlds-indigenous-peoples.html

[71] Hickman C, Marks E, Pihkala P, Clayton S, Lewandowski E, Mayall E, et al. Climate anxiety in children and young people and their beliefs about government responses to climate Change: A global survey. The Lancet Planetary Health [Internet]. 2021;**5**(12):863-873. Available from: https://www.thelancet.com/journals/lanplh/article/PIIS2542-5196(21)00278-3/fulltext

[72] Kagire L. Global Citizen Prize Winner on Steering the Environment Protection ship, Getting Young People Aboard [Internet]. The New Times. 2023. Available from: https://www.newtimes.co.rw/article/7781/news/rwanda/global-citizen-prize-winner-on-steering-the-environment-protection-ship-getting-young-people-aboard#:~:text=Eco%2Dfeminist%20Ineza%20Umuhoza%20Grace [Accessed: August 8, 2023]

[73] Ponder J, Cox-Peterson A. Action-based science instruction: Service learning, stewardship and civic involvement. In: Cultural Studies and Environmentalism: The Confluence of Ecojustice, Place-Based (Science) Education, and Indigenous Knowledge Systems. Dordrecht: Springer; 2010. pp. 287-302

[74] UN. Transforming Our world: the 2030 Agenda for Sustainable Development. Resolution A/Res/70/1. [Internet]. UN. 2015. Available from: https://sdgs.un.org/2030agenda

[75] European Parliament. Texts Adopted - Climate and Environmental Emergency - Thursday, 28 November 2019 [Internet]. www.europarl.europa.eu. 2019. Available from: https://www.europarl.europa.eu/doceo/document/TA-9-2019-0078_EN.html

[76] Cordero E., Centeno D., Todd A. M. The role of climate change education on individual lifetime carbon emissions. Pausata F, editor. PLoS One [Internet]. 2020;15(2):e0206266. Available from: https://journals.plos.org/plosone/article?id=10.1371/journal.pone.0206266

[77] Sinnes AT, Kyle WC Jr, Alant B, Kazima M, Nampota D, Onwu GOM. Editorial. African Journal of Research in Mathematics, Science and Technology Education. 2011;**15**(3):1-4

[78] Onwu G, Kyle WC Jr. Increasing the socio-cultural relevance of science education for sustainable development. African Journal of Research in Mathematics, Science and Technology Education. 2011;**15**(3):5-26

[79] Narayanasamy N. Participatory Rural Appraisal: Principles, Methods and Application. New Delhi: Sage; 2009

[80] Schabort FA. Can Science Education Be Empowering to Girls in Rural South Africa? Contextualizing Science Education through Action Research [Doctoral Thesis]. Ås Norway: Department of Mathematical Sciences and Technology Norwegian University of Life Sciences; 2013

[81] Kapp D. Girls Who Green the World: Thirty-Four Rebel Women out to Save our Planet. New York: Delacorte Press; 2022

[82] Martinez X. We rise: The Earth Guardians Guide to Building a Movement that Rescues the Planet. Rodale; 2017

[83] Méndez M. Climate Change from the Streets: How Conflict and Collaboration Strengthen the Environmental Justice Movement. New Haven, CT: Yale University Press; 2020

[84] S S, E B. Implications of sense of place-based education for ecological integrity and cultural sustainability in diverse places. In: Cultural Studies and Environmentalism: The Confluence of Ecojustice, Place-Based (Science) Education, and Indigenous Knowledge Systems. Dordrecht, Netherlands: Springer; 2010. pp. 287-302

[85] Callaghan CT, Mesaglio T, Ascher JS, Brooks TM, Cabras AA, Chandler M, et al. The benefits of contributing to the citizen science platform iNaturalist as an identifier. PLoS Biology. 2022;**20**(11):2-5

[86] Hann C. Evaluation of the Marine Debris Tracker App on the U.S. West Coast [Internet]. West Coast Marine Debris Alliance. 2017. Available from: https://pub-data.diver.orr.noaa.gov/marine-debris/pacificnorthwest/20170330%20Hann.C_MDT%20App%20Report_FINAL.pdf [Accessed: August 6, 2023]

[87] Shizha E. Reclaiming our indigenous voices: The problem with

postcolonial sub-Saharan African school curriculum [internet]. Journal of indigenous Social Development. 2013;**2**:1-18. Available from: https://scholarspace.manoa.hawaii.edu/server/api/core/bitstreams/24a9b689-1a2a-4f41-88b2-c41c748c465f/content

[88] Delgado M, Staples L. Youth-Led Community Organizing: Theory and Action. New York: Oxford University Press; 2008

[89] Alkholy S, Gendron F, McKenna B, Dahms T, Ferreira M. Convergence of indigenous science and Western science impacts students' interest in STEM and identity as a scientist. Ubiquitous Learning: An International Journal. 2017;**10**(1):1-13

[90] United Nations Climate Change Conference (COP 28). Event highlights. Available from: https://www.weforum.org/events/united-nations-climate-change-conference-cop28-2023/

[91] Kyle W, Jr C. The way forward: Science is imperative for human survival. APEduc Journal [Internet]. 2020;**1**(1):179-184 Available from: https://apeducrevista.utad.pt/index.php/apeduc/article/view/79/15

[92] NASA Earth Observatory. (2020). Airborne Nitrogen Dioxide Plummets Over China.. Avaialble from: https://earthobservatory.nasa.gov/images/146362/airborne-nitrogen-dioxide-plummets-over-china

[93] NASA Earth Observatory. (2020). How to Find and Visualize Nitrogen Dioxide Satellite Data: Find Research Quality Nitrogen Dioxide Data to Assess Regional Air Quality and Potential Trends in Industrial Activity. Available from: https://www.earthdata.nasa.gov/learn/articles/find-no2-data

[94] McDonald B. The Future Is Now: Solving the Climate Crisis with Today's Technologies. New York: Penguin Random House; 2022

Chapter 2

Introduction to the STEM Student Success Model

Leander Kwabena Brown

Abstract

This chapter is based on a literature review to gain insight into the critical components of a STEM Student Success Model for underrepresented minority (URM) undergraduate (UG) science, technology, engineering, and mathematics (STEM) students since many articles reviewed focused on the equity, inclusion, and diversity of URM STEM students who are most likely to need help in their studies. The salient components of the model are learning with peers, interactions with faculty, STEM employability skills, and a supportive campus environment. The success of URM STEM students is necessary to address the preparation of URM graduates for the STEM workforce which in turn, will help address the workforce shortage in STEM fields if the United States is to maintain a competitive advantage in STEM disciplines.

Keywords: STEM education, underrepresented minority student, learning with peers, interactions with faculty, STEM employability skills, overcoming barriers to academic success, supportive campus environment

1. Introduction

The United States of America (USA) is a technologically advanced nation and a world leader in STEM because of its many research universities, fundamental technologies, and connections between science, technology, and business. However, China is poised to become the world leader in STEM graduates and competes with the USA for technological supremacy which has triggered a national STEM crisis [1, 2]. The crisis results from the demands of the workforce being greater than the supply of STEM graduates in the right STEM fields at any given time. The shortage has resulted in a need and an opportunity for unrepresented minority STEM students to help bridge the gap through higher education institutions. To aid in these goals, a STEM Student Success Model with the most relevant components for success is proposed.

STEM education is an approach to teaching and learning that integrates the content and skills of STEM and other disciplines using an interdisciplinary and applied approach that focuses on the development of STEM skills. STEM education facilitates economic development, international competitiveness, and job creation [3]. The U.S. Department of Homeland Security [4] maintains the STEM-designated degree program list of classifications of instructional programs (CIPs) that fall within the regulatory definition of STEM fields. See the Appendix for a listing of the two-digit primary and additional CIP codes.

IntechOpen

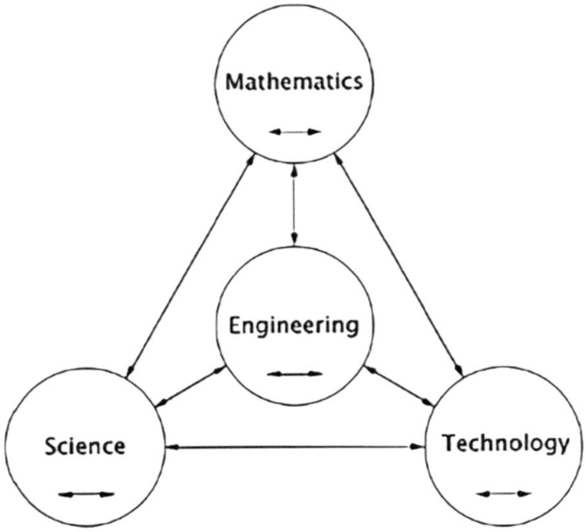

Figure 1.
STEM translation model. Note. The STEM translation model shows the connections within and between the four main STEM disciplines based on the original model [5].

During their first two UG years, STEM students must complete 100- and 200-level gateway STEM classes before advancing to the upperclassman level. In the meantime, STEM students regardless of their major must learn how to translate between STEM disciplines which are inherently interdisciplinary. The STEM Translation Model in **Figure 1** depicts how STEM students must translate within and between STEM disciplines [5]. Also, to help in the translation, STEM students engage in convergent cognition when they take classes that complement each other such as using an algebra function with a programmatic function which demonstrates the interdisciplinary nature of STEM disciplines [6]. However, those students who are underprepared for the challenge or are unable to progress in their gateway classes may either switch to non-STEM majors or drop out of college [7]. Higher education institutions (HEIs) can provide UG STEM students with access to support activities with peers, faculty, and other stakeholders so they can enroll, persist, and graduate with STEM degrees timely [8].

This study seeks to address the single research question "What critical elements make up a STEM Student Success Model?". Since the model is exploratory, there are no hypotheses or propositions rather the National Survey on Student Engagement [9] and other reliable sources will be used to address the research question.

2. Components of the STEM student success model

2.1 Learning with peers

Learning with peers in the STEM community includes those academic activities in which STEM students can participate and/or lead a student group with faculty oversight and training. It is characterized by small groups of STEM students learning together with a peer student leader or facilitator. They include such activities as, peer tutoring, peer supplemental instruction, peer cooperative learning, peer-led team learning, and related small group learning activities.

2.1.1 Peer tutoring

While learning with peers comes in many forms, one of the most well-known forms is tutoring in one-on-one sessions or in group sessions led by peers but faculty or professionals can tutor students. Peer tutoring programs offered to all students in a course have been widely found to help STEM students in introductory gateway STEM courses [10], especially in large course courses with hundreds of students and with limited time for students to receive individualized instruction. Students are able to get the tutoring support they need to be an academic success. Struggling students who regularly attended peer tutoring increased exam performance, increased knowledge of biology concepts, and increased course persistence relative to their struggling peers who were not attending the peer tutoring sessions.

Peer tutoring is student-centered where the students seek help and set up appointments with their tutors in one-on-one or group sessions. A qualitative study of 10 tutors and 10 tutees in seven high risk STEM courses over one semester was performed at a university in Malaysia [11] while a quantitative analysis of 213 participants using new model, the Weekly Tutoring Group (WTG) at the University of Rhode Island was available on multiple days and times throughout the week [12]. While the studies both focused on peer tutoring, the first study was in two focus groups and participant observations but actual tutoring sessions were one tutor with three to nine tutees while the second study had groups of one peer leader of two to six students who were former supplemental instruction (SI) participants. Participants in WTG and tutoring groups saw an increase in regular attendance and a significant difference in proficiency and in actual grades, particularly when the students attended seven or more sessions. The tutoring sessions in both studies were similar to supplemental instruction (SI) but different because SI targets specific high-risk courses while tutors assist high risk students who seek help [11, 12].

2.1.2 Peer supplemental instruction

Supplemental instruction (SI) was designed by Deanna Martin of the University of Missouri-Kansas City in 1974. SI is designed to foster learning and improve the academic performance of low-achieving students but focuses on identifying high-risk courses, such as those with significant grades of D, F, Withdrawal, or Incomplete (DFWI) or other students who traditionally are less prepared for difficult STEM courses. A major problem for STEM disciplines is students who achieve grades of C- or lower and are unable to progress in their STEM field either switch majors or drop out of college [13].

Supplemental instruction is characterized by increased passage rates in gateway courses, higher course grades, early engagement, collaborative learning, small group problem solving, and study skills with sessions facilitated by a trained peer group leader. As attendance in SI sessions increased so did positive student outcomes [13–15]. A study on the importance of SI at San Francisco State University showed that it benefits to STEM students who choose to make use of it and URM students who traditionally are less prepared for difficult STEM courses. While the impact of SI in introductory STEM courses has seen increased passage rates, the impact in upper-level courses was associated with higher proportions of STEM students obtaining As and Bs. Also, SI users performed better than non-users and were more likely to take subsequent STEM courses while URM SI students had significantly increased passage rates [14]. An SI study at California State University, San Marcos was conducted of STEM students enrolled in four biology courses and participated in online SI. They received increases in academic performance similar to traditional SI participants, had higher course grades, and had lower fail rates as compared to students who did not participate in either form of SI [15].

A study of peer SI (PSI) was conducted at Georgia Gwinnett College, a public college in Lawrenceville, Georgia. PSI incorporates two high-impact practices in improving STEM student education and retention that include early engagement and exposure to college survival skills and collaborative learning so that students can discover an array of study styles and perspectives while learning and problem-solving in groups. Model student leaders with faculty training and oversight facilitate PSI sessions, such as addressing STEM skills, metacognitive skills, and STEM course concepts as shown in **Figure 2** [14].

As PSI attendance for students in STEM classes at Georgia Gwinnett College increased to 4 or more sessions per term so did course averages and exam grades [14].

2.1.3 Peer cooperative learning

Peer cooperative learning is a type of group learning where students work together on class projects or other academic activities toward a common goal and each member sets and accomplishes his/her learning goal(s) concurrent with other members and students may work in groups on assignments in the classroom. Two studies were reviewed: one was a mixed methods study of an introductory algebra course with 20 final participants of 30 enrolled students at a community college in the southwestern United States [16] while the second one was a quantitative

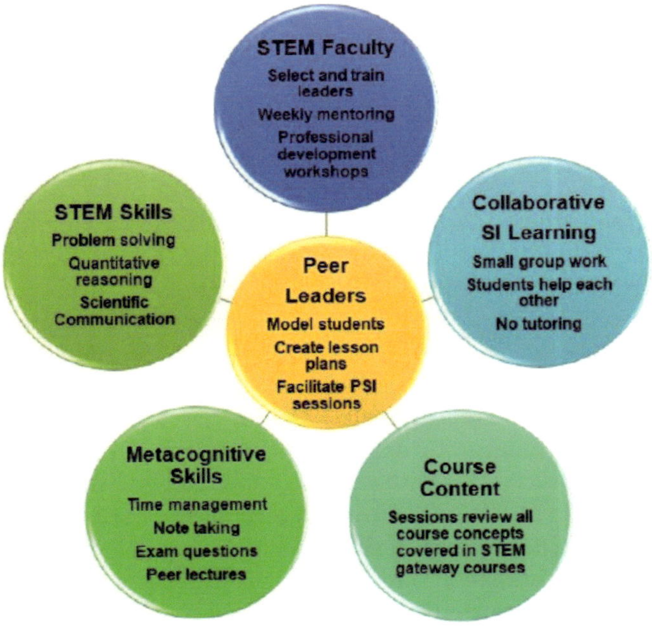

Figure 2.
The STEM peer supplemental instruction model. Note. The STEM peer supplemental instruction model shows peer student leaders at the center of PSI activities while STEM faculty and students play an active role in STEM student success and is based on an adaption of the supplemental instruction model [14].

analysis of 456 students before implementation and 552 students after implementation across five STEM departments of peer cooperative learning techniques at Bridgewater State University [17]. Cooperative learning sessions resulted in an increase in student attendance, a decrease in student withdrawals, an increase in student motivation to work with peers in and outside of class sessions, and an increase in retention rates [16, 17].

2.1.4 Peer led team learning

In peer-led team learning (PLTL), students attend workshops that are led by a trained peer facilitator outside of class. The PLTL workshops are aligned with class topics in the curriculum as coordinated by the instructor or an educational specialist. Three quantitative studies were reviewed: Both studies were of PLTL and non-PLTL students in gateway biology courses. The first study was conducted at a private university in eastern Puerto Rico [18] and the two other studies were conducted at Syracuse University in New York [18, 19]. The researchers in each study concluded that the PLTL model contributes to the academic success of PLTL students and encourages student engagement and critical thinking resulting in academic achievement, retention, and graduation. However, students noted that they needed more than one PLTL weekly workshop [18, 19] which supports other peer-led learning studies in which sessions were more frequent than weekly or students attended multiple sessions of the same PTLT course.

2.1.5 Other interactions with peers

Students can learn from interacting with peers with different viewpoints or who come from different backgrounds. In a NSSE 2022 survey [9], among first year students, 47% frequently had discussions with people with different political views, 47% frequently had discussions with people from a different economic background, and 59% frequently had discussions with people from a different race or ethnicity. This is significant because it contributes to inclusion and diversity in first-year students at the university sampled but no information was provided on senior students.

2.2 Interactions with faculty

STEM faculty can provide positive faculty support to students including intellectual challenge and stimulation, opportunities to discuss coursework outside of class and feedback about academic work, advice about educational programs and graduate study, and other activities. STEM student interactions with faculty can prove valuable in such activities as course goals and requirements, research projects, career plans, and STEM exam wrappers. As noted in **Figure 2**, faculty is involved with STEM students and leaders beyond the classroom. Specifically, as part of faculty oversight of peer supplemental instruction faculty select and train peer leaders, provide weekly mentoring, and conduct professional development workshops. There is a connection between faculty mentorship and research such that STEM students are able to establish a science identity to enroll, persist, and graduate timely. In STEM exam wrappers, after exams, students assess how they did in problem areas so that instructors will know any problems and respond to these evaluations so that it helps the students on the next exams during the course [20, 21].

2.3 STEM employability skills

There is a distinct gap in skill levels between the competencies gained in STEM programs and what employers are seeking in new graduates as confirmed by human resource professionals [22]. To close the skills gap, several studies noted that employers should communicate their employability skill requirements to the HEI including soft skills, and the HEI should embed employability skills and career-preparedness into the STEM curricula [23–25]. However, there are many other ways students can be prepared for the workforce. As noted in the prior learning with peers section, students learn how to develop problem-solving skills, analytical abilities, and soft skills during the group sessions. Likewise, students can seek career-related campus employment, internships, and research work with faculty in STEM laboratories.

2.4 Supportive campus environment

A supportive campus environment with a STEM focus is one that is designed to improve the retention, persistence, graduation, and overall academic success of STEM students and prepares them for STEM careers [26]. STEM students have access to a STEM-focused supportive campus environment which includes a wide variety of functions, services, and assistance. Common interventions available to all students include tutoring, mentoring, counseling, academic advising, academic coaching, social interaction, scholarships, internships, and career services to boost successful student outcomes [27–29]. STEM students can engage in co-curricular activities to supplement their STEM learning experiences outside the classroom. Co-curricular activities include participation in clubs, organizations, and associations. Positive academic and social experiences in this supportive environment may lead to STEM students persisting and graduating [30].

2.4.1 STEM intervention programs

STEM Intervention Programs (SIPs) have emerged on college campuses in the United States to foster, support, and sustain the interest of students in STEM, broaden the participation of URM students in STEM fields, enrich STEM student experiences, and respond to external influences and opportunities [31]. SIP studies tended to be conducted at large research universities, directed across multiple SIPs, and focused on URM students but did not address long term challenges such as continued funding.

Institutions offer SIPs that are tailored to meet the specific needs of targeted, diverse recipients resulting in a wide variety of program designs, purposes, and services for STEM students. The Program for Excellence in Education and Research in the Sciences is an academic support program at the University of California, Los Angeles established in 2003. It was designed to support first- and second year URM STEM students with an emphasis on research engagement, mentoring, STEM persistence, and entry into doctoral programs. While the study was of 141 STEM students in 2009 and 2010 with an average time to graduation of 4.2 years and found a correlation between student success and UG research [32] it was not specifically noted that any sessions were peer-led but has proven successful for STEM students.

DOI: http://dx.doi.org/10.5772/intechopen.112614

2.4.2 STEM learning community

A STEM learning community (SLC) is a type of SIP in which STEM students join a learning community to accomplish their goals with a group goal and share learning experiences or activities in and outside of classes and are often organized around a common theme. Students connect with peers, faculty, administrators, and industry, conduct research, and participate in community service events. SLCs may include a residential component called a living-learning community. HEIs use learning communities (LCs) to promote the academic and social integration of entering students, especially within STEM majors so that entering STEM freshmen have a support base for their college transition and beyond. A study showed that by adding friendships to the same biological sciences freshman cohort it led to a support base for STEM students but had a negative impact when students were segmented based on their academic records [33]. In a second study, a case study of 119 student narratives of an SLC found that the psychosocial or learning factors for students identified were improved as a result of their participation in a STEM learning community [34]. The studies found that the SLCs resulted in friendship support and other support, such that the participants' grades and GPAs increased.

2.4.3 Co-curricular activities

Although not directly tied to the curriculum, STEM students can engage in co-curricular activities to supplement their STEM learning experiences outside the classroom. Co-curricular activities include participation in clubs, organizations, associations, student government, community service, committee membership, career exploration, mentorship, and research. Integrating research and teaching exposes students to the process of knowledge creation through mentored or faculty led UG research initiatives. These activities give the STEM students the full complement of UG research experience to enter the STEM workforce [35].

2.5 Removing academic barriers

There are barriers to STEM education that need to be addressed as part of the strategic plan. The removal of institutional academic barriers for undergraduate STEM students may begin as early as secondary school where prerequisite science and math courses may be offered. The potential barriers include math proficiency, course load, online classes and testing, technically complex materials, and equity and diversity of URM students. In removing these barriers, HEIs seek favorable student outcomes, especially for increased retention and graduation rates of STEM students [36]. In removing academic barriers, the student seeks help or is provided help through tutoring, supplemental instruction, cooperative learning, peer led telelearning, and through many other ways as noted above. There is evidence that systematically pairing a core STEM subject with another complementary subject may lead to greater overall learning in both subjects which is called convergent cognition. An example is when a core algebra class (algebra function) is paired with an introductory computer science class (programmatic function) such that the student uses algebra in writing an algorithm for a computer program which demonstrates the interdisciplinary nature of STEM disciplines [6] as shown in **Figure 3**.

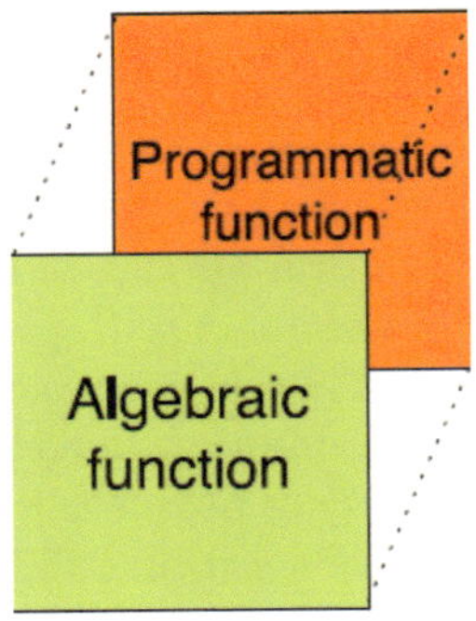

Figure 3.
Convergent cognition arising from complementary functions. Note. Convergent cognition is shown as an example when a STEM student pairs a core algebra function with a programming function so that they complement each other and the student's knowledge is enhanced in both classes [6].

2.6 Underrepresented minority STEM students

There is a category of STEM students for which academic barriers need to be removed as shown by disparities in STEM retention and graduation rates between URM and non-URM students.

URM students who identify as African-American/Black, American Indian/Alaska Native, or Hispanic/Latino made up 33.2% of the U.S. population but represented only 22% of STEM undergraduate degrees and 9% of doctoral STEM degrees in 2016. The 6-year graduation rate for URM STEM majors was 33.8% as compared to 53.1% for non-URM STEM majors [37]. Consequently, URM students must navigate the challenges that disproportionately affect them so ultimately their representation in STEM fields will be equal to or better than their percentage of the USA population.

Early intervention efforts, particularly in the areas of mathematics through summer bridge and related peer learning programs help ensure URM students are adequately prepared to succeed in STEM fields. Likewise, network-based mentoring allows students to receive both professional and peer mentoring services as a mentor and a mentee that can alleviate barriers to success among URM groups in STEM fields including faculty mentoring programs. Also, HEIs can track successes and failures at the institutional level and collect data to help explain existing trends so there can be a stronger focus on removing institutional barriers. Finally, HEIs can capitalize on known successes, recognize the need for accountability, and facilitate future progress for increased persistence and graduation of URM STEM students [37–40].

2.7 The STEM student success model

The STEM Student Success Model is an introductory model for STEM students and other stakeholders to use in determining what are the most likely critical components based on the NSSE and other reliable sources for STEM students to enroll, persist, and graduate timely. The STEM Student Success Model outlines in pictorial form the basic overlapping critical elements for which URM STEM students may enroll, persist, and graduate timely. The most salient stakeholders that support the model are peer students, faculty, and the institution. The main elements are listed in **Figure 4** and detailed as follows:

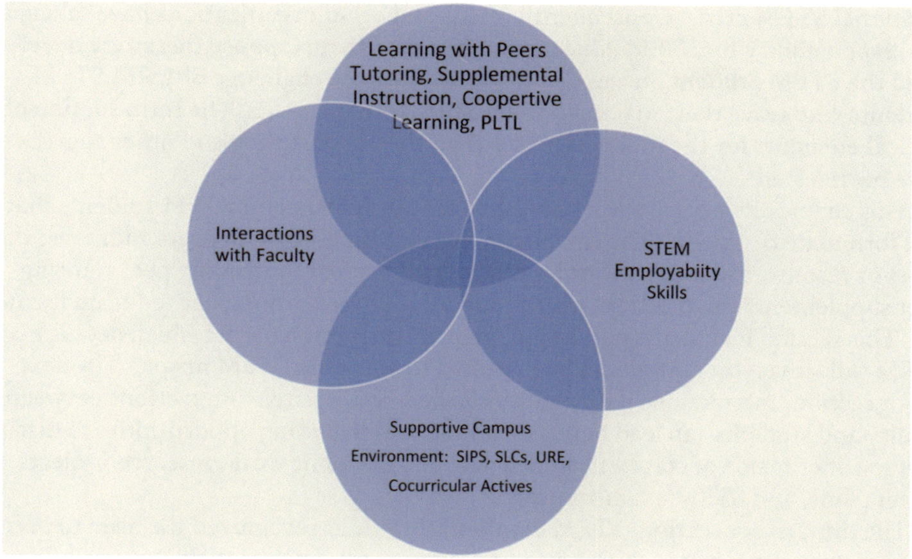

Figure 4.
The STEM student success model. Note. The STEM student success model is based on major elements of the NSSE survey and other sources.

- Learning with peers is significant because many programs identified peers as leaders or participants in tutoring, supplemental instruction, cooperative learning, and peer-led team learning. The STEM Peer Supplemental Instruction Model as adopted presented STEM concepts covered in PSI sessions that include problem-solving, analytical skills/quantitative reasoning, metacognition, course content, and collaborative SI learning.

- Interactions with faculty as a positive base is important in and outside of the classroom, including faculty-led research and peer leader training and oversight.

- STEM employability skills are critical as these skills prepare students for the workforce with potential employer input and should be embedded into the curriculum whenever possible. Students can gain practical experience through research and internships.

- A supportive campus environment is critical as it offers STEM intervention programs, STEM learning communities, undergraduate research experience, and co-curricular activities to supplement their academic careers.

The model depicts all four main sections but users would also need to access the details for more information on how STEM students were successful in their studies.

3. Conclusion

The United States is a technologically advanced nation and a world leader in STEM, but it competes with China for technological supremacy which has triggered a national STEM crisis. Consequently, HEIs and other organizations have taken on the responsibility for STEM education. As part of that response this study developed the STEM Student Success Model to address the challenge of URM STEM students who leave their majors in the first two critical years. The introduction of critical elements for the model was based on the NSSE survey and other sources that have been tested.

This chapter identified the most likely critical elements for STEM students that will formulate the model. The first major section, learning with peers addressed the types of resources that STEM students may have access to including peer tutoring, peer supplemental instruction, peer cooperative learning, and peer-led team learning. This section included a peer supplemental instruction model which details major STEM skills that students would be expected to master in STEM majors. The next major section, interactions with faculty showed how positive interactions between faculty and students can lead to positive outcomes including opportunities to discuss coursework outside of class and feedback about academic work, research projects, career plans, and STEM exam wrappers.

The third major section, STEM employability skills recognized the need to prepare students for the workforce by having employers communicate their employability skill requirements to the HEI, participating in internships and research, and embedding employability skills and career-preparedness into the STEM curricula. The last section, a supportive campus environment addressed what the HEI can offer to students including STEM intervention programs, STEM learning communities, and co-curricular activities. Also, the subsection on URM STEM students addressed disparities between URM and non-URM students and how the disparities can be overcome through early intervention programs, mathematics skills development, mentoring network usage, and data tracking to identify what works for URM students. The model depicts all four main sections but users would also need to access the details for more information on how STEM students were successful in their studies.

Acknowledgements

The author would like to thank all my family members who encouraged me to pursue a writing career.

Conflict of interest

The author declares no conflicts of interest.

Notes/thanks/other declarations

The author makes no other declarations.

Nomenclature

The most important nomenclature regarding undergraduate STEM education in an unusual or highly specific way are:

Twenty-first-century skills. Skills or core competencies developed to help STEM students keep pace with the evolving modern world include digital learning, critical thinking, problem-solving, and many other useful skills that STEM students may be required to demonstrate [41].

Active learning. An approach to instruction that involves actively engaging students with course material through discussions, problem solving, case studies, role plays, and other methods also known as student centered learning [42].

Gateway STEM courses. A set of introductory STEM courses typically taken in the first 2 years of college at the 100- and 200-level course sequences in such courses as biology, chemistry, physics, mathematics, engineering, and related disciplines. While successfully passing gateway courses does not guarantee degree completion in the sciences, previous research has identified these courses as among the greatest obstacles [43].

Persistence and retention rates. The persistence rate is a measure of the percentage of students who return to college at *any* institution for their second year while the retention rate represents the percentage of students who return to *the same* institution [44] but can also apply to an upper-class cohort.

Professional development. Any effort to influence and advance faculty attitudes, practices, and skills in academia including both student and faculty development and educational development. It is a critical component of ongoing work to improve student learning outcomes in higher education through positive faculty and student interactions, especially STEM education [45].

Science identity. The establishment of the student's interest and consistent work in science or STEM discipline as a researcher and future scientist and enroll, persist, and graduate timely with a STEM degree.

Underrepresented minority. A group whose percentage of the population in a group is lower than their percentage of the population in the country, such as in STEM disciplines. In the USA, the group may have been denied access and/or suffered past institutional discrimination. According to the Census and other federal measuring tools, URM groups include African Americans, Asian Americans, Hispanics or Chicanos/Latinos, and Native Americans.

Acronyms and abbreviations

CIPs	Classification of Instructional Programs
DFWI	D, F, Withdrawal, Incomplete
DHS	Department of Homeland Security
HEI	Higher Education Institution
NSSE	National Survey of Student Engagement
PLTL	Peer-Led Team Learning
PSI	Peer Supplemental Instruction
STEM	Science, Technology, Engineering, and Mathematics
SIP	STEM Intervention Program
SLC	STEM Learning Community
SI	Supplemental Instruction
UG	Undergraduate
URM	Underrepresented Minority
USA	United States of America
WTG	Weekly Tutoring Group

A. Overview of U.S. Department of Homeland Security STEM Designated Degree Program List

The U.S. Department of Homeland Security (DHS) STEM Designated Degree Program List is a complete list of fields of study that DHS considers to be STEM fields of study under *8 CFR 214.2(f)*. A *STEM field of study* is one that is included in the Department of Education's Classification of Instructional Programs (CIPs) taxonomy. STEM-related fields include fields involving research, innovation, or development of new technologies in STEM disciplines under *8 CFR 214.2(f)(10)(ii)(C)(2),* The four primary two-digit CIP codes are:

- Engineering (14).
- Biological and biomedical sciences (26).
- Mathematics and statistics (27).
- Physical sciences (40).

There are 18 additional STEM fields of study in the program list with two-digit CIP codes as follows:

- Agriculture, agriculture operations, and related sciences (01).
- Natural resources conservation (03).
- Architecture and related services (04).
- Communications, journalism, and related programs (09).
- Communications technologies/technicians and support services (10).
- Computer and information sciences and support services (11)
- Education (13).
- Engineering technologies and engineering-related fields (15).
- Military science, leadership, and operational art (28).
- Military technologies and applied sciences (29).
- Multi/interdisciplinary studies (30).
- Science technologies/technicians (41).
- Psychology (42).
- Homeland security, law enforcement, firefighting, and related protective services (43).

- Social sciences (45).
- Transportation and materials moving (49).
- Health professions and related programs (51).
- Business, management, marketing, and related support services (52).

The 2-digit codes are further divided into 6-digit codes that are available at https://www.ice.gov/sites/default/files/documents/stem-list.pdf which contains a complete STEM-designated degree program listing.

Author details

Leander Kwabena Brown
Washington DC Metro Alumni Chapter, Mississippi Valley State University National Alumni Association, Riverdale, MD, United States

*Address all correspondence to: leanderb32@gmail.com

References

[1] Dollar D. The US-China tech rivalry shapes the economic relationship [Internet]. In: Dollar & Sense Podcast. Washington, DC: The Brookings Institution; 2020. Available from: https://www.brookings.edu/podcast-episode/the-us-china-tech-rivalry-shapes-the-economic-relationship/

[2] Herman A. America's STEM crisis threatens our national security. American Affairs Journal. 2019;**III**(1):127-148

[3] University of Education, Winneba. FSE to Promote STEM Education in Ghana [Internet]. 2019. Available from: https://uew.edu.gh/news/fse-promote-stem-education-ghana

[4] U.S. Department of Homeland Security. 2020 STEM Designated Degree Program List [Internet]. 2021. Available from: https://www.ice.gov/sites/default/files/documents/stem-list.pdf

[5] Glancy AW, Moore TJ. Theoretical Foundations for Effective STEM Learning Environments. West Lafayette, Indiana: Purdue University, Purdue e-Pubs, School of Engineering Education Working Papers, Paper 1. 2013

[6] Rich PJ, Leatham KR, Wright GA. Convergent cognition. Instructional Science. 2012;**41**(2):431-453. DOI: 10.1007/s11251-012-9240-7

[7] Langie G, Pinxten M. The transition to STEM higher education: Policy recommendation – Conclusions of the readySTEMgo-project. International Journal of Engineering Pedagogy. 2018;**8**(10):10-13. DOI: 10.3991/ijep.v8i2.8286

[8] Sithole A, Chiyaka ET, McCarthy P, Mupinga DM, Bucklein BK, Kibirige J. Student attraction, persistence and retention in STEM programs: Successes and continuing challenges. Higher Education Studies. 2017;7(1):46-59. DOI: 10.5539/hes.v7n1p46

[9] National Survey on Student Engagement. A Pocket Guide to Choosing a College: NSSE 2022 Answers from Students [Internet]. Available from: https://docs.udc.edu/irap/NSSE22-Pocket-Guide-Report-UDC.pdf

[10] Batz Z, Olsen BJ, Dumont J, Dastoor F, Smith MK. Helping struggling students in introductory biology: A peer-tutoring approach that improves performance, perception, and retention. Cbe—Life Sciences Education. 2015;**14**(2):1-12

[11] Chan YF, Sidhu GK, and Lee LF. Active learning in higher education: A case study. In: Tang SF & Logonnathan L, editors. Taylor's 7th Teaching and Learning Conference 2014 Proceedings. Singapore: Springer; 2015. pp. 519-526. DOI: 10.1007/978-981-287-399-6_47.

[12] Mendes SH, Fede JH, Wilks MB. Data-based program reform: A shift from supplemental instruction to weekly tutoring groups. Learning Assistance Review. 2017;**22**(2):75-96

[13] Peterfreund AR, Rath KA, Xenos SP, Bayliss F. The impact of supplemental instruction on students in STEM courses: Results from san Francisco State University. Journal of College Student Retention Research Theory and Practice. 2007;**9**(4):487-503

[14] Achat-Mendes C, Anfuso C, Johnson C, Awong-Taylor J. Learning, leaders, and STEM skills: Adaptation of the supplemental instruction model to improve STEM success and build transferable skills

in undergraduate courses and beyond. Journal of STEM Education: Innovations and Research. 2019;**20**(2):14-23

[15] Hizer SE, Schultz PW, Bray R. Supplemental instruction online: As effective as the traditional face-to-face model? Journal of Science Education and Technology. 2017;**26**(1):100-115. DOI: 10.1007/s10956-016-9655-z

[16] Rivera N. Cooperative Learning in a Community College Setting: Developmental Coursework in Mathematics [doctoral dissertation]. Tempe, Arizona: Arizona State University; 2013

[17] Salomone M, Kling T. Required peer-cooperative learning improves retention of STEM majors. International Journal of STEM Education. 2017;**4**(1):1-19. DOI: 10.1186/s40594-017-0082-3

[18] Dreyfuss AE, Fraiman A. Peer-led team learning: An active learning method for the 21st century. In: Paper Presented at Proceedings of ICERI2015 Conference. Seville, Spain: International Academy of Technology, Education, and Development; 2015. pp. 0698-0706

[19] Snyder JJ, Carter BE, Wiles JR. Implementation of the peer-led team-learning instructional model as a stopgap measure improves student achievement for students opting out of laboratory. CBE Life Sciences Education. 2015;**14**(2):1-6. DOI: 10.1187/cbe.13-08-0168

[20] Hodges LC, Beall LC, Anderson EC, Carpenter TS, Cui L, Feeser E, et al. Effect of exam wrappers on student achievement in multiple, large STEM courses. Journal of College Science Teaching. 2020;**50**(1):69-79

[21] Kim YK, Sax LJ. The effect of positive faculty support on mathematical self-concept for male and female students in STEM majors. Research in Higher Education: Journal of the Association for Institutional Research. 2018;**59**(8):1074-1104. DOI: 10.1007/s11162-018-9500-8

[22] McGunagle D, Zizka L. Employability skills for 21st century STEM students: The employers' perspective. Higher Education, Skills and Work-based Learning. 2020;**10**(3):591-606. DOI: 10.1108/HESWBL-10-2019-0148

[23] Clinkscales-Verdell A, Keith JM, Warnock J, White VM. Best practices for underrepresented minority students in an engineering summer bridge program. In: Paper Presented at ASEE 123rd Annual Conference and Exposition. New Orleans, LA: American Society for Engineering Education. 2016 Paper ID #17173

[24] Turner BD. STEM employability skills: Employer and faculty expectations and perceptions of skill competencies in STEM graduates [doctoral dissertation]. New Castle, Delaware: Wilmington University; 2019

[25] University of Houston. Scholar Enrichment Program [Internet]. Available from: https://www.uh.edu/nsm/scholarenrichment/about/

[26] Florida State University. Student Support Services-STEM [Internet]. Available from: https://care.fsu.edu/college-programs/student-support-services-stem

[27] Martin JP, Choe NH, Halter J, Foster M, Froyd J, Borrego M, et al. Interventions supporting baccalaureate achievement of Latinx STEM students matriculating at 2-year institutions: A systematic review. Journal of Research in Science Teaching. 2019;**56**(4):440-464. DOI: 10.1002/tea.21485

[28] Rincon BE, George-Jackson CE. STEM intervention programs: Funding

practices and challenges. Studies in Higher Education. 2016;**41**(3):429-444. DOI: 10.1080/03075079.2014.927845

[29] Estrada M. Ingredients for Improving the Culture of STEM Degree Attainment with Co-curricular Supports for Underrepresented Minority Students [Commissioned Paper]. 2014. Available from: https://sites.nationalacademies.org/cs/groups/dbassesite/documents/webpage/dbasse_088832.pdf

[30] Dika SL, D'Amico MM. Early experiences and integration in the persistence of first-generation college students in STEM and non-STEM majors. Journal of Research in Science Teaching. 2016;**53**(3):368-383. DOI: 10.1002/tea.21301

[31] George CE, Castro EL, Rincon B. Investigating the origins of STEM intervention programs: An isomorphic analysis. Studies in Higher Education. 2019;**44**(9):1645-1661. DOI: doi 10.1080/03075079. 2018.1458224

[32] Sellami N, Toven-Lindsey B, Levis-Fitzgerald M, Barber PH, Hasson T. A unique and scalable model for increasing research engagement, STEM persistence, and entry into doctoral programs. Cbe—Life Sciences Education. 2021;**20**(1):1-11

[33] Jeffrey W, Schaefer DR, Xu D, McPartlan P, Solanki S. STEM learning communities promote friendships but risk academic segmentation. Scientific Reports. 2022;**12**(1):1-9. DOI: 10.1038/s41598-022-15,575-y

[34] Carrino SS, Gerace WJ. Why STEM learning communities work: The development of psychosocial learning factors through social interaction. Learning Communities Research and Practice. 2016;**4**(1):1-22

[35] El-Haggar N, Mezhoudi N, Alrawjih F. The impact of co-curricular activities supported by generic skills on students' performance at university level. International Journal of Innovative Technology and Exploring Engineering. 2019;**8**(12):981-986. DOI: 10.35940/ijitee.K1290.1081219

[36] De S, Arguello G. STEM education in college: An analysis of stakeholders' recent challenges and potential solutions. Florida Distance Learning Association Journal. 2020;**5**(9):1-13

[37] Markle RS, Williams TM, Williams KS, deGravelles KH, Bagayoko D, Warner IM. Supporting historically underrepresented groups in STEM higher education: The promise of structured mentoring networks. Frontiers in Education. 2022;**7**:1-14. DOI: 10.3389/feduc.2022.674669

[38] Ghazzawi D, Pattison D, Horn C. Persistence of underrepresented minorities in STEM fields: Are summer bridge programs sufficient? Frontiers in Education. 2021;**6**:1-12. DOI: 10.3389/feduc.2021.630529

[39] Whitcomb KM, Singh C. Underrepresented minority students receive lower grades and have higher rates of attrition across STEM disciplines: A sign of inequity? International Journal of Science Education. 2021;**43**(7):1054-1089. DOI: 10.1080/09500693.2021.1900623

[40] Estrada M, Burnett M, Campbell AG, Campbell PB, Denetclaw WF, Gutiérrez Carlos G, et al. Improving underrepresented minority student persistence in STEM. Cbe Life Sciences Education. 2016;**15**(3):1-10. DOI: 10.1187/cbe.16-01-0038

[41] Singh M. Acquisition of 21st century skills through STEAM education. Academia Letters. 2021:1-7. DOI: 10.20935/AL712

[42] Freeman S, Eddy SL, McDonough M, Smith MK, Okoroafor N, Jordt H, et al. Active learning increases student performance in science, engineering, and mathematics. Proceedings of the National Academy of Sciences. 10 June 2014;**111**(23):8410-8415. Available from: https://www.pnas.org/content/111/23/8410

[43] Ferrare JJ. A multi-institutional analysis of instructional beliefs and practices in gateway courses to the sciences. CBE - Life Sciences Education. 2019;**18**(2):1-16. DOI: 10.1187/cbe.17-12-0257

[44] National Student Clearinghouse Research Center. Term enrollment estimates fall 2020. 2020. Available from: https://nscresearchcenter.org/wp-content/uploads/CTEE_Report_Fall_2020.pdf

[45] Emery N, Maher JM, Ebert-May D. Studying professional development as part of the complex ecosystem of STEM higher education. Innovative Higher Education. 2019;**44**(6):469-479. DOI: 10.1007/s10755-019-09475-9

Chapter 3

Integrating STEM: An Interdisciplinary Approach to PreK-12 Education

Teresa J. Kennedy and Michael R.L. Odell

Abstract

STEM education employs an interdisciplinary pedagogical approach that merges the subjects of science, technology, engineering, and mathematics into traditional disciplines covered across all levels within primary and secondary school curricula. STEM topics span the United Nation's Sustainable Development Goals (SDGs) and naturally weave STEM disciplines throughout, focusing on the importance of developing a global STEM-literate workforce capable of responding to the worldwide challenges presented today and into the future. STEM-focused frameworks provide students with the knowledge and skills they need to be successful in the twenty first century and beyond, preparing them for the rapidly changing technological and scientific landscapes. This chapter explores STEM as a meta-discipline, documenting the origins and emergence of STEM education in the United States, provides definitions of STEM literacy and includes primary and secondary implementation models.

Keywords: science, technology, engineering, mathematics, STEM, STEM education, interdisciplinary frameworks, interdisciplinarity, twenty first century skills, primary STEM pedagogy, secondary STEM pedagogy, sustainable development goals (SDGs)

1. Introduction

The origins of STEM Education in the United States can be traced back to the launch of Sputnik by the Soviet Union in 1957. There was concern in the Western World that national security required investing in science and mathematics education to maintain safety and remain competitive economically. Over time, concerns about STEM competitiveness have increasingly focused more heavily on economic competitiveness and technological innovation in a global economy. This trend can be seen in reports such as *A Nation at Risk* [1], which served as a reaction to the U.S. perception that it was losing its competitive edge as new economic powers were rising in Asia and Europe. Additionally, the release of the results of the *Trends in International Mathematics and Science Study* (TIMMS) beginning in 1995 showed that U.S. students were not achieving at high levels when compared to students in other developed nations. These same achievement trends had also been presented in the Program for International Student Assessment (PISA) results since 1997, inspiring the 2007 National Academy of Sciences report *Rising above the gathering storm: Energizing and employing America for a brighter economic future* (RAGS), and its sequel, *Rising Above the Gathering Storm Revisited: Approaching Category 5* (RAGSR), in 2010 which once again sounded the alarm concerning U.S. student achievement in relation to other countries internationally [2, 3]. However, compared to earlier alarmist reports, the RAGS report provided suggestions for educational interventions that were practical at scale. These recommendations, referred to as *actions,* included the following:

1.1 Action A-1: annually recruit 10,000 science and mathematics teachers by awarding 4-year scholarships and thereby educating 10 million minds

The U.S. federal government and individual state governments have invested heavily in STEM education since the 1990s. Several initiatives, such as the National Science Foundation Robert Noyce Teacher Scholarship Program initiated in 2002, provided college students with significant scholarships if they sought a teaching credential and joined the K-12 teaching force. Private initiatives such as 100Kin10, referring to training 100,000 new Math and Science Teachers in 10 years, from 2011 to 2021, provided funding and support through partnerships to prepare new science and mathematics teachers. The U.S. Department of Education and almost every U.S. state also provided student loan relief for college graduates who chose to teach STEM disciplines in "high needs schools" (public schools located in areas where 30% or more of the student population comes from families with incomes below the poverty line). Additionally, large university networks, such as the UTeach Program, created teacher pathways that allow college students to earn a 4-year degree and a teaching certificate concurrently, supporting the preparation of preservice science teachers [4]. UTeach is currently implemented in 55 universities across the United States and has prepared over 8,300 new STEM teachers since 1997 [5]. While there is still a critical shortage of STEM K-12 teachers in the U.S., progress is being made.

1.2 Action A-2: strengthen the skills of 250,000 teachers through training and education programs at summer institutes, in master's programs, and in advanced placement (AP) and international baccalaureate (IB) training programs

Since the 1960s, there has been significant investment in teacher professional learning, with a significant focus on mathematics beginning in the 1990s. The U.S. Department of Education provides funds to all states for teacher professional development to support effective instruction under a section of the U.S. education law within the *Every Student Succeeds Act* (ESSA, Title II-A). In addition, teachers are required to engage in professional learning as a requirement of recertification. Many states also reimburse teachers who enroll and complete AP and IB training programs to make advanced coursework available to their students since high school teachers who seek to teach dual credit courses must earn a master's degree with 18 credit hours in the STEM discipline to qualify as a university instructor.

In the 2000s, the emergence of dual credit/dual enrollment opportunities allowed secondary students to enroll in college level courses during their high school years. According to the Texas Higher Education Coordinating Board [6], roughly 12,000 high school students enrolled in dual credit courses in the state of Texas alone in 1999. In 2020, that number exceeded 370,000 high school students, and during the 2021–2022 academic year, over 300,000 Texas students enrolled in AP coursework. Additionally, THECB documented 181 International Baccalaureate (IB) Schools in

Texas. This trend is not unique to Texas. Since the 1990s, there has been an increase in access to advanced coursework in the U.S., although access is not evenly distributed.

1.3 Action A-3: enlarge the pipeline of students who are prepared to enter college and graduate with a degree in science, engineering, or mathematics by increasing the number of students who pass AP and IB science and mathematics courses

Two recommendations specifically described how to engage students in STEM in a manner that would prepare them to enter college and graduate with a STEM degree. The first of these called for states to develop *Statewide specialty high schools*, based on the rationale that specialty secondary education could foster leaders in science, technology, and mathematics since they immerse students in high-quality STEM education, serve as a mechanism to test teaching materials, and also provide a training ground for teachers. School models that were developed to address Action A-3 included Residential State STEM Academies, Schools within a School, STEM Magnet Schools, Early College High Schools, and University Affiliated Laboratory Schools. **Table 1** provides a description of each model as well as an example school in the state of Texas, USA.

The five models highlighted prepare students to transition to a STEM degree program of their choice upon completion. These models showcased K-12 or secondary schools collaboratively designed through the guidance of their partnering institution of higher education or educational organization.

K-12 laboratory schools have a long history in the United States as well as worldwide. Just as their educational policies and practices can differ, so do their admission and enrollment criteria. While some schools follow requirements or restrictions that determine whether students are eligible to attend, others are "open enrollment" schools, allowing students to enroll on a first-come, first-serve basis without strict admission requirements, regardless of the student's location or district. Typically, a lottery system, a random selection process that determines which students can enroll, is implemented when more students apply than space allows. For example, the University of Texas at Tyler University Academy is an open enrollment public school focused on K-12 pathways leading to engineering and/or biomedical sciences. Similar to the Early College High School (ECHS) model, students can earn college credits while enrolled in high school, but also have the advantage of being located on a university campus which allows for participation in unique research opportunities. The University Academy also serves as a research and clinical teaching platform to provide authentic STEM teaching and learning experiences.

STEM Pathway programs provide innovative school improvement science models [7]. Research has shown that students, who attend schools focusing on STEM or pursue a STEM Pathway while enrolled in a traditional school, have a higher likelihood to pursue a STEM degree in college [8]. For context, the most recent graduation class of the University of Texas at Tyler University Academy, 100% of graduates are pursuing postsecondary education opportunities with 87% of graduates enrolling in a STEM degree program. Although STEM schools can have a positive impact on college enrollment in STEM fields, it should be noted that schools are not the only factor influencing students' decisions. Personal interests, family support, socioeconomic factors, and individual motivations also play a role in determining whether students pursue STEM degrees in college.

The RAGS report recommended that middle and high school students engage in inquiry-based learning including laboratory and real-world experiences and research.

K-12 STEM School Models in the United States	
K-12 School Model	**Description and Example School**
Residential State STEM Academies	K-12 schools seeking to bring the best and brightest students from across a state to study STEM in a residential boarding school setting. Students enroll in college degree programs and receive additional supports and experiences as they pursue their selected STEM pathway. Prominent RSSA program in Texas: The Texas Academy of Mathematics and Science at the University of North Texas (Denton, Texas). https://tams.unt.edu/
Schools within a School	K-12 schools offering multiple specialty programs in diverse areas while sharing common courses and electives that are operated under the larger umbrella of the school. Students are provided with diverse opportunities that cater to their differing learning needs and interests within a single facility. Prominent SWS program in Texas: School of Science and Engineering (SEM) at Townview Magnet Center (Dallas, Texas). https://www.dallasisd.org/semagnet
STEM Magnet Schools	Secondary schools operating within an existing school district as schools of choice, focusing on a specific area such as STEM, Fine Arts, or Leadership, or a combination of areas. Additionally, all subjects are taught through the lens of the school focus. Prominent SMS program in Texas: Liberal Arts and Science Academy (Austin, Texas). https://www.austinisd.org/schools/lasa
Early College High Schools	Secondary schools partnering with an institution of higher education to enable students to earn college credits concurrently while completing their high school course requirements. In some cases, students earn an associate degree from a community college or complete up to 2 years toward a STEM degree from the partner university. Prominent ECHS program in Texas: Houston Community College Associate of Science in Engineering Science (ASES) Program (Houston, Texas). https://www.hccs.edu/programs/areas-of-study/science-technology-engineering--math/engineering/ut-tyler-engineering/course-sequence-chart/
University Affiliated Laboratory Schools	K-12 schools partnering with an institution of higher education to provide a unique learning environment for students within a structured environment for experimenting with curriculum designs, educational technologies, and new teaching methods. Laboratory Schools have the added mission of preparing new teachers. While there are currently over 100 U.S. lab schools, only a few focus exclusively on STEM. Prominent UALS in Texas: University of Texas at Tyler University Academy (Campuses in Tyler, Longview and Palestine, Texas). https://www.uttua.org/

Table 1.
K-12 U.S. school models developed to address RAGS report action A-3.

However, many programs recognize the importance of including younger students, as can be seen in **Figure 1**. Research has shown that early childhood settings (including pre-primary school experiences) support STEM learning since "young children are capable of engaging in, at developmentally appropriate levels, the scientific practices that high school students carry out" ([10], p. 2; [11], p. 16). This makes sense since STEM capabilities begin in the initial early years with the development of manipulative skills, learning about safety and following safety rules and instructions, recognizing a problem, planning and identifying the necessary items to solve the problem, realizing

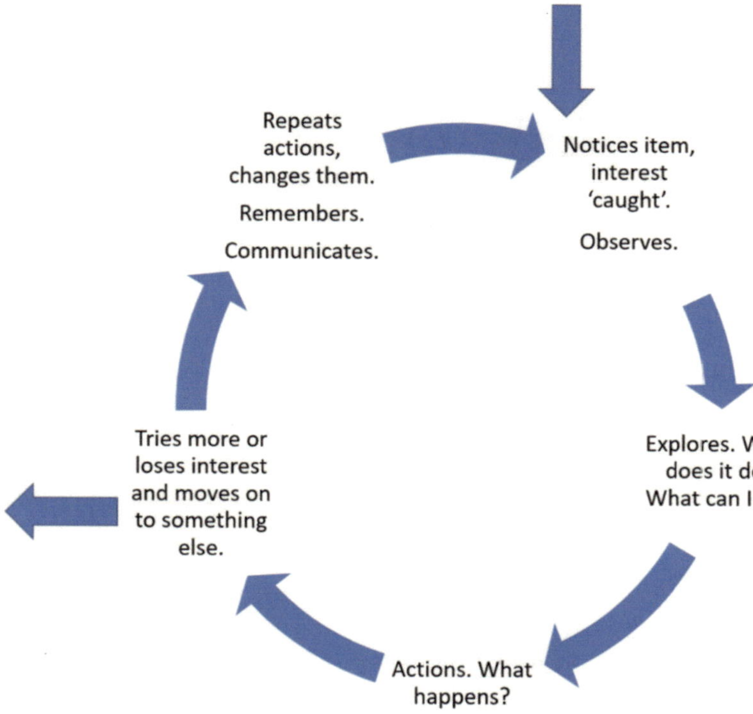

Figure 1.
The STEM play cycle [9].

a design, evaluating outcomes, and changing paths; a process that celebrates successful discovery and allows for redirection to the next problem encountered [12, 13].

Young children constantly attempt to solve problems they encounter in their environment, regardless of whether they are engaged in their own discoveries, involved in formal education activities (occurring during academic schooling), or informal education activities (occurring outside of the classroom; at a playground, camp, museum, etc.). **Figure 1** shows the STEM Play Cycle, promoting creativity as children observe, question, explore, investigate, and construct meaning through problem-solving and applying previous experiences. The cycle is continuous until the child moves on to a different variation of inquiry or losses interest and moves on to something completely new.

According to the National Science Teaching Association ([14], p. 1), "learning science and engineering practices in the early years can foster children's curiosity and enjoyment in exploring the world around them and lay the foundation for a progression of science learning in K-12 settings and throughout their entire lives." STEM education naturally brings the concepts of investigation and design together through all four disciplines it subsumes since scientific inquiry involves the formulation of a question that could be answered through *investigation*, and engineering design involves the formulation of a problem that could be solved through *design* ([15], p. 247).

2. K-12 STEM education and twenty first century skills

The National Education Association of the United States developed the *Framework for 21st Century Learning* in 2002 through a collaboration of educators, education experts, and business leaders, with the goal of defining the skill set needed for success in work, life, and citizenship [16]. The framework consisted of 18 skills, which were further refined into four learning and innovation skills in 2012. Referred to as the Four C's, these skills include critical thinking, communication, collaboration, and creativity [17].

The Four C's are highlighted in many twenty first century skills (21CS) policy documents from around the globe, often including social and emotional intelligence, technological literacy, and problem-solving skills [18]. Education systems, both formal and informal, play a critical role in fostering development of 21CS. Joynes et al. [19] cited a broad range of literature discussing 21CSs, concluding that "there is evidence of general agreement across commentators on the need for new forms of learning to tackle global challenges" (p. 40). The challenge then becomes combining *hard skills* (scientific content) with *soft skills* (effective verbal written communication, career readiness, emotional I.Q., collaboration, creativity, work ethic) to build greater connections between the two. Within the context of STEM, essential skills include rigorous core content (biology, chemistry, Earth sciences, engineering, mathematics, physics, and technology) combined with critical thinking skills [20]. Curriculum that focuses on cognition of core constructs within the framework of 21CS provides opportunities for students to apply critical thinking and problem-solving in perspective with real-world scenarios.

Specific 21CS connections with STEM education include:

- Critical Thinking and Decision Making: Students identify and analyze complex problems, break them down into manageable parts, and develop innovative solutions through problem-solving.

- Collaboration and Communication: Through different variations of project-based learning (PBL), students work in teams to tackle real-world challenges. This collaborative approach facilitates the development of strong communication skills, helps students learn to work effectively with others, and promotes appreciation of diverse perspectives.

- Creativity and Innovation: STEM fields naturally foster a culture of creativity as students design new technologies, conduct experiments, or invent novel solutions.

3. STEM education and connections to UNESCO's education goals

To this point, this chapter has provided examples from the United States. As presented, the U.S. has responded to several critical reports about the status of STEM Education. There is another series of reports from the Intergovernmental Panel on Climate Change (IPCC) that has many in the U.S. thinking more globally. These reports have sounded the alarm about climate change and its impact on all facets of life. Beyond testing comparison such as PISA, what is the role of STEM Education now that there is a crisis that impacts everyone regardless of nationality? The United Nations Educational, Scientific and Cultural Organization (UNESCO) aims to address this crisis through STEM education, which is closely related to the education goals set by UNESCO, as outlined in its Education 2030 agenda. The goals emphasize the importance of quality education for all, lifelong learning opportunities, and the promotion of sustainable development. STEM education aligns with several key aspects of UNESCO's education goals.

The Sustainable Development Goals (SDGs), established by the United Nations in 2015 in support of the 2030 Agenda for Sustainable Development, are a set of independent yet interconnected goals created with the twenty first century skills framework in mind. **Figure 2** lists the 17 SDGs that weave STEM disciplines throughout with the collective goal of ending poverty, protecting the health of our planet, and providing equitable educational opportunities to ensure that, by 2030, all members of civil society can engage in prosperous and fulfilling lives.

Quality Education (SDG 4) is dedicated to ensuring inclusive and equitable quality education for everyone on a global scale. STEM education plays a significant role in achieving this goal by providing students with the knowledge and skills needed for success in our changing world. SDG 4 specifically employs 21CS through the development of critical thinking, problem-solving, and analytical skills, which are essential for lifelong learning and personal development. Additionally, one of UNESCO's key priorities is the promotion of gender equality in education. STEM fields have historically been male-dominated [21], and UNESCO aims to address this gender gap by empowering girls and women to pursue STEM education and careers (SDG 5) as well as competitively enter all areas of the global workforce. SDG 4 also emphasizes the importance of lifelong learning, from informal and formal education opportunities to continuous skill development and knowledge acquisition throughout one's lifetime. STEM education encourages a mindset of curiosity, questioning, inquiry, and adaptability. It promotes lifelong learning and active engagement in civil society.

In addition to education, the SDGs cover a wide range of global challenges, including poverty, inequality, climate change, life on land and below water, as well as environmental sustainability in all areas. STEM education plays a crucial role in addressing every one of these challenges. For example, STEM subjects provide the foundation for advancing healthcare and technology to improve people's lives (SDG 9), which address environmental issues and the development of clean energy solutions (SDG 7) and smart and sustainable urban solutions (SDG 11). STEM enables us to better understand and protect our ocean, which regulates our climate, generates

Figure 2.
Sustainable development goals. (Graphic used with permission by the division of science policy and capacity building, UNESCO natural sciences sector: https://en.unesco.org/sustainabledevelopmentgoals).

over half of the world's oxygen, and absorbs 50 times more carbon dioxide than our atmosphere. STEM education is vital for understanding and conserving marine and terrestrial ecosystems (SDGs 14 and 15), and it prepares students to better understand how to contribute to climate mitigation and adaptation efforts. Addressing climate change (SDG 13) requires a strong understanding of science and technology. STEM education prepares students to engage in climate-related research and the development of green technologies, as well as equips students with the knowledge and skills needed for environmental research, conservation, and sustainable resource management. From SDG 1 (ending poverty) to promoting interdisciplinary collaboration and partnerships (SDG 17), STEM education advances essential 21CS that cross all areas of civil society and equip students with the skills they need to drive future innovation in our rapidly evolving world, preparing the next generation to address global challenges and actively contribute to a more equitable and sustainable future.

4. Inquiry-based STEM teaching and learning

Over the past 20 years, STEM has been increasingly viewed as an integrated meta-discipline due to the natural relationships among science, technology, engineering, and mathematics. This shift removes the traditional barriers between individual STEM content areas and focuses on innovation and the applied process of designing solutions to complex contextual problems, challenging students to innovate and invent, while also promoting problem-solving and critical thinking skills that can be applied to their academic as well as everyday lives ([22], pp. 37–38). Although it is still common for STEM disciplines to be viewed and taught as distinct content areas, most of them acknowledge that they are both interconnected and interdependent. Schools that focus on STEM are adopting and implementing inquiry-based interdisciplinary approaches to engage students in authentic STEM learning that provides relevance in the real world.

As outlined in **Table 1**, specific STEM school models or approaches can vary by location, educational system, and student interest. Some STEM schools may focus on a particular STEM discipline, such as computer science or environmental science, while others take a more holistic approach to STEM education. However, the end goal of each of these school models is to prepare students for careers in STEM fields as well as create STEM-literate citizens by providing a strong foundation in science, technology, engineering, and mathematics.

According to Bybee [23], STEM education can take on various forms depending on the way STEM subjects are presented in the classroom. However, most STEM learning experiences do have one thing in common—they seek to provide students with opportunities to remove the artificial boundaries between STEM disciplines so that students can better understand the connected nature of knowledge using critical skills, leading to success in the twenty first century economy through applying the skills and knowledge that they have learned or are in the process of learning ([18], p. 479).

STEM specialty school models tend to engage students in inquiry-based learning such as the 5E Model of instruction. The five phases of the 5E Model are Engage, Explore, Explain, Elaborate, and Evaluate which provides "a 'common sense' value; it presents a natural process of learning" ([24], p. ix). **Figure 3** shows this process.

STEM specialty school models tend to implement different variations of project-based learning (PBL), problem-based learning (PrBL), and phenomena-based learning (PhBL) through implementation of the 5E Model. Summaries of each follow:

Figure 3.
Evidence-based practices: The 5E model of instruction [25]. (Graphic used with permission from the San Diego County Office of Education, 2018. https://ngss.Sdcoe.Net/evidence-based-practices/5E-model-of-instruction).

Project-based learning (PBL) is a student-centered instructional approach that focuses on students actively engaging in real-world, meaningful projects to gain knowledge and develop skills. PBL emphasizes hands-on, experiential learning, rather than passive learning through lectures and memorization. While PBL generally deemphasizes traditional lectures in favor of more active and collaborative learning, it does not necessarily exclude lectures altogether. Lectures, when included, are integrated into the overall PBL framework in a targeted and strategic manner. In this sense, lectures do not serve as the primary mode of content delivery and instead serve specific purposes such as providing an introduction to concepts prior to the beginning of the project; skill development and techniques needed to successfully complete the project; guidance and clarification on instructions and common challenges that arise during the project; reflection and debriefing after the completion of project phases; and when experts deliver a guest lecture to provide real-world perspectives or inspire students. In a PBL setting, students typically work on a project over an extended period, during which they explore and investigate a specific topic or problem resulting in a product that is presented publicly.

Key characteristics of project-based learning include:

1. Real-World Relevance: Students are presented with a problem or challenge and develop projects to address a real-world problem or question. This is designed to make the learning experience more meaningful and applicable to students' lives.

2. Inquiry: Students are encouraged to ask questions, conduct research, and explore topics collaboratively. This is designed to promote deeper learning and critical thinking.

3. Collaboration: PBL typically involves students working in collaborative groups. This approach mirrors real-life work environments where teamwork is required and provides the students with the opportunity to share ideas, engage in problem-solving, and learn from each other.

4. Ownership: PBL facilitates students taking ownership of their own learning by providing a level of voice in choice in the learning process. Students have a say in the direction of their project. Students utilize soft skills to collaborate, manage time, and meet their project goals.

5. Interdisciplinary: In many cases, PBL requires an interdisciplinary approach requiring students to connect and apply knowledge and skills from multiple disciplines.

6. Assessment: Assessment in PBL focuses on the quality of the final project. Students are typically provided with requirements and rubrics as a standard or aspirations that the final product must meet.

7. Reflection: In PBL, students are encouraged to reflect on their learning. Teachers provide students with checkpoints along the way so that students can gauge their progress. In some cases, traditional assessments are included so that students are prepared for high stakes tests.

8. Public Presentation: Students present their final product to their peers, teachers, and experts when practical, to help them develop critical presentation skills and enhance communication skills.

Project-based learning is seen as an effective way to foster critical thinking, creativity, problem-solving, and a deeper learning of subject matter. It also prepares students for the challenges and demands of the real world by giving them opportunities to apply their knowledge in authentic contexts. Furthermore, the implementation of PBL has been shown to be an effective STEM school reform model [26] that can be applied in various educational settings, from K-12 classrooms to higher education and professional development programs. While it is an engaging instructional approach, it is quite different from traditional teaching methods and requires teacher training to ensure understanding of PBL principles, how to facilitate student inquiry and manage classroom dynamics involving group work and collaboration, and assessment of the learning process, collaboration, and critical thinking.

4.1 Problem-based learning

Problem-based learning (PrBL) is similar to project-based learning; however, it begins with a real-world complex problem or challenge. In PrBL, the learning process begins with the presentation of a problem or a challenge, and students are tasked with exploring, analyzing, and solving that problem over a specified period. This approach has often been utilized in medical schools, but is also implemented in K-12 settings, especially in mathematics classrooms.

Unique features of PrBL in contrast to PBL include:

1. Teacher as Facilitator: Instead of traditional lecturing, instructors using PrBL serve as facilitators. They provide support and resources to help students navigate the problem-solving process but do not dictate solutions.

2. Assessment in PBL is often based on the quality of the solutions proposed. Assessment focuses on depth of understanding demonstrated and the process of learning. Student products may involve presentations, written reports, discussions, or other forms of evaluation.

3. Reflective thinking is an essential component of PBL. Students are encouraged to assess their own learning process, identify gaps in their knowledge, and set goals for further learning.

Problem-based learning targets students developing a deep understanding of the content covered, utilize critical thinking skills and self-directed learning abilities, and focus on the application of knowledge in real-world contexts. It encourages students to become self-learners, capable of addressing the complex problems they may face in future careers.

4.2 Phenomena-based learning (PhBL)

Phenomena-based learning (PhBL) is a variant of PBL that revolves around students exploring complex, real-world phenomena that are of interest to them [27]. This educational movement was initiated by Finland's educational system in 2016 striving to expand traditional project-based learning (PBL) and problem-based learning (PrBL) approaches into learning experiences that immerse students deeper into contextual situations aligned with real-life issues while also applying knowledge and skills from multiple disciplines. As with PBL and PrBL, students are encouraged to share their innovative solutions widely.

The unique aspects of PhBL include:

- Phenomena-based learning (PhBL) starts with the identification of a central phenomenon, often chosen based on its relevance to students' lives or the curriculum. This phenomenon serves as the anchor for student learning.

- Phenomena-based learning (PhBL) emphasizes the application of knowledge and skills in authentic contexts. Students learn by addressing real-world problems related to the chosen phenomenon, making the learning experience more meaningful.

Phenomena-based learning (PhBL) projects often begin when students are personally impacted by a specific phenomenon. For example, the COVID-19 Engineering Design Challenge, spanning 2020–2023, engaged over 1000 students from 14 countries to design solutions to challenges and worldwide problems related to the pandemic [28]. Students were encouraged to identify areas of need, brainstorm and design solutions, and communicate their ideas to their classmates, families, and community members via online environments or following physical distancing guidelines established by their teachers and/or schools.

PBL, PrBL, and PhBL are effective methods for promoting deeper learning. The success of these approaches depends on the quality of instruction, the relevance of the chosen problems or phenomena, and the support provided to students as they learn. These methodologies are especially suited to STEM.

4.3 PBL/PrBL/PhBL products

STEM-focused schools engage in a wide range of real-world projects that are designed to apply and deepen student understanding of STEM concepts and skills. These inquiry-based projects emphasize hands-on learning, problem-solving, and collaboration. Specific projects can vary depending on the curriculum, resources, and student interest.

Samples include:

- Engineering Design Projects
- Scientific Investigations
- Computer Coding
- Software Development
- Environmental Studies
- Computer Modeling (Mathematics)
- Robotics Challenges
- Aerospace Design
- Biomedical and Health Science Projects
- Innovation Challenges
- Local Community Service Project
- Competitions

Projects are designed to help students learn STEM content, but they also serve the function of helping students develop critical thinking, problem-solving, teamwork, and communication skills. Where applicable, projects can provide opportunities for students to collaborate with industry professionals, mentors, and experts in the field, which can often be a barrier for many schools due to location and lack of opportunities to interact with professionals, especially in rural areas. However, the pandemic illustrated the value of web-based conferencing to bring experts into the classroom remotely. Projects can span across schools and grade levels and include virtual participation as well.

4.4 The role of the teacher in PBL, PrBL, and PhBL

In PBL, PrBL, and PhBL, teachers play a crucial role in guiding and facilitating the learning process. While these pedagogical frameworks shift some of the responsibilities for learning to students, teachers remain essential to create a supportive and structured environment.

Key roles include:

- Design and planning. Teachers select or approve appropriate projects, problems, or phenomena, align them with curriculum standards, establish learning goals, and define the scope of the project.

- Introduce the projects to students. Teachers frame the problem or challenge, explain its relevance by making connections between the project and real-world applications, and assist students to understand the project's purpose, objectives, and overall expectations. They may invite guest speakers or arrange field trips to enhance this connection.

- Model specific skills or processes relevant to the project. Teachers demonstrate how to approach specific tasks or challenges and foster a growth mindset by encouraging students to embrace challenges, persevere through difficulties, and view failures as opportunities for learning and improvement.

- Create a collaborative learning environment. Teachers facilitate group dynamics by building student teamwork skills (assisting students to navigate conflicts or challenges within their teams) and build a positive and supportive classroom culture (celebrate students' success and achievements).

- Monitor students' progress throughout the project. Teachers schedule check-ins and conferences to discuss work progress and ensure students stay on track and meet deadlines.

- Scaffold instruction to support students' learning. Teachers offer advice and guidance, provide relevant resources and supports to help students successfully navigate projects, and suggest alternate approaches when needed.

- Facilitate the inquiry process. Teachers encourage students to ask questions, explore topics, seek additional sources of information, and conduct research.

- Design assessment rubrics and criteria for evaluating student work. Teachers provide formative (ongoing) feedback to help students improve their project work and meet learning objectives and promote metacognition by asking students to assess their progress, identify areas for improvement, and set goals for further learning.

Through the process of leading PBL, PrBL, and PhBL activities, the teacher's role evolves from being the sole source of information to becoming a guide, mentor, role model, and facilitator of student learning. The teacher empowers all students to take personal ownership of their learning and develop essential skills and knowledge through their own research.

5. Conclusion

According to Nobel Laureate Physicist Leon Lederman, "STEM literacy" in a knowledge-based economy is defined as the ability to adapt to and accept changes driven by new technology and work with others, often across borders, to anticipate the multilevel impacts of their actions, communicate complex ideas to a variety of audiences, and perhaps most importantly, find "measured yet creative solutions to problems which are today unimaginable" ([29], p. 3).

Integrating STEM in preK-12 classrooms naturally promotes 21CS development, especially when implemented through different variations of project-based learning

(PBL, PrBL, and PhBL) that are aligned with the 5E Model. Implementing STEM curricula requires careful planning, interdisciplinary collaborations, and a commitment to engage students in active learning. French [30] emphasized that desirable learning experiences often occur when students are engaged in something hands-on and of interest. Personalized inquiry-based experiences promote student agency shaped around student choice, kinesthetic experiences, and discussions that lead to activism, and support a lifelong learning agenda aligned with the needs of the twenty first century and beyond [31]. PBL, PrBL, and PhBL are engaging instructional approaches. Students who participate in projects tend to retain information and skills longer because they have a deeper understanding of the content through active exploration and application, and practice presenting their research. This outcome is particularly beneficial for STEM schools as it facilitates retention of students' interest in STEM subjects and encourages them to pursue STEM-related careers. STEM education raises awareness of critical issues and challenges, serving as the driver for sustainable development.

STEM disciplines are closely tied to every UN Sustainable Development Goal (the SDGs) as they equip individuals with the knowledge and skills required to address a wide range of global challenges, from education and gender equality to clean energy, innovation, and environmental sustainability. STEM education is a critical component of achieving the SDGs and creating a more sustainable and equitable world. SDG integration naturally addresses complex global challenges in a collaborative manner, leading to solutions that better address the root causes of an issue. Achievement of the SDGs is dependent upon global implementation of STEM education and the development of the essential skills needed for success in the twenty first century workplace and beyond (problem-solving, critical thinking, creativity, improved communication skills, collaboration, perseverance, information and digital literacy, and entrepreneurial skills).

Author details

Teresa J. Kennedy* and Michael R.L. Odell
The University of Texas at Tyler, Texas, USA

*Address all correspondence to: tkennedy@uttyler.edu

References

[1] Gardner DP, Larsen YW, Baker WO, Campbell A, Crosby EA, Foster CA, et al. A Nation at Risk: The Imperative for Educational Reform. Washington, D.C.: National Commission on Excellence in Education; 1983. Available from: https://files.eric.ed.gov/fulltext/ED226006.pdf

[2] National Academy of Sciences, National Academy of Engineering, and Institute of Medicine. Rising above the Gathering Storm: Energizing and Employing America for a Brighter Economic Future. Washington DC, USA: National Academy Press; 2007. Available from: https://www.nap.edu/catalog/11463/rising-above-the-gathering-storm-energizing-and-employing-america-for

[3] National Academy of Sciences, National Academy of Engineering, and Institute of Medicine. Rising above the Gathering Storm Revisited: Approaching Category 5. Washington DC, USA: National Academy Press; 2010. Available from: https://nap.nationalacademies.org/catalog/12999/rising-above-the-gathering-storm-revisited-rapidly-approaching-category-5

[4] Odell MRL, Kennedy TJ, Simmons J, Pedersen J. Laboratory schools to support the preparation of UTeach preservice teachers. In: Goodell J, Koc S, editors. Preparing STEM Teachers: A Replication Model. Chapter 9. Charlotte, NC, USA: Information Age Publishing, Inc.; 2020. Available from: https://www.infoagepub.com/products/Preparing-STEM-Teachers

[5] UTeach Institute. Shaping national STEM success: Quality teaching + quality student outcomes. 2019. Available from: https://institute.uteach.utexas.edu/sites/default/files/UTeach-Shaping-National-Success-Fall-2019.pdf

[6] THECB. Dual credit data. Texas higher education coordinating board. 2023. Available from: http://www.txhighereddata.org/index.cfm?objectId=AEE9A640-D971-11E8-BB650050560100A9

[7] Pedersen J, Odell MRL, Kennedy TJ, Dyer K, Simmons JA, Oliveras-Ortiz Y. Improvement science models used for achievement in mathematics. In: Peterson DS, Carlile SP, editors. Improvement Science: Promoting Equity in Schools. Chapter 9. Gorham, ME: Myer Education Press; 2021. pp. 132-146. Available from: https://myersedpress.presswarehouse.com/pages/contact-us

[8] National Research Council. Successful K-12 STEM Education: Identifying Effective Approaches in Science, Technology, Engineering, and Mathematics. Committee on Highly Successful Science Programs for K-12 Science Education. Board on Science Education and Board on Testing and Assessment, Division of Behavioral and Social Sciences and Education. Washington, DC: The National Academies Press; 2011. Available from: http://www.stemreports.com/wp-content/uploads/2011/06/NRC_STEM_2.pdf

[9] Tunnicliffe SD, Kennedy TJ. The foundation of STEM through play. In: Japan Society for Science Education (JSSE) STEM Education in Early Childhood Conference Proceedings. Kyoto, Japan: Nakanishi Printing Co., Ltd.; 21 Aug 2021. pp. 191-194. Available from: https://www.jstage.jst.go.jp/article/jssep/45/0/45_191/_article/-char/ja

[10] McClure ER. More than a foundation: Young children are capable STEM learners. Young Children. 2017;**72**(5):83-89. Available from: https://www.naeyc.org/resources/pubs/yc/nov2017/STEM-learners

[11] McClure ER, Guernsey L, Clements DH, Bales SN, Nichols J, Kendall-Taylor N, et al. STEM Starts Early: Grounding Science, Technology, Engineering, and Math Education in Early Childhood. New York: The Joan Ganz Cooney Center at Sesame Workshop; 2017. Available from: https://joanganzcooneycenter.org/wp-content/uploads/2017/01/jgcc_stemstartsearly_final.pdf

[12] Kennedy TJ, Tunnicliffe SD. Introduction: The role of play and STEM in the early years. In: Tunnicliffe SD, Kennedy TJ, editors. Play and STEM Education in the Early Years: International Policies and Practices. Chapter 1. Switzerland: Springer Nature; 2022. pp. 3-37. Available from: https://link.springer.com/book/10.1007/978-3-030-99830-1

[13] Prieto AB, Kennedy TJ. GLOBE, STEM and citizen science in the early years in Argentina. In: Tunnicliffe SD, Kennedy TJ, editors. Play and STEM Education in the Early Years: International Policies and Practices, Chapter 19. Switzerland: Springer Nature; 2022. pp. 383-423. Available from: https://link.springer.com/chapter/10.1007/978-3-030-99830-1_19

[14] National Science Teaching Association (NSTA). NSTA Position Statement: Early Childhood Science Education. 2014. Available from: https://www.nsta.org/about/positions/earlychildhood.aspx

[15] Kennedy TJ, Odell MRL. Engaging students in STEM education. Science Education International. 2014;**25**(3):246-258. Available from: https://eric.ed.gov/?id=EJ1044508

[16] National Education Association. Preparing 21st century students for a global society: An educator's guide to the "four Cs.". 2012. Available from: https://dl.icdst.org/pdfs/files3/0d3e72e9b873e0ef2ed780bf53a347b4.pdf

[17] Battelle for Kids. Framework for 21st century learning. 2019. Available from: https://static.battelleforkids.org/documents/p21/P21_Framework_Brief.pdf

[18] Kennedy TJ, Sundberg CW. 21st century skills. In: Akpan B, Kennedy TJ, editors. Science Education in Theory and Practice: An Introductory Guide to Learning Theory. Switzerland: Springer Nature. Chapter 32. Print ISBN 978-3-030-43619-3; eBook ISBN 978-3-030-43620-9; 2020. pp. 479-496. DOI: 10.1007/978-3-030-43620-9

[19] Joynes C, Rossignoli S, Fenyiwa Amonoo-Kuofi E. 21st Century Skills: Evidence of Issues in Definition, Demand and Delivery for Development Contexts (K4D Helpdesk Report). Brighton, UK: Institute of Development Studies; 2019. Available from: https://assets.publishing.service.gov.uk/media/5d71187ce5274a097c07b985/21st_century.pdf

[20] Bybee RW, Fuchs B. Preparing the 21st century workforce: A new reform in science and technology education. Journal of Research in Science Teaching. 2006;**43**(4):349-352. DOI: 10.1002/tea.20147/epdf

[21] Kennedy TJ, Sundberg CW. International perspectives and recommendations on equity and gender: Development studies in science education. In: Akpan B, editor. Science Education: A Global Perspective (Chapter 15). Switzerland: Springer Nature; 2017. pp. 295-311. Available from: https://www.springer.com/gp/book/9783319323503

[22] Kennedy TJ, Odell MRL. STEM education as a meta-discipline. In: Akpan B, Çavaş B, Kennedy TJ, editors.

Contemporary Issues in Science and Technology Education. Chapter 4. Switzerland: Springer Nature; 2023. pp. 37-52. Available from: https://link.springer.com/chapter/10.1007/978-3-031-24259-5_4

[23] Bybee RW. The Case for STEM Education: Challenges and Opportunities. Arlington, Virginia: National Science Teachers Association, NSTA Press; 2013

[24] Bybee RW. The BSCS 5E Instructional Model: Creating Teachable Moments. Arlington: NSTA Press; 2015. Available from: https://static.nsta.org/pdfs/samples/PB356Xweb.pdf

[25] San Diego County Office of Education. Evidence-Based Practices: The 5E Model of Instruction. 2018. Available from: https://www.sdcoe.net/ngss/evidence-based-practices/5e-model-of-instruction

[26] Odell MRL, Kennedy TJ, Stocks E. The impact of PBL as a STEM school reform model. Interdisciplinary Journal of Problem-Based Learning. 2019;**13**(2). Available from: https://docs.lib.purdue.edu/ijpbl/vol13/iss2/4/

[27] Kennedy TJ, Fields DL. Augmenting upper and early learning STEM lessons through phenomenon-based learning. Journal of Higher Education Theory and Practice. 2023;**23**(18):170-182

[28] Kennedy TJ. Student innovations related to COVID-19: The international engineering design challenge. American Journal of Biomedical Science & Research. 2021;**13**(6):592-595. Available from: https://biomedgrid.com/pdf/AJBSR.MS.ID.001921.pdf

[29] National Governors Association. Innovation America: Building a Science, Technology, Engineering, and Mathematics Agenda. Washington DC, USA: NGA; 2007. p. 3. Available from: https://files.eric.ed.gov/fulltext/ED496324.pdf

[30] French D. Student Agency and Personalized Learning. Somerville, MA: Center for Collaborative Education; 2016. Available from: https://www.theteachercollaborative.org/contact

[31] Kennedy TJ, Cherry A. Sustainable development and science and technology education. In: Akpan B, Çavaş B, Kennedy T, editors. Contemporary Issues in Science and Technology Education. Chapter 10. Switzerland: Springer Nature; 2023. pp. 131-150. Available from: https://link.springer.com/chapter/10.1007/978-3-031-24259-5_10

Chapter 4

Addressing the Gender Gap in the STEM Field

Rubie Ibuola

Abstract

The Science, technology, engineering, and mathematics (STEM) field has gained more traction in the last decades because of the digitalization and technological and economic advancement of society. Amidst the rapid growth in the field, there has been a consistent gender gap, with males making up more than 50% of the STEM field. The gap is often started to be seen as early as secondary school when students start to choose courses. This chapter explores some of the major factors causing this gap from primary school level to career level, drawing on past studies and case studies, and poses possible solutions to bridge the gap. It has a major focus on the United Kingdom but draws examples and case studies from other countries.

Keywords: STEM, gender, gap, female, STEM education, gender gap, gender disparity

1. Introduction

The first introduction of the Science, Technology, Engineering, and Mathematics (STEM) acronym was in 2001 by some science administrators at the U.S. National Science Foundation (NSF). SMET was the initial acronym when referring to the career fields in those disciplines. Later, an assistant director of education and human resources at NSF, named Judith Ramaley, rearranged the words to form the STEM acronym. Since then, the acronym has been adopted by several countries beyond the United States [1].

Stohlmann [2] defined STEM education as curricula that attempt to integrate science, technology, engineering, and mathematics into one category based on interdependence and connections in the natural world. Over time, because of the natural overlap between science, technology, engineering, and mathematics, the curricula in schools started evolving to reflect the relationships among the fields [3]. It became important to develop curricula that support these disciplines as time went on because of the impact STEM education and STEM-related careers had on society at large. According to Bell-young [4], the STEM curriculum in schools became important in the United Kingdom (UK) for various reasons; the backbone it is to society, the solid intellectual foundation it offers to students who take this path, the practical learning experience it offers, and the problem-solving skills it fosters. With the digitalization and technological focus of society in the last decades, it became imperative to focus on the fields that support technological advancement, which largely happens to be STEM.

To understand better the STEM landscape in the UK, the Royal Academy of Engineering undertook a study, in conjunction with the Lloyd's Register Foundation, to provide stakeholders with an accurate picture of the STEM education landscape. The UK recognizes STEM subjects as critically important to its economic success. For example, according to [5], engineering accounts for 25% of gross value added for the UK economy and manufactured goods account for 50% of UK exports. Science, engineering, and technology underpin the whole economy in one way or the other, including power generation and electricity distribution, utilities, the food chain, healthcare, and our physical, transportation, and information and communications infrastructure [5].

According to [6], countries that focus on the production of information and technology attach more importance to STEM education to improve students' skills accordingly. In the study conducted, it was found that countries such as the USA, England, and Japan that integrate STEM into the education system have grown economically and there has been an increase in students' achievement in international exams such as the Programme for International Student Assessment (PISA) and the trends in international mathematics and science study (TIMSS).

Most countries design STEM education from as early as primary school and have a curated curriculum to support them through secondary and tertiary education [7]. STEM education is regarded as a priority in most high-, middle- and low-income countries [8, 9]. However, most countries in Africa are still catching up with this development. The gender gap in STEM education is prevalent in most developed and developing countries [10] and is attributed to a number of factors ranging from political factors to economic and social factors.

Although women are currently more likely than men to hold a college degree in the vast majority of The Organization for Economic Co-operation and Development (OECD) countries, their choices of college major have been and persistently continue to be different from those of men. According to Granato [11], across all OECD countries, women were greatly over-represented among bachelor's degree graduates in the education field (in which the females share was 77%), but they represented only 26% of engineering graduates and 22% of graduates in the Information and Communication Technology (ICT) field in the 2020 statistics. This eliminates the idea that there are more men getting educated than women, the focus just needs to be given to understanding why more women tilt more to other courses outside the STEM field and how we can encourage their involvement.

This chapter sheds light on the gender gap in the STEM field by exploring some of the major reasons for this gap and also posing practical solutions across all levels of society, so as to enable stakeholders in the field to pay closer attention to how they can contribute to bridging the gap, starting from their organizations.

The chapter utilizes data from secondary sources, all from peer-reviewed journals, limited to the last 15 years to highlight the most recent work done in the field.

2. Gender statistics in the STEM field

In spite of the progress women have made in education and the workforce over the past 5 decades, we still continue to see men dominate science, technology, engineering, and mathematics (STEM) fields [12]. From 1970 to 2019, the proportion of women in the U.S. workforce increased from 38 to 48%, yet women's representation in STEM fields increased from 8% to only 27% in the same time period [12]. The same negligible growth is seen in the UK, between 2016 and 2019.

Figure 1 shows the gender statistics of STEM workers in the UK, and while there has been an increase since 2016, it is still comparatively low overall compared to the male gender.

There is growing literature as to the reasons why more men end up in STEM careers compared to women. Granato [11] poses that the issue can be traced to the early foundation years of children. If females acquire lower science and maths ability and knowledge prior to choosing their major this could potentially translate into a lower probability of choosing a STEM degree. Tian et al. [14] gives more context to this, showing in the study that if more females engaged in spatial activities, it could translate to an interest in mathematics, which is found to be a primary reason why males end up in STEM fields. Granato [11] highlights more external factors that could be the reason for the gender gap, the study poses that external and environmental influences, which can originate from a close environment, such as the family, or from the broader social setting in which students grow up and make their choices, which can influence their decision on the career path to take. The study by Granato [11] follows a five-year investigation into the gender gap in STEM fields in a higher institution in Italy, highlighting the possible reasons for the gender gap in STEM and focusing on governmental policies as a factor to be considered by countries to close the gap.

2.1 Possible reasons for the existing gender gap

Effort has been made to identify the various factors across cognitive, motivational, sociocultural, and environmental reasons that are contributing to women's lower participation in STEM college majors [15]. Studies on cognitive factors have traditionally centered around science and mathematics ability in childhood and adolescence. Strong science and mathematical competence is found to be fundamental for STEM

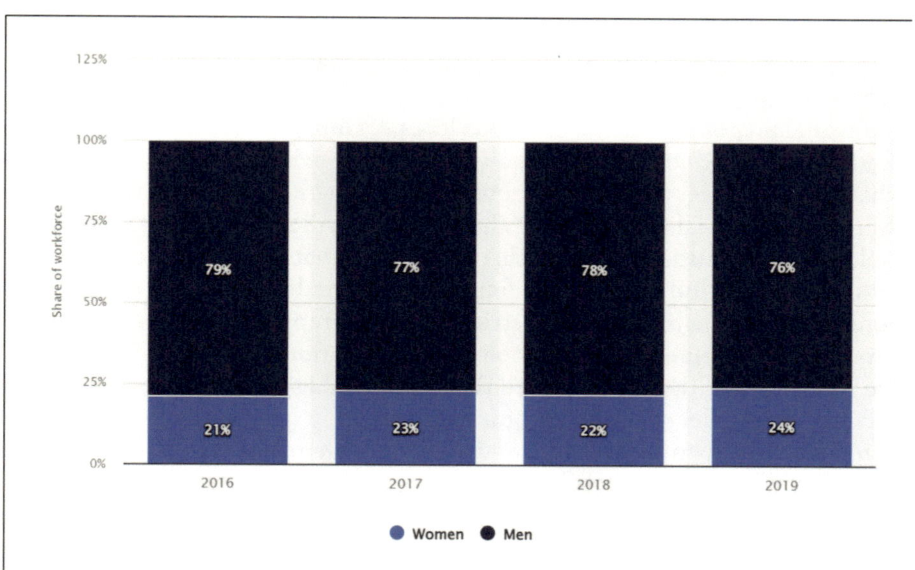

Figure 1.
Distribution of STEM (science, technology, engineering, mathematics) workforce in the United Kingdom from 2016 to 2019, by gender [13].

achievement, and high-school students with higher science and math achievement are more likely to choose STEM when it comes to college majors [16]. However, an analysis in the US showed that though boys have long been believed to have stronger math abilities than girls, assessments showed no gender difference in math performance among boys and girls in grades two through grade eleven [17], which makes that factor a probability and not an absolute on the impact it has on student's decision to go into STEM. Another study conducted in the U.S. went beyond the known cognitive factors and abilities like giftedness in math and calculus and aimed to trace the origins of the gender gap in STEM college majors to spatial skills in middle childhood. The study found that enhancing spatial skills in childhood such as through spatial play at home and spatial activities in schools could set more children, especially girls, on a pathway toward STEM achievement in adulthood [14].

Granato [11] breaks down the determinants of the major choice by students to go into STEM courses into three main factors: pre-college education, family and social background, and higher education supply. To explain these factors further, the study poses that a student's experience in primary and secondary school influences their skills, interests and subsequently their career. Mathematics was also found to be the highest predictor of the decision to go into STEM courses [17], further evidence from more developed countries like the UK and the US indicates that taking maths courses in high school is a strong predictor of a later STEM major choice [18, 19].

A child's environment and family also show high correlations to their choice of course in college. Some women with high ability may choose to exert lower effort and select less difficult majors with lower monetary returns when identity enters their choices because it is expected from them under the prevailing gender identity norms in their environment [20, 21]. Parents' educational achievement is also highlighted under family influence that impacts the decision of what course to study in college and other studies have shown the existence of a gendered pattern in the influence of parents' education or occupation on their children's educational and occupational choices [22]. **Figure 2** highlights the share of women in STEM who also have parents in the STEM field.

Women are more likely to end up in STEM courses and occupations if either or both parents have a similar occupation.

The availability of higher education in close proximity to college students was also found to influence their course choices [11]. If there is a supply of higher education that a student faces upon high school exit. If higher education institutions are not evenly geographically distributed, then students will settle for courses readily available by the institution closest to them or the one that grants them admission. For example, [24] found that Spanish female college graduates have stronger family relations than males, which largely restricts their geographical mobility and has a negative effect on their educational choices and aspirations.

The conclusion of the study showed that students' high school experience explains up to half of the gender gap in STEM graduation rates in Italy and young girls were found to be less likely to choose courses with a focus on maths and technical skills if they came from homes where both parents were not in similar fields.

2.2 Solutions to bridging the gender gap

Existing research poses that Women's under-representation in science, technology, engineering, and mathematics (STEM) studies and careers is seen as a barrier to an

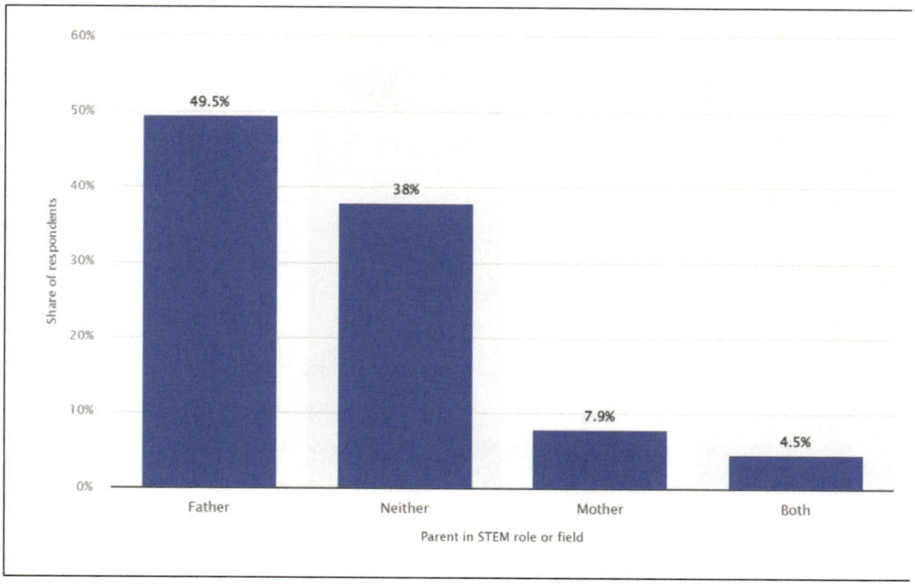

Figure 2.
Share of women in STEM (science, technology, engineering, mathematics) fields with STEM parents worldwide, as of 2016 [23].

extensive and sufficient supply of STEM skills which is considered crucial to boosting economic and technological innovation and growth [25]. There has been an effort made to address this gap and under-representation. In addressing the gender gap from a policy perspective, [11] suggests early interventions targeting both female students and their parents because parents need awareness of the influence they have on their children's career choices. Intervention for female students alone might not be sufficient to fully address the STEM gender gap.

Mishel [26] highlights the existence of discrimination against women in the workforce and society's expectations of women to prioritize caring for the family above pursuing careers, therefore interrupting their work for family and household issues. With the demands of most STEM careers, most women might choose other careers outside of STEM to accommodate the flexibility of catering to family matters. If more organizations and institutions accommodated the demands of family life, more women might be encouraged into STEM fields. In addition to this, women are generally believed to be less likely to succeed, get promoted, and occupy leadership positions in sectors where there is male dominance [27]. It is also important to highlight that women who succeed despite these stereotypes usually experience negative reactions for not performing their socially prescribed duties which are usually family-related. These can lead women into low self-efficacy and self-imposed isolation which could later have more adverse effects on their personal and career journey [28].

In a study conducted across five European countries, 39 female staff in STEM fields were interviewed to know the best methods of bridging the gender gap in STEM, the content analysis provided 5 categories which were: modeling, encouragement, change of mindset, campaigns, teaching, policies, and preventing gender discrimination [29]. **Table 1** below shows the sub-category of each category.

	Categories	Sub-categories
1	Modeling	• Highlighting the figure of women in STEM areas • Imprinting success stories in girls' minds • Introducing role models • Offering women experiences • Presenting successful careers • Providing behavior patterns • Showing examples • Showing success stories • Telling stories to girls about successful women in STEM • Getting to know the trajectory ofrelevant women in STEM areas
2	Encouragement	• Encouraging girls to STEM areas at school • Encouraging girls' participation in scientific events • Giving examples about facing challenges • Helping girls to believe in themselves • Highlighting the obstacles • Inspiring girls • Introducing female students to science at a younger age • Promoting female students in rural areas • Providing training and courses • Raising awareness in girls • Scaffolding girls
3	Change of mindset	• Changing the attitudes of others • Changing the beliefs of parents • Changing the beliefs of the society • Stop thinking in boxes • Talking about students' prejudices they bring from home, TV
4	Campaigns	• Conducting campaigns • Gaining public support • Informing society • Inviting role models to schools • Organizing events to encourage female students
5	Teaching	• Arousing curiosity • Creating an environment like laboratories • Creating different and interesting experiences for girls • Encouraging project-based learning • Encouraging students to do practical experiments • Encouraging students to express opinions clearly • Helping girls explore technology in an interactive and intuitive way • Providing girls with practical knowledge • Providing hands-on activities

	Categories	Sub-categories
6	Policies	• Financial support to STEM programs • Managing HR effectively • Promoting mentoring programs • Promoting student career discovery programs
7	Preventing gender discrimination	• Eliminating gender discrimination • Ensuring equality of genders • Eradicating macho stereotypes • Preventing societal prejudices • Treating boys and girls equally

Source: Adapted from Kaplan-say et al. [3].

Table 1.
Categories and sub-categories of ways to bridge the gender gap in STEM fields.

The categories and subcategories show that a collective effort is required to bridge the digital gap in STEM fields. It is not left to organizations alone, but educational institutions, parents, and government policies also have an impact on a student's decision to study a course or work in a field.

3. Conclusion

Despite the effort to bridge the gender gap, existing research still shows more work is still required to encourage more women into the STEM field.

More young girls and women need to be mentored because seeing people like them in similar fields encourages their own participation. A lot of stereotypes need to be addressed from home as well as in the workplace. Women should not have to choose family over work, some countries like Sweden and Norway have put policies in place that ensure job security for the women even through different seasons of family life, some of these include maternity and paternity leave, and the option to work virtually and shorter hours after childbirth. If more countries adopt these measures, more women might see a reason to enter the field. Research also calls for reorientation in the minds of young girls and women as to their roles in society. They should know they have options and should not be shamed or discriminated against for whatever option they choose, regardless of the norm seen in their society.

Campaigns have been one of the most popular means used in a lot of countries, but research shows they are often short-lived without addressing the core of the issue, so they should be supplemented with other means highlighted above, to encourage more women into STEM fields.

This chapter should bring all shareholders to collaborate and act together to support and encourage female students to pursue a career in STEM areas.

Conflict of interest

The authors declare no conflict of interest.

Author details

Rubie Ibuola
University of Portsmouth, Portsmouth, United Kingdom

*Address all correspondence to: targematakema_rubie@yahoo.com

References

[1] Hallinen J. STEM Education Curriculum. 2023. Available from: https://www.britannica.com/topic/STEM-education

[2] Stohlmann M, Moore T, Roehrig G. Considerations for teaching integrated STEM education. Journal of Pre-College Engineering Education Research. 2012;**2**(1):28-34. DOI: 10.5703/1288284314653

[3] AlMuraie A, Algarni A, Alahmad N. Upper-Secondary School Science Teachers' Perceptions of the Integrating Mechanisms and Importance of Stem Education. 2021. Available from: https://files.eric.ed.gov/fulltext/EJ1315750.pdf

[4] Bell-young L. What is STEM Education and Why is it Important? 2020. Available from: https://www.reagent.co.uk/blog/what-is-stem-education/

[5] Royal Academy of Engineering. The UK STEM Education Landscape. 2016. Available from: https://raeng.org.uk/media/bcbf2kyb/112408-raoe-uk-stem-education-landscape_final_lowres.pdf

[6] Acar D. The Effect of STEM Education on the Academic Success, Critical Thinking, and Problem-Solving Skills of the Elementary 4th-Grade Students. Unpublished PhD thesis. Ankara: Gazi University Department of Primary Education; 2018

[7] Ohize EJ. Repositioning science, technology, engineering and mathematics education for economic recovery in Nigeria. In: Conference Proceedings. "Reposition Science, Technology, Engineering & Mathematics (STEM) Education for Economic Recovery in Nigeria." Kontagora, Niger State. 2017. Available from: www.sciencesfcekg.sch.ng/downloads/ConferenceProceedings_11th_2017.pdf

[8] Burnett N, Jayaram S. Innovative Secondary Education for Skills Enhancement: Skills for Employability in Africa and Asia. Washington, DC: Results for Development Institute; 2012. Available from: https://www.r4d.org/.../InnovativeSecondaryEducationSkillsEnhancement-PhaseI-Syn

[9] Ostler E. 21st century STEM education: A tactical model for long-range success. International Journal of Applied Science and Technology. 2012;**2**(1):28-31

[10] UNESCO. Cracking the Code: Girls' and Women's Education in Science, Technology, Engineering, and Mathematics (STEM). UNESCO; 2017. Available from: unesdoc.unesco.org/images/0025/002534/253479e.pdf

[11] Granato S. Early Influences and the Choice of College Major: Can Policies Reduce the Gender Gap in Scientific Curricula (STEM)? 2023. Available from: https://www.sciencedirect.com/science/article/pii/S0161893823000352

[12] Martinez A, Christnacht C. Women Making Gains in STEM Occupations But are Still Underrepresented. 2021. Available from: https://www.census.gov/library/stories/2021/01/women-making-gains-in-stem-occupations-but-still-underrepresented.html

[13] Statista. Distribution of STEM (Science, Technology, Engineering, Mathematics) Workforce in the United Kingdom from 2016 to 2019, by Gender. 2023. Available from: https://www.statista.com/statistics/1251340/united-kingdom-stem-workforce-gender/

[14] Tian J, Ren K, Newcombe S, Weinraub M, Vandell D, Gunderson E. Tracing the Origins of the STEM Gender

Gap: The Contribution of Childhood Spatial Skill. 2022. Available from: https://onlinelibrary.wiley.com/doi/epdf/10.1111/desc.13302?saml_referrer

[15] Wang M-T, Degol JL. Gender gap in science, technology, engineering, and mathematics (STEM): Current knowledge, implications for practice, policy, and future directions. Educational Psychology Review. 2017;**29**(1):119-140. DOI: 10.1007/s10648-015-9355-x

[16] Guo J, Parker PD, Marsh HW, Morin AJS. Achievement, motivation, and educational choices: A longitudinal study of expectancy and value using a multiplicative perspective. Developmental Psychology. 2015;**51**(8):1163-1176. DOI: 10.1037/a0039440

[17] Hyde JS, Lindberg SM, Linn MC, Ellis AB, Williams CC. Gender similarities characterize math performance. Science. 2008;**321**(5888):494-495. Available from: https://doi.org/10.1126/science.1160364

[18] Tchuente G. High school human capital portfolio and college outcomes. Journal of Human Capital. 2016;**10**(3):267-302

[19] Philippis MD. STEM graduates and secondary school curriculum: Does early exposure to science matter? Journal of Human Resources. 2021;**59**:1219-10624R1

[20] Cheng A, Kopotic K, Zamarro G. Can parentsà growth mindset and role modelling address STEM gender gaps? SSRN Electronic Journal. 2017:4-8

[21] Bordón P, Canals C, Mizala A. The gender gap in college major choice in Chile. Economics of Education Review. 2020;**77**(May):2-4. Article 102011

[22] Pascual M. Intergenerational income mobility: The transmission of socio-economic status in Spain. Journal of Policy Modeling. 2009;**31**(6):835-846

[23] Statista. Share of Women in STEM (Science, Technology, Engineering, Mathematics) Fields with STEM Parents Worldwide, As of 2016. 2016. Available from: https://www.statista.com/statistics/671746/worldwide-women-in-stem-with-stem-parents/

[24] Farré L, Ortega F. Family ties, geographic mobility, and the gender gap in academic aspirations. SSRN Electronic Journal. 2021;**14561**:3-8

[25] Peri G, Shih K, Sparber C. STEM workers, H-1B visas, and productivity in US cities. The Economics of International Migration. 2016;**49**(3):277-307

[26] Mishel E. Discrimination against queer women in the US workforce: A résumé audit study. Socius. 2016;**2**:2378023115621316

[27] Meyer M, Cimpian A, Leslie SJ. Women are underrepresented in fields where success is believed to require brilliance. Frontiers in Psychology. 2015;**6**:235

[28] Picho K. The psychosocial experience of high school girls highly susceptible to stereotype threat: A phenomenological study. The Journal of Educational Research. 2016;**109**(6):608-623

[29] Kaplan-Say A, Yurtseven N, Karadeniz S, Vatanartıran S. Unraveling the Situation of Women in STEM Areas from the European Teacher Perspective: Insights from FEMALES Project. 2023. Available from: https://web.s.ebscohost.com/ehost/pdfviewer/pdfviewer?vid=16&sid=a4477e89-a42a-4b2b-8770-7e9ce5dabc39%40redis

Chapter 5

Bridging the Gap: Progressive Teaching Strategies for Gender Equity in STEM Education

Umar Daraz, Younas Khan, Muhammad Azeem Ashraf and Samson Maekele Tsegay

Abstract

This study explores the impact of progressive teaching methods on gender equity in STEM education in Malakand Division, Pakistan. Using a cross-sectional exploratory design, it engages educators, students, parents, community leaders, and policymakers through purposive sampling. Through in-depth interviews and focus groups, the research uncovers region-specific challenges and opportunities. Thematic analysis highlights the community's call for progressive teaching to combat barriers, stereotypes, and promote inclusivity. It stresses the need for inclusive learning spaces, challenging societal norms, and enhancing female representation in STEM leadership. The study advocates for innovative pedagogies, policy changes, and community involvement as catalysts for change. While acknowledging local obstacles like socio-cultural norms and resource constraints, it underscores the potential of community engagement and robust policies to overcome these challenges. Cultural influences on STEM participation, such as familial expectations, are recognized, underscoring the importance of addressing these factors. The study concludes with a plea for interdisciplinary action to dismantle systemic barriers and cultivate a culture of inclusivity and mentorship. Policy recommendations focus on inclusive curricula, targeted recruitment, mentorship initiatives, and community outreach to foster an equitable STEM environment in Malakand Division, Pakistan.

Keywords: STEM education, progressive teaching strategies, STEM women, gender equity, Pakistan

1. Introduction

STEM education has become a cornerstone of global progress, innovation, and societal development, transcending geographical boundaries [1]. Its significance extends beyond the mere imparting of technical knowledge, encompassing the cultivation of critical thinking, problem-solving skills, and a broader understanding of the world [2]. Stemming from the Enlightenment era, which emphasized reason and scientific inquiry, the foundations of modern STEM education were laid centuries ago [3].

IntechOpen

Throughout history, STEM education has evolved in response to societal and technological advancements. The Industrial Revolution of the nineteenth century propelled the need for skilled workers proficient in mathematics, engineering, and scientific principles [4]. This period witnessed the establishment of technical schools and universities offering specialized education in STEM fields, marking a significant milestone in the formalization of STEM education [5, 6].

The twentieth century saw unprecedented advancements in science and technology, catalyzed by groundbreaking discoveries in fields such as quantum mechanics, relativity, and computing. The aftermath of World War II further accelerated technological progress, leading to the emergence of disciplines like computer science and biotechnology [7]. The Space Race between the United States and the Soviet Union in the 1950s and 1960s epitomized the global significance of STEM education, spurring investments in research and education [8].

As the world entered the digital age in the late twentieth century, STEM education became even more pronounced, with the rise of the internet, personal computing, and telecommunications revolutionizing communication and commerce [9]. In the twenty-first century, STEM education has emerged as a linchpin for global competitiveness and economic prosperity, addressing challenges such as climate change, healthcare, and cybersecurity [10].

Amidst these advancements, the issue of gender equity in STEM education has garnered significant attention. Despite the historical contributions of women in STEM fields, they have often faced systemic barriers to full participation. Efforts to address this imbalance have intensified in recent decades, with initiatives aimed at dismantling barriers and fostering inclusivity [11].

This study aims to explore progressive teaching strategies tailored to address gender equity in STEM education within the context of Malakand Division, Pakistan. By aligning with global aspirations while acknowledging historical developments and local nuances, we aim to empower future innovators both globally and regionally.

2. Literature review: gender gap in STEM fields

Numerous international studies have established a persistent gender gap in STEM fields. A comprehensive meta-analysis conducted by Wang and Degol [12] reveals that, globally, women are underrepresented in STEM occupations. For instance, 18% of girls in tertiary education are enrolled in STEM education globally in comparison to 35% of boys. This underrepresentation is attributed to various factors, including societal stereotypes, lack of role models, and subtle biases in educational practices. However, the literature consistently emphasizes the transformative potential of progressive teaching strategies in mitigating these barriers and fostering gender equity [13]. Research by Archer-Kuhn and MacKinnon [14] highlights the effectiveness of innovative pedagogies, such as project-based learning and collaborative activities, in promoting inclusivity and engagement in STEM subjects. These strategies not only enhance academic performance but also contribute to breaking down gender stereotypes by emphasizing the collaborative and interdisciplinary nature of STEM disciplines [15]. Policy frameworks play a pivotal role in shaping the educational landscape. Laursen et al. [16] emphasize the importance of gender-inclusive policies in STEM education. This includes targeted initiatives to encourage girls' participation in STEM activities, provision of mentorship programs, and the integration of gender perspectives into STEM curricula [17]. Such policies aim to create an environment conducive to the empowerment of all students, irrespective of gender.

Shifting the lens to Malakand Division, Pakistan, the existing literature paints a nuanced picture of the challenges faced in promoting gender equity in STEM education. A study conducted by Sadia [18] identifies socio-cultural norms and limited access to resources as key challenges hindering girls' participation in STEM subjects. Socio-cultural factors significantly impact educational choices in the region. Khan et al. [19] discuss the influence of cultural norms and familial expectations on students' decisions to pursue STEM careers. Empirical evidence suggests that interventions need to be culturally sensitive and address community-specific concerns to be effective in promoting gender equity in STEM education [20]. The importance of community engagement is underscored by research conducted by Haque and Doberstein [21], emphasizing the need for collaborative efforts involving educators, parents, and community leaders. This collaborative approach is seen as instrumental in creating awareness, dispelling stereotypes, and garnering support for progressive teaching strategies that promote gender equity in STEM education [22].

3. Current discourse and the present study

The pursuit of gender equity in STEM education is crucial globally and particularly urgent in regions like Malakand Division, Pakistan. Challenges such as socio-cultural norms and limited resources contribute to a gender gap in STEM enrollment. Existing teaching methods may not effectively address this issue, highlighting the need for tailored strategies.

This study aims to investigate the effectiveness of progressive teaching strategies in promoting gender equity in STEM education in Malakand Division. By examining local challenges and trends, it seeks to inform policy, educational practices, and community engagement to bridge the gender gap in STEM fields.

While global literature acknowledges the need for gender equity in STEM, there is a gap in understanding the context of Malakand Division. This research fills this void by exploring local nuances and providing empirical evidence on innovative pedagogies' effectiveness.

Researchers combine international insights with local challenges, conducting surveys and engaging stakeholders to offer context-specific recommendations. The study's significance lies in its potential to transform the educational landscape, fostering inclusivity in STEM and driving local economic development.

The study's novelty lies in its localized approach to progressive teaching strategies, providing practical solutions tailored to the challenges of Malakand Division. It offers a blueprint for educators, policymakers, and community leaders, empowering the next generation of innovators and advancing gender equity in STEM.

4. Methodology

The qualitative methodology for "Bridging the Gap: Progressive Teaching Strategies for Gender Equity in STEM Education" encompasses Research Design, Participant Selection, Data Collection Methods, Data Analysis, Ethical Considerations, and Limitations with Mitigations. It focuses on developing inclusive pedagogical approaches to address gender disparities in STEM, ensuring rigorous research practices and ethical standards are maintained throughout the study.

4.1 Research design

This study, conducted in Malakand Division, Pakistan, employs an exploratory descriptive design to delve into gender equity in STEM education [23]. It introduces a novel angle by examining the impact of progressive teaching strategies on mitigating gender disparities, a fresh approach within qualitative research. By focusing on the local context, it aims to uncover nuanced insights into the challenges faced by women in STEM fields due to cultural constraints. This method offers a platform for in-depth exploration, bridging the gap between theory and practice, and providing actionable recommendations for promoting inclusivity and empowerment within the region.

4.2 Participants selection

This study employs purposive sampling to select participants vital to the dynamics of STEM education in Malakand Division, Pakistan—educators, students, parents, community leaders, and policymakers. Utilizing the saturation point concept [24], 40 participants were chosen, with 8 from each group, ensuring a diverse and comprehensive representation. This approach aligns with established methodologies, optimizing the collection of varied perspectives on gender equity and progressive teaching strategies in STEM education within the local context [25].

What's new in this research is its emphasis on exploring progressive teaching strategies' impact on gender equity within a specific cultural setting. Unlike previous studies, which may have focused solely on gender disparities, this research delves deeper by examining how innovative teaching methods can address these disparities. By including a diverse demographic mix, comprising educators, students, parents, community leaders, and policymakers, this study ensures a holistic exploration of progressive teaching strategies' efficacy in promoting gender equity in STEM education in Malakand Division, Pakistan (**Table 1**).

Appendix A below encapsulates the key demographic details, providing a clear overview of the participants' diversity and relevance to the study on gender equity in STEM education.

Participants group	Gender	Age range	Educational level	Qualification	Work experience
Educators (8)	M/F	25–55	Education/STEM	Varied degree	5–20 years
Students (8)	M/F	18–24	STEM courses	Varied levels	—
Parents (8)	M/F	30–60	Varied backgrounds	Varied degrees	—
Community leaders (8)	M/F	35–65	Varied occupations	Varied degrees	—
Policymakers (8)	M/F	35–60	Education/Policy	Advanced degrees	Extensive

Table 1.
Demographic characteristics of participants.

4.3 Data collection methods

In our qualitative study on gender equity in STEM education in Malakand Division, Pakistan, we employ in-depth interviews and structured Focus Group Discussions (FGDs) as our primary data collection methods. What sets our approach apart is the meticulous design of the interview guide, inspired by established thematic coverage frameworks by DiCicco-Bloom and Crabtree [26] and Boateng [27], ensuring a nuanced exploration of stakeholders' experiences and perceptions. Additionally, we orchestrate six strategically structured FGDs, catering to different stakeholder sections and including a combined session. This mixed-method approach, drawing from established methodologies, allows for the capture of both individual insights and collective perspectives. By innovatively combining these methods, our study enriches the exploration of gender equity in STEM education, offering a holistic understanding of challenges and opportunities specific to Malakand Division, Pakistan.

4.4 Data analysis

After data collection and transcription, our study undertook systematic thematic analysis to uncover patterns within qualitative data, following methodologies like Braun and Clarke [28] and Bryman [29]. This rigorous approach aimed to explore nuances of gender equity in STEM education in Malakand Division, Pakistan. Additionally, we integrated aspects of grounded theory to allow emergent themes to surface organically, distinguishing our analysis and offering a deeper understanding beyond existing frameworks.

4.5 Ethical considerations

Before commencing data collection, participants were thoroughly informed about the study's purpose, procedures, and their rights. Informed consent was diligently obtained from each participant, ensuring their voluntary participation. Anonymity and confidentiality were paramount, with all participant identities strictly protected. The collected data were securely stored to uphold participant privacy. To ensure credibility, findings were shared with participants for member checking, allowing them to verify the accuracy of interpretations.

4.6 Limitations and mitigations

While confined to the Malakand Division, potentially limiting generalizability, the study maximized depth through a purposive sample and cultural sensitivity. A sample size of 40 was mitigated by employing a saturation point strategy. Cultural biases were addressed through local researchers. Reflexivity tackled researcher bias, and time constraints were managed by prioritizing depth and critical reflection. External factors' impact was monitored, and member checking, though prone to social desirability bias, sought honest participant feedback, reinforcing the study's credibility within its contextual constraints.

5. Results

The results highlight key themes: Perspectives on STEM Education and Gender Equity, Innovative Pedagogies, Inclusive Learning Environments, Policy

Interventions for Gender Equity in STEM, Challenges and Opportunities in Promoting Equity, Cultural Factors on STEM Participation, and Community Engagement. These findings contribute insights into progressive teaching strategies for bridging gender gaps in STEM education, emphasizing the importance of inclusive practices, policy interventions, and collaborative efforts with stakeholders to enhance equity.

6. Perspectives on STEM education and gender equity

Global perspectives on STEM education and gender equity emphasize the critical need for inclusivity, highlighting the imperative to bridge gender gaps. Recognizing diverse voices and fostering equal opportunities are pivotal for a thriving, equitable future in science, technology, engineering, and mathematics [30]. Respondents underscored the importance of fostering an inclusive environment that encourages and supports the participation of women in STEM disciplines. In this regard, a student-3, explored:

> *"In recognizing gender disparities in STEM, I advocate for dismantling barriers and promoting equal opportunities for all, irrespective of gender. By fostering inclusivity, we create a space where everyone can thrive in STEM education and careers."*

Respondents highlighted the role of societal expectations and stereotypes in steering individuals toward or away from STEM fields based on their gender. Overcoming these stereotypes emerged as a crucial step in promoting gender equity in STEM education. An educator-4, expressed:

> *"Societal expectations often limit career choices based on gender. Personally, I am committed to challenging these stereotypes to ensure that individuals, regardless of gender, have the freedom to pursue their interests and potential in STEM."*

The lack of representation of women in STEM leadership positions emerged as a significant concern. Respondents pointed out that having more female role models and mentors in STEM can inspire and guide the next generation of women in the field, fostering a sense of belonging and empowerment. However, parent-8, viewed:

> *"Visibility matters. As someone passionate about STEM, I believe in the importance of being a role model and mentor to empower the future generation of female scientists and innovators."*

The importance of inclusive teaching methodologies was also highlighted. Respondents stressed the need for progressive teaching strategies that promote diversity and inclusivity, addressing different learning styles and ensuring that STEM education is accessible and engaging for everyone. In this context, an educator-7 explained:

> *"Progressive teaching strategies, to me, mean catering to diverse learning styles and making STEM education accessible and engaging for all students, irrespective of gender or background."*

7. Innovative pedagogies and inclusive learning environments

Innovative pedagogies and inclusive learning environments redefine education by embracing diverse approaches that cater to varied learning styles. Creating inclusive spaces ensures equitable participation, fostering a dynamic educational landscape that empowers learners and promotes collaboration, preparing them for a globally interconnected world [31]. Respondents expressed the importance of integrating digital tools and platforms to enhance engagement and facilitate personalized learning experiences. In this regard, Educator-5, explained:

> *"Embracing technology in education opens new avenues for innovative pedagogies. Personally, I find that incorporating digital tools not only enhances student engagement but also allows for a more personalized and adaptive learning environment."*

The concept of active learning emerged as a central theme, emphasizing the shift from traditional lecture-based approaches to more hands-on and collaborative learning experiences. Respondents highlighted the benefits of active learning in promoting critical thinking, problem-solving skills, and a deeper understanding of the subject matter. Moreover, Educator-6, demonstrated:

> *"Active learning methodologies empower students to actively participate in the learning process. As an educator, I witness firsthand how these methods foster critical thinking and a deeper understanding of concepts, creating a more enriching learning environment."*

The theme of inclusivity in education was also prevalent, with respondents stressing the importance of creating learning environments that cater to diverse needs and backgrounds. Inclusive pedagogies were seen as essential in ensuring that every student feels valued and supported in their educational journey. Likewise, Educator-4, explained:

> *"Inclusivity is at the heart of effective teaching. Personally, I am committed to creating an inclusive learning environment where diverse perspectives are not only acknowledged but celebrated, fostering a sense of belonging among students."*

Respondents discussed the role of project-based and experiential learning in enhancing student engagement and preparing them for real-world challenges. The hands-on nature of these approaches was seen as a way to bridge the gap between theoretical knowledge and practical application. In this connection, Educator-1, reviewed:

> *"Project-based and experiential learning bring theory to life. I believe in providing students with opportunities to apply their knowledge in real-world scenarios, preparing them for the challenges they will face beyond the classroom."*

8. Policy interventions for gender equity in STEM

Policy interventions for gender equity in STEM are crucial for dismantling systemic barriers. Implementing targeted strategies, such as mentorship programs and inclusive policies, can mitigate gender disparities. Robust policies empower individuals, fostering an environment where all genders contribute meaningfully to science, technology, engineering, and mathematics [32]. Respondents stressed the importance of creating policies that actively encourage and support the recruitment and retention of women in STEM fields, ensuring equal opportunities for career progression. Similarly, Policy Maker-3, viewed:

> *"Targeted recruitment and retention policies are essential. Personally, I advocate for strategies that actively promote the inclusion of women in STEM, ensuring equal opportunities for career advancement and recognition."*

Respondents discussed the need for policies that facilitate mentorship programs and support networks for women in STEM, providing guidance and fostering a sense of community that can help overcome challenges and barriers. Likewise, Policy Maker-1, explained:

> *"Mentorship is crucial. I believe in the power of policies that promote mentorship and support networks, creating environments where women in STEM can thrive, share experiences, and navigate their career paths successfully."*

Respondents emphasized the role of policies in ensuring that educational materials and curricula are designed to be inclusive and representative, challenging stereotypes and providing diverse role models for students. In this context, Educator-4, expressed:

> *"Policy interventions should shape a gender-inclusive curriculum. Personally, I advocate for policies that ensure educational resources and curricula promote diversity, challenge stereotypes, and offer a variety of role models for aspiring STEM students."*

Workplace policies promoting flexibility and work-life balance were discussed as crucial for gender equity. Respondents stressed the importance of policies that allow for flexible work arrangements, parental leave, and supportive workplace cultures to address the challenges women may face in balancing career and family responsibilities. Moreover, Policy Maker-4, visualized:

> *"Workplace policies should support work-life balance. I am an advocate for policies that create a supportive workplace culture, offering flexibility and parental leave, addressing the unique challenges women may encounter in balancing career and family."*

9. Challenges and opportunities in promoting equity in STEM education

Promoting gender equity in STEM education confronts local challenges and taps into unique opportunities. Identifying barriers specific to the community, while leveraging local strengths, guides targeted interventions [11]. Respondents discussed how traditional expectations and societal norms can discourage girls from pursuing STEM education. Cultural biases, perceptions about gender roles, and limited exposure to STEM opportunities were identified as barriers hindering the active involvement of girls in STEM fields. In this regard, Community Leader-3, opinioned:

> *"Socio-cultural norms influence STEM participation. Personally, I recognize the challenges posed by traditional expectations and biases. Addressing these norms is vital to fostering an environment where girls feel empowered to pursue STEM education in Malakand Division."*

Respondents echoed the literature's findings, emphasizing how the lack of adequate infrastructure, educational materials, and technology hampers the quality of STEM education. The scarcity of resources disproportionately affects girls, limiting their opportunities to explore and engage in STEM subjects fully. However, Parent-5, viewed:

> *"Resource limitations hinder STEM quality. I advocate for addressing the scarcity of infrastructure and technology. Ensuring equal access to resources is essential to providing a conducive environment for girls to excel in STEM education in Malakand Division."*

Respondents discussed the importance of collaborative efforts between communities, schools, and policymakers to raise awareness about the benefits of STEM education for girls. Strengthening local educational policies to prioritize gender equity and promote inclusivity in STEM was seen as a crucial step toward overcoming barriers. Moreover, Community Leader-8, expressed:

> *"Community engagement and policy support matter. Personally, I believe in collaborative efforts to raise awareness and strengthen local educational policies. Empowering communities and schools to prioritize gender equity in STEM is essential for breaking barriers in Malakand Division."*

10. Cultural factors on STEM participation

Cultural factors significantly influence STEM participation. Recognizing and understanding cultural nuances is essential for fostering inclusivity. By bridging cultural gaps and promoting diverse perspectives, we can create environments that encourage broader and more equitable engagement in science, technology, engineering, and mathematics [20]. Respondents highlighted how cultural norms and expectations can influence individuals, particularly women, steering them towards or away from STEM disciplines based on perceived gender roles and stereotypes. In this regard, Student-2, reviewed:

> *"Societal expectations shape STEM choices. Personally, I recognize the influence of cultural norms and expectations, particularly on women, impacting career choices in STEM. We must address these biases to encourage a more diverse STEM participation."*

Respondents discussed how cultural beliefs about innate abilities in STEM subjects can either encourage or discourage individuals from pursuing STEM careers. Overcoming stereotypes related to who is deemed naturally suited for STEM is essential for fostering inclusivity. Likewise, Educator-6, expressed:

"Cultural perceptions affect STEM confidence. I advocate for challenging stereotypes around innate STEM abilities. Personally, I believe we must emphasize that anyone can excel in STEM, breaking down cultural barriers and fostering confidence in pursuing STEM fields."

Respondents shared how familial and community pressures can play a crucial role in shaping educational and career choices, affecting individuals' decisions to pursue or avoid STEM fields. Creating awareness and support systems within families and communities is crucial for overcoming these barriers. Moreover, Community Leader-2, viewed:

"Familial and community expectations matter. Personally, I emphasize the need for awareness and support systems within families and communities. Breaking down these expectations can open doors for more individuals to pursue and succeed in STEM."

Respondents discussed the importance of diverse role models in STEM who reflect various cultural backgrounds. The lack of representation can limit the visibility of STEM opportunities for individuals from underrepresented cultural groups, hindering their participation. In this context, Parent-7, explained:

"Representation matters in STEM. I advocate for diverse role models to inspire future generations. Personally, I believe that showcasing individuals from various cultural backgrounds succeeding in STEM is crucial for breaking down barriers and encouraging wider participation."

11. Community engagement and stakeholder collaboration

Promoting gender equity in STEM education relies on robust community engagement and stakeholder collaboration. By involving communities and key stakeholders, we cultivate a shared commitment to dismantling barriers [33]. Respondents emphasized the need for community outreach programs that educate parents about the opportunities in STEM and dispel gender stereotypes, encouraging families to actively support their daughters' interest in STEM fields. In this regard, Parent-3, demonstrated:

"Involving parents is crucial for girls in STEM. Personally, I advocate for community outreach programs that educate parents, dispelling gender stereotypes. Engaging families actively in supporting their daughters in STEM is vital for gender equity."

Respondents stressed the importance of collaborative efforts between educational institutions, STEM professionals, and community organizations to provide mentorship programs for girls. Having visible role models in STEM helps inspire and guide young girls, fostering a sense of belonging and empowerment. Likewise, Educator-4, augmented:

"Mentorship is a powerful tool for gender equity in STEM. I believe in collaborative efforts between institutions, STEM professionals, and community organizations to provide mentorship programs. Visible role models inspire and guide girls in their STEM journey."

Respondents discussed the importance of collaboration between schools, businesses, and industries to create opportunities for girls to engage in real-world STEM experiences, internships, and mentorship programs. These collaborations bridge the gap between classroom learning and practical application, enhancing girls' interest and confidence in STEM. In this context, Educator-1, explained:

> *"Industry partnerships enhance STEM opportunities for girls. Personally, I emphasize collaboration between schools, businesses, and industries to create real-world STEM experiences. These partnerships bridge the gap between learning and application, boosting girls' interest and confidence in STEM."*

Respondents shared examples of successful community-led STEM events, workshops, and clubs that actively involve girls and create a supportive community. These programs provide hands-on experiences and opportunities for collaboration, nurturing girls' passion for STEM. Moreover, Community Leader-5, opinioned:

> *"Community-led STEM programs make a difference. I advocate for initiatives that actively involve girls in STEM, creating a supportive community. These programs provide hands-on experiences, fostering passion and collaboration among girls in STEM."*

12. Discussions

Respondents emphasized the critical need for inclusivity in STEM education and the importance of dismantling barriers to ensure equal opportunities for all genders. This aligns with BrckaLorenz et al. [34], who stress the importance of recognizing diverse voices and fostering equal opportunities in STEM. Moreover, the commitment of one respondent to challenging societal expectations regarding gender and career choices echoes the findings of Andrews and Boklage [35], who emphasize the role of societal expectations in influencing STEM career paths. The concern over the lack of representation of women in STEM leadership positions is supported by Killpack and Melón [36] and Leslie et al. [37], who highlight the underrepresentation of women in STEM leadership roles [38]. However, the emphasis on visibility and role modeling from a parental perspective adds a nuanced dimension, suggesting a grassroots approach to fostering empowerment in STEM.

Respondents advocated for progressive teaching strategies that cater to diverse learning styles and promote inclusivity in STEM education. This aligns with Lincoln et al. [39], and Moriarty [40], who emphasize the importance of inclusive spaces in fostering equitable participation. Additionally, the endorsement of active learning methodologies resonates with Freeman et al. [41], who highlight the benefits of active learning in promoting critical thinking and deeper understanding. While the emphasis on inclusivity and project-based learning is consistent with previous research, the specific mention of digital tools and platforms to enhance engagement reflects a contemporary approach to pedagogy, suggesting an adaptation to modern learning environments.

Respondents stressed the importance of policy interventions to promote gender equity in STEM, including targeted recruitment and retention policies and mentorship programs. This aligns with Laursen et al. [16] and Brush et al. [42], who highlight the effectiveness of targeted policies in increasing women's participation in STEM fields. However, while previous research emphasizes the importance of policies, this study adds a focus on workplace policies promoting flexibility and work-life balance, acknowledging the intersectionality of gender and other factors in STEM careers.

Respondents discussed local challenges such as socio-cultural norms and resource limitations, highlighting the importance of community engagement and policy support. This aligns with Tanwir & Khemka [43] and [44], who stress the significance of identifying barriers specific to the community while leveraging local strengths. The emphasis on collaborative efforts and grassroots initiatives adds depth to the discussion, suggesting a bottom-up approach to addressing gender equity in STEM education. This differs from previous research by highlighting the importance of community-led initiatives in addition to policy interventions.

Respondents acknowledged the influence of cultural factors on STEM participation, particularly regarding societal expectations and perceptions of intelligence. This aligns with the findings of Williams and Shipley's [45], who recognized the impact of cultural norms on career choices in STEM, echoing the work of Kricorian et al. [46]. While previous research has addressed cultural factors in STEM participation, this study adds a focus on familial and community pressures, highlighting the importance of support systems within families and communities. This suggests a more holistic approach to addressing cultural barriers in STEM education.

Respondents emphasized the importance of community engagement and stakeholder collaboration in promoting gender equity in STEM education. This aligns with Allen et al. [47] and Appel et al. [48], who stress the need for robust community engagement to dismantle barriers. However, while previous research emphasizes the importance of community engagement, this study adds a focus on collaborative efforts between schools, businesses, and industries, suggesting a broader approach to stakeholder collaboration in promoting gender equity in STEM education.

13. Conclusion

The study reveals a collective call for progressive teaching strategies to dismantle barriers, challenge stereotypes, and foster inclusivity, emphasizing the need for a holistic and collaborative approach. The findings underscore the critical importance of inclusivity in STEM education, with a focus on bridging gender gaps and providing equal opportunities for all. Respondents advocate for the creation of inclusive learning environments that empower individuals, irrespective of gender, aligning with empirical evidence supporting the positive outcomes of such approaches. Addressing societal expectations and stereotypes emerges as a key theme, emphasizing the necessity of overcoming these barriers to promote gender equity in STEM. The commitment to challenging stereotypes and fostering an environment where individuals have the freedom to pursue their interests aligns with research highlighting the impact of societal expectations on career choices. The lack of representation of women in STEM leadership positions is identified as a significant concern, prompting a call for more female role models and mentors. This aligns with empirical evidence highlighting the underrepresentation of women in STEM leadership roles and the positive impact of visible role models on inspiring the next generation [38]. Innovative pedagogies and inclusive learning environments are recognized as transformative elements in STEM education. The integration of technology, active learning methodologies, and a commitment to inclusivity are deemed essential for empowering students and preparing them for the future, consistent with research supporting the positive outcomes of such approaches. Policy interventions are highlighted as crucial for achieving gender equity in STEM, with a focus on targeted recruitment and retention policies, mentorship programs, gender-inclusive curricula, and supportive workplace policies. This aligns with empirical evidence demonstrating the effectiveness of policy interventions in addressing gender disparities in STEM fields [42].

Local challenges in Malakand Division, including socio-cultural norms and resource limitations, are acknowledged, but the study identifies community engagement and robust educational policies as opportunities to overcome these hurdles. This resonates with research emphasizing the importance of collaborative efforts and community engagement in addressing challenges to gender equity in STEM [47, 48]. Cultural factors are recognized as significant influencers of STEM participation, with expectations, perceptions of intelligence, and familial pressures shaping individuals' choices. Overcoming these cultural barriers is identified as crucial for fostering a more diverse and equitable engagement in STEM fields. Community engagement and stakeholder collaboration emerge as vital elements in promoting gender equity in STEM education. Involving parents, establishing mentorship programs, fostering industry partnerships, and supporting community-led STEM initiatives are seen as integral to creating a supportive and empowering environment, although effective strategies are suggested for a successful outcome.

14. Policy implications

Policy implications for empowering tomorrow's innovators in STEM education in Malakand Division, Pakistan, involve multifaceted strategies. Implementing inclusive STEM curricula is crucial, addressing societal expectations and fostering diversity. Targeted recruitment and retention policies should be prioritized, promoting equal opportunities and breaking leadership barriers. Mentorship programs and visible role models are essential for inspiring the next generation. Workplace policies supporting work-life balance, flexible arrangements, and parental leave can address gender disparities. Community engagement should dispel stereotypes, garner parental support, and establish industry partnerships for real-world experiences. Educators and policymakers must collaboratively prioritize innovative pedagogies, technological integration, and hands-on learning. Addressing socio-cultural barriers requires localized educational policies; and ensuring gender-inclusive policies at all levels creates an equitable, diverse, and supportive STEM environment. A comprehensive approach to policy interventions is necessary for dismantling barriers and fostering an environment conducive to gender equity in STEM education in Malakand Division, Pakistan.

15. Gap for the future research

Future research should explore the long-term impact of implemented policies on gender equity in Malakand Division's STEM education. Investigating the effectiveness of inclusive curricula, mentorship programs, and workplace policies is crucial. Additionally, understanding the evolving role of community engagement and cultural factors in shaping STEM participation can inform targeted interventions. Longitudinal studies assessing the career trajectories of individuals benefiting from these policies can provide insights into sustained gender equity. Exploring the scalability and adaptability of these strategies in diverse contexts will guide the development of tailored policies for fostering inclusive STEM education in similar regions.

Appendix A

See **Table A1**.

Participants	Gender	Age range	Educational level	Specialization	Experience related to STEM education
Educator-1	M	37	PhD Degree (22 Years)	Math	18 Years
Educator-2	F	54	Master's Degree (16 Years)	Science	6 Years
Educator-3	M	48	Post-Doctorate Degree (22 and Above)	Science	8 Years
Educator-4	M	41	PhD Degree (22 Years)	Technology	12 Year
Educator-5	F	34	Master's Degree (16 Years)	Technology	12 Year
Educator-6	F	46	PhD Degree (22 Years)	Engineering	17 Years
Educator-7	M	55	Master's Degree (16 Years)	Engineering	20 Year
Educator-8	F	36	Master's Degree (16 Years)	Math	20 Years
Student-1	F	22	University Student	Math Student	—
Student-2	F	18	Middle School Student	Math Student	—
Student-3	M	20	Secondary School Student	Math Student	—
Student-4	M	21	Middle School Student	Science Student	—
Student-5	M	24	College Student	Engineering Student	—
Student-6	M	19	College Diploma Student	Technology Student	—
Student-7	F	19	College Graduate	Technology Student	—
Student-8	F	24	University Student	Engineering Student	—
Parent-1	M	30	Master's Degree (16 Years)	Science	—
Parent-2	M	40	PhD Degree (22 Years)	Science	—
Parent-3	M	58	BSC Degree (14 Years)	Math	—
Parent-4	M	60	Master's Degree (16 Years)	Math	—

Participants	Gender	Age range	Educational level	Specialization	Experience related to STEM education
Parent-5	M	55	Master's Degree (16 Years)	Engineering	—
Parent-6	M	34	Master's Degree (16 Years)	Technology	—
Parent-7	F	39	BSC Degree (16 Years)	Science	—
Parent-8	F	41	College Degree (12 Years)	Science	—
Community Leader-1	M	60	MS Degree (18 Years)	Science	3 Years
Community Leader-2	M	58	Master's Degree (16 Years)	Science	2 Years
Community Leader-3	M	61	Master's Degree (16 Years)	Technology	2 Years
Community Leader-4	M	40	BSC Degree (14 Years)	Technology	4 Years
Community Leader-5	M	43	BSC Degree (14 Years)	Technology	3 Years
Community Leader-6	M	38	Master's Degree (16 Years)	Math	2 Years
Community Leader-7	M	35	MS Degree (18 Years)	Math	2 Years
Community Leaders-8	F	56	BSC Degree (14 Years)	Engineering	2 Year
Policymakers-1	M	40	PhD Degree (22 Years)	Science	3 Years
Policymakers-2	M	38	PhD Degree (22 Years)	Science	3 Years
Policymakers-3	M	60	Post-Doctorate Degree (22 Years and Above)	Science	4 Years
Policymakers-4	M	52	Post-Doctorate Degree (22 Years and Above)	Technology	4 Years
Policymakers-5	M	54	PhD Degree (22 Years)	Technology	3 Years
Policymakers-6	M	60	MS Degree (22 Years)	Engineering	4 Years
Policymakers-7	F	45	PhD Degree (22 Years)	Math	4 Years
Policymakers-8	F	55	PhD Degree (22 Years)	Math	3 Years

Table A1.
Demographic characteristics of participants.

Author details

Umar Daraz[1], Younas Khan[2], Muhammad Azeem Ashraf[3]*
and Samson Maekele Tsegay[4]

1 Department of Sociology, University of Malakand, Pakistan

2 Department of Sociology, Kohat University of Science and Technology, Kohat, Pakistan

3 Institute of Educational Sciences, Hunan University, China

4 School of Education, Anglia Ruskin University, Cambridge, UK

*Address all correspondence to: azeem20037@gmail.com

References

[1] Jamali SM, Ale Ebrahim N, Jamali F. The role of STEM education in improving the quality of education: A bibliometric study. International Journal of Technology and Design Education. 2023;**33**(3):819-840

[2] Chiu TK, Li Y. How can emerging technologies impact STEM education? Journal for STEM Education Research. 2023;**6**:375-384

[3] Kuhn TS. The relations between history and history of science. Daedalus. 1971;**100**(2):271-304

[4] McClellan JE, Dorn H. Science and Technology in World History: An Introduction. Baltimore, Maryland, USA: JHU Press; 2015

[5] Kavak Ş. The evolution and global significance of STEM education in the 21st century. The Journal of International Scientific Researches. 2023;**8**(3):410-415

[6] Suhirman S, Prayogi S. Overcoming challenges in STEM education: A literature review that leads to effective pedagogy in STEM learning. JurnalPenelitian Pendidikan IPA. 2023;**9**(8):432-443

[7] Baker DP, Crist JT, Zhang L, Powell JJW, Shima K, Stock M. Science Productivity, Higher Education Development and the Knowledge Society (SPHERE Project) Final Report. Education City, Qatar: Georgetown University School of Foreign Service in Qatar; 2015

[8] Idris R, Govindasamy P, Nachiappan S, Bacotang J. Revolutionizing STEM education: Unleashing the potential of STEM interest career in Malaysia. International Journal of Academic Research in Business and Social Sciences. 2023;**13**(7):1741-1752

[9] Thornburg D. The New Basics: Education and the Future of Work in the Telematic Age. Alexandria, USA: ASCD; 2002

[10] El Nagdi M, Roehrig GH. Gender equity in STEM education: The case of an Egyptian girls' school. In: Theorizing STEM Education in the 21st Century. London, UK: IntechOpen; 2019. p. 89

[11] Ullah R, Ullah H, Bilal M. Biological versus feminists perspectives on Girls' underperformance in STEM subjects in Pakistan. Liberal Arts and Social Sciences International Journal (LASSIJ). 2020b;**4**(1):10-18

[12] Wang M-T, Degol JL. Gender gap in science, technology, engineering, and mathematics (STEM): Current knowledge, implications for practice, policy, and future directions. Educational Psychology Review. 2017;**29**:119-140

[13] CohenMiller A, Saniyazova A, Sandygulova A, Izekenova Z. Gender equity in STEM higher education in Kazakhstan. In: Gender Equity in STEM in Higher Education. New York: Routledge; 2021. pp. 140-157

[14] Archer-Kuhn B, MacKinnon S. Inquiry-based learning in higher education: A pedagogy of trust. Journal of Education and Training Studies. 2020;**8**(9):1

[15] Valverde-Berrocoso J, Garrido-Arroyo MdC, Burgos-Videla C, Morales-Cevallos MB. Trends in educational research about e-learning: A systematic literature review (2009-2018). Sustainability. 2020;**12**(12):5153

[16] Laursen SL, Austin AE, Soto M, Martinez D. Advancing the agenda for

gender equity. Change: The Magazine of Higher Learning. 2015;**47**(4):16-24

[17] Kong S, Carroll K, Lundberg D, Omura P, Lepe B. Reducing gender bias in STEM. MIT Science Policy Review. 2020;**1**:55-63

[18] Sadia S. Factors Influencing Female Intention to Participate in Science Technology Engineering Mathematics (STEM) Education in Pakistan. [Master dissertation/thesis] UTAR (Universiti Tunku Abdul Rahman); 2022

[19] Khan R, Khan S, Khan M. Impact of socio-cultural factors on women's higher education. Pakistan Review of Social Sciences (PRSS). 2020;**1**(2):36-46

[20] Ullah R, Ullah H, Allender T. Girls underperforming in science: Evidences from Khyber Pakhtunkhwa, Pakistan. Journal of Elementary Education. 2020a;**29**(2):1-14

[21] Haque CE, Doberstein B. Adaptive governance and community resilience to cyclones in coastal Bangladesh: Addressing the problem of fit, social learning, and institutional collaboration. Environmental Science & Policy. 2021;**124**:580-592

[22] Yan CT, Haque S, Chassler D, Lobb R, Battaglia T, Martinez LS. "It has to be designed in a way that really challenges people's assumptions": Preparing scholars to build equitable community research partnerships. Journal of Clinical and Translational Science. 2021;**5**(1):e182

[23] Rose KR. An exploratory cross-sectional study of interlanguage pragmatic development. Studies in Second Language Acquisition. 2000;**22**(1):27-67

[24] Francis JJ, Johnston M, Robertson C, Glidewell L, Entwistle V, Eccles MP, et al. What is an adequate sample size? Operationalising data saturation for theory-based interview studies. Psychology and Health. 2010;**25**(10):1229-1245

[25] Ritchie J, Lewis J, Nicholls CM, Ormston R. Qualitative Research Practice. Vol. 757. London: Sage; 2003

[26] DiCicco-Bloom B, Crabtree BF. The qualitative research interview. Medical Education. 2006;**40**(4):314-321

[27] Boateng W. Evaluating the efficacy of focus group discussion (FGD) in qualitative social research. International Journal of Business and Social Science. 2012;**3**(7):54-57

[28] Braun V, Clarke V. Thematic analysis. In: Cooper H, Camic PM, Long DL, Panter AT, Rindskopf D, Sher KJ, editors. APA Handbook of Research Methods in Psychology. Research Designs: Quantitative, Qualitative, Neuropsychological, and Biological. Vol. 2. American Psychological Association; 2012. pp. 57-71

[29] Bryman A. Social Research Methods. Oxford, UK: Oxford university press; 2016

[30] Ro HK, Fernandez F, Ramon EJ. Gender Equity in STEM in Higher Education: International Perspectives on Policy, Institutional Culture, and Individual Choice (Edition 1). New York: Taylor & Francis; 2021

[31] Mulcahy D, Morrison C. Re/assembling 'innovative' learning environments: Affective practice and its politics. Educational Philosophy and Theory. 2017;**49**(8):749-758. DOI: 10.1080/00131857.2016.1278354

[32] van den Brink M, Stobbe L. Gender equality interventions in the STEM fields: Perceptions, successes and dilemmas. In: Women in STEM Careers: International Perspectives on Increasing Workforce

Participation, Advancement, and Leadership. Cheltenham, UK: Edward Elgar Publishing; 2014. pp. 187-203

[33] Chapman S, Vivian R. Engaging the Future of STEM; 2017. Available from: https://apo.org.au/sites/default/files/resource-files/2017-03/apo-nid75558.pdf

[34] BrckaLorenz A, Haeger H, Priddie C. An examination of inclusivity and support for diversity in STEM fields. Journal for STEM Education Research. 2021;**4**(3):363-379

[35] Andrews ME, Boklage A. Supporting inclusivity in STEM makerspaces through critical theory: A systematic review. Journal of Engineering Education. 2023;Early view:1-31. DOI: 10.1002/jee.20546

[36] Killpack TL, Melón LC. Toward inclusive STEM classrooms: What personal role do faculty play? CBE—Life Sciences Education. 2016;**15**(3):es3

[37] Leslie S-J, Cimpian A, Meyer M, Freeland E. Expectations of brilliance underlie gender distributions across academic disciplines. Science. 2015;**347**(6219):262-265

[38] González-Pérez S, Mateos de Cabo R, Sáinz M. Girls in STEM: Is it a female role-model thing? Frontiers in Psychology. 2020;**11**:2204

[39] Lincoln AE, Pincus S, Koster JB, Leboy PS. The Matilda effect in science: Awards and prizes in the US, 1990s and 2000s. Social Studies of Science. 2012;**42**(2):307-320

[40] Moriarty MA. Inclusive pedagogy: Teaching methodologies to reach diverse learners in science instruction. Equity & Excellence in Education. 2007;**40**(3):252-265

[41] Freeman S, Eddy SL, McDonough M, Smith MK, Okoroafor N, Jordt H, et al. Active learning increases student performance in science, engineering, and mathematics. Proceedings of the National Academy of Sciences. 2014;**111**(23):8410-8415

[42] Brush CG, De Bruin A, Welter F. A gender-aware framework for women's entrepreneurship. International Journal of Gender and Entrepreneurship. 2009;**1**(1):8-24

[43] Tanwir M, Khemka N. Breaking the silicon ceiling: Gender equality and information technology in Pakistan. Gender, Technology and Development. 2018;**22**(2):109-129

[44] Lonchar CM. Barriers to Girls' Education in the Developing World. Ohio University; 2022

[45] Williams DH, Shipley GP. Cultural taboos as a factor in the participation rate of native Americans in STEM. International Journal of STEM Education. 2018;**5**:1-8

[46] Kricorian K, Seu M, Lopez D, Ureta E, Equils O. Factors influencing participation of underrepresented students in STEM fields: Matched mentors and mindsets. International Journal of STEM Education. 2020;**7**:1-9

[47] Allen PJ, Chang R, Gorrall BK, Waggenspack L, Fukuda E, Little TD, et al. From quality to outcomes: A national study of afterschool STEM programming. International Journal of STEM Education. 2019;**6**:1-21

[48] Appel DC, Tillinghast RC, Winsor C, Mansouri M. STEM outreach: A stakeholder analysis. In: 2020 IEEE Integrated STEM Education Conference (ISEC). Princeton, NJ, USA; 2020. pp. 1-9. DOI: 10.1109/ISEC49744.2020.9280723

Chapter 6

Thinking, Alternatively: The Application of Art, Philosophy and Holistic Practice in the Teaching of Science, Technology, Engineering and Mathematics (STEM)

Christine Challen and Kay Sidebottom

Abstract

The challenge for education in contemporary times is how to combine rich subject knowledge with conceptualization, whilst also supporting curiosity, criticality and creative problem-solving through application. More importantly is the embedding of well-being in curriculums to ensure the evolution of being, becoming, and thriving subjects towards enhancing life-long learning. This chapter will explore the potential of interdisciplinary working across philosophy and the humanities to develop the kinds of critical education required. Using practical examples, we will demonstrate how educators can think in interdisciplinary ways, within the constraints of formal curricula activity. We will explore opportunities for 'thinking, alternatively' across STEM subjects, focusing particularly on the role of art and philosophy. Drawing on posthuman thinking and the work of Deleuze, Guattari, Dewey and others, we consider philosophy to be an active and creative practice which can work to open new ontological positions and modes of thought. We consider how art can work with science, not as a tool in the service of understanding scientific concepts, but to reimagine what science is, what it can do, and to break down nature/culture binaries. We aim to show how transdisciplinary educational practice can aid human and non-human well-being, through reframed relations within and without the classroom.

Keywords: STEM, conceptualisation, arts, philosophy, interdisciplinarity

1. Introduction

"Imagination is more important than Knowledge" Albert Einstein

"Where is the wisdom we have lost in knowledge?"

"Where is the knowledge we have lost in information?" (T.S. Eliot, The Rock)

The complexity of our current worldly condition - an entangled, relational set of human and environmental predicaments such as climate change, digital revolution, environmental degradation and the rise of zoonotic pandemics - requires new conceptual understandings of the world and consequently, different framings of education and schooling. Braidotti ([1], p. 1181) terms this contemporary phenomenon as the 'posthuman predicament'; a convergence of multiple technological, ecological and democratic challenges which, although often dealt with separately, need to be understood as connected and intra-acting. Posthumanism as referred to here is essentially a final call to '...mark the end of the self-reverential arrogance of a dominant Eurocentric notion of the human, and to open new perspectives' ([2], p. 3). The *post* means 'after' humanism, acknowledging the benefits of humanism (particularly around struggles for equity via identity-related civil action), but recognising the limitations of it regarding the reification of what is typically encoded as 'Man': that is, white, male, European, able-bodied, neurotypical, straight and so on. Posthuman thinking thus requires us to go beyond, or *after* humanism as we augment and reposition the voices of those overlooked and oppressed by Enlightenment ideas of 'humanity.' It also turns towards the body as an important and often overlooked site of learning, calling us to start our understanding from our own situated, embodied contexts in acts of the politics of location [3] which do not universalise but understand both teachers and students as individuals with facing different, often intersectional, modes of oppression.

Despite the irrefutable complexities of current times, the compartmentalised way in which educational disciplines are currently taught within English education continues to reflect humanistic, Cartesian binaries and dualisms which fail to account for our predicament, or to create new systems of knowledge which allow us to deal with the various issues differently. The siloed nature of education is further entrenched via neo-liberal practices of measurement, comparison and competition which do not allow for complexity and nuance in either practice or judgement. The emphasis is on pedagogical activities which will have a significant impact on students' measurable outcomes; not those that impact individuals emotionally or relationally [4]. Within these spaces it is difficult (and in fact counter-cultural) for teachers to find room for consideration of deeper ontological or epistemological questions around education in the current times. Alongside this sits a culture of anti-intellectualism, whereby subjects such as art, literature and science are downgraded into the kind of knowledges that can be memorised and regurgitated to suit prescribed tests rather than studied deeply and meaningfully. Thinking, in an age of academic and expert distrust, is seen as the practice of the elite (that is, the preserve of those at the higher echelons of the educational system); yet we are in a time that calls for new ideas and approaches to complex ethical dilemmas more than ever. 'Think, we must' as said Virginia Woolf ([5], p. 60); but in accelerated educational consumer cultures of product, customer and service, time and space for these activities is increasingly eroded.

Practices such as interdisciplinarity, which work towards the integration of differing knowledges and draw on diverse ontological approaches to embrace multiple perspectives can be helpful in allowing new ideas and worldly understandings to be foregrounded. However, as a practice interdisciplinary education can also be seen as pre-supposing and cementing the very disciplinary differences it purports to disrupt. Nina Lykke therefore suggests that we need to move beyond disciplines entirely by embracing *postdisciplinarity*, a term which '... refers to more transgressive ways of producing academic knowledge which destabilise, deconstruct and disrupt the hegemony of distinct disciplines and the classic academic divides between human, social, technical, medical and natural sciences'. (Nina Lykke [6], p. 333). Whilst the current constraints of 'education-as-usual' may restrict teachers' ability to truly move beyond disciplines in this way, we argue in this chapter that trans-disciplinarity - the synthesis of subjects and removal of boundaries through the transversal merging of science, arts and humanities - can help us to reimagine 'subjects.'

In this chapter we explore the challenges and opportunities for working in trans-disciplinary ways across STEM subjects, focusing particularly on the role of art and philosophy. Drawing on posthuman thinking and the work of Deleuze, Guattari, Dewey and others, we consider philosophy to be an active and creative practice which can work to open new ontological positions and modes of thought. We then consider how art can work with science, not being employed as a tool in the service of understanding scientific concepts and ideas, but to reimagine what science is, what it can do, and to break down nature/culture binaries. In doing so we aim to show how trans-disciplinary educational practice can aid human and non-human well-being, through reframed relations both within and without the classroom. We draw on a range of practical examples from practice in order to illustrate our ideas, not in a prescriptive sense but as a jumping-off point for educators' own innovations and experiments.

Also embedded across this chapter are quotations, stories and reflections from creative narratives, poetry and biography. In this way we emulate the practices we espouse by breaking the disciplinary silos entrenched within academic writing. This approach is diffractive, in that it allows us to read things through other things; in the same way that a pebble when thrown into water with a second pebble will cause the waves to intersect, so interrupting academic writing with creative texts will allow new readings, ideas and patterns of thought to emerge [7]. Diffractive writing, in a similar way to interdisciplinary pedagogy, allows us to be affected in different ways. In adopting this approach, we also encourage readers to take 'lines of flight' [8] in their thinking; we therefore pose the following questions to you, our reader. What other connections can you make, and what other texts or work of art can you connect to the ideas shared here?

2. Teaching in complex times

The challenges of working within an educational system driven by data, testing and results render it difficult to teach outside the boundaries of subjects and disciplines. Standardised tests limit the scope for innovation and moving outside of prescriptive lesson plans. The reality of testing culture is pervasive and damaging; recently a teacher told me (Christine) the following about her daughter: "*She was driven by exam results and exam result day was her Christmas day.*" Meanwhile, outside the artificial process of examinations lie real socio/economic problems that society and the world is facing, and these require deeper, innovative and inter-connective thinking.

An example of the lack of such connections in schools was highlighted recently in my (Christine's) teaching. In the year 10 Science end-of-term assessment paper was a question which asked the pupils to measure the area around each circle of antibiotic resistance in an agar plate (a required practical that children undertake in Biology). Not one of them attempted this question, yet when asked in class if they had learnt how to calculate the area of a circle in maths and what the formula is, they were all able to answer the question correctly. Despite this mathematical knowledge they failed to make this connection during the test, and in failing to see Maths and Science as inextricably connected, lost valuable marks as well.

Further illustrating the siloed understandings of knowledge in education, on August 3rd, 2023, Katherine Birbalsingh (Head Teacher of Michaela School, London and former adviser to the UK government on social mobility) tweeted the following:

> *You want kids to read? Make them read. Constantly. Break it down. You want kids to add and subtract? Do maths. Constantly. Means no phones. No yoga. No gardening. No chatter and lack of attention.*

Birbalsingh, known as '@MissSnuffy' on the social media platform X (previously known as Twitter) is well-known for her provocative stance regarding the promotion of traditional education. The tweet here exemplifies in many ways the emphasis in contemporary English teaching practice on Cartesian dualisms between mind and body. Rather than seeing activities such as gardening as having educational, as well as therapeutic and environmental value, they are seen as extraneous to the 'real' and valid practices of maths and reading. By establishing such false binaries however, we risk students being unable to think divergently across disciplinary boundaries; struggling to integrate maths within other subjects, or failing to understand maths as a life skill which could be equally applied in calculating gaps when planting seeds, measuring wood for planters, or estimating growth and harvesting times. The examples provided here demonstrate that there is a real need for strategies to be embedded which allow for deeper thinking, so that knowledge can be adapted to different situations and subjects. As Eliot states in The Rock ([9], p. 147) '*Where is the wisdom we have lost in knowledge? Where is the knowledge we have lost in information?* So, how can this be done effectively in practice?

Rainbows, Bikes, and Lego

What is a rainbow? This is a simple question that often begins with very young nursery/early primary-aged children producing pictures of colourful semi-circles. As they move up into higher primary, or secondary school, this visualisation can be developed into connections with physics and the diffraction of white light, understandings of meteorology/weather, and maths through an appreciation of the angles of diffraction of the light colours. English Literature and History can also be embedded through the introduction of songs, poems and stories, mythology and magic around the concept of the rainbow and the infamous "pot of gold" at its end. By drawing together nature, science and culture, multiple understandings of a common phenomenon can help children to understand the ways in which knowledge is not siloed but relational, contextual and mediated.

Another example which I (Christine) undertook as part of a STEM transition from year 6 primary to Year 7 secondary was titled "The Beat of Life." In this session the pupils created a heart out of clay using straws to illustrate the arteries' veins. Some decided to sketch and label a heart from the power-point picture and one of these is framed on my living room wall as it was so good (**Figure 1**).

The children then had to create a storyboard of how the heart relates to health, exercise, poetry, poems and literature and if time allowed write a rap called 'The Beat of Life'. The pupils thoroughly enjoyed these sessions and were enthused about being able to think differently about science. One of the parents said to me after it "I hope the teachers at my son's secondary are as engaging and passionate about their subject as you. He has learnt a lot today".

I have also used sweets to get the pupils/students to create different body systems. This is easy to implement in practice a selection of sweets and get the pupils/students to be creative

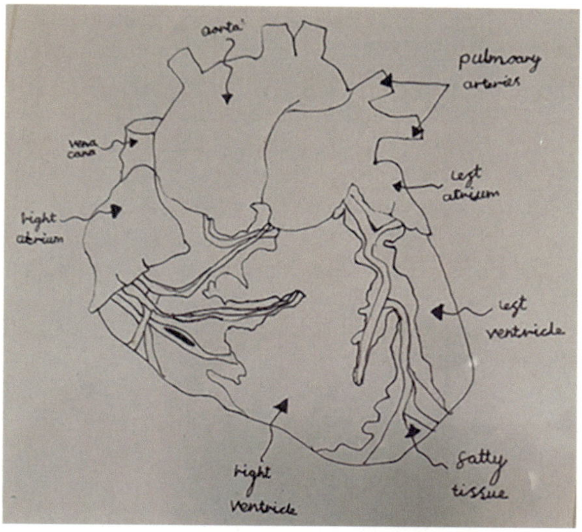

Figure 1.
Framed pupil heart sketch from "beat of life session."

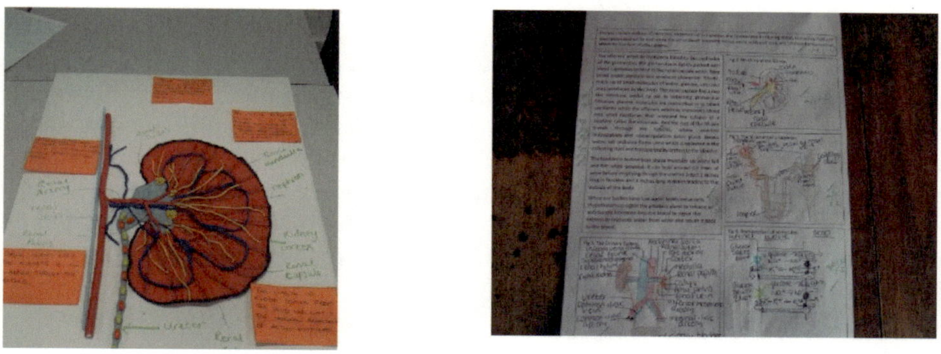

Figure 2.
Candy anatomy of the kidney and student assessment showing how this has been used to assess learning.

as shown in the diagrams below. Further it has huge successful results as the students used this in assessments to recall the different organ systems and their parts to obtain high grades of merits and distinctions (**Figure 2**).

In more recent times I have also used Lego to support the engagement of pupils in science/medical topics. In February I attended a workshop session which involved building a mini heart from Lego. This was suitable for all ages, young and old, and it made me think of adapting and building intersubject connections in this activity not just for school but University students as well. While being a creative, artistic process, this activity also instilled a spatial awareness of the heart structure; the red and blue colours representing the oxygenated and deoxygenated blood and opening room for comprehending the reality of blood flow between the heart and lungs (**Figure 3**).

These lung miniature kits can be bought from companies such as Brick This (https://www.brickthis.co.uk/), they have full instructions to build and support useful conversations around the blood flow of oxygenated and deoxygenated blood in the heart. This process simultaneously enhances spatial awareness and kinaesthetic learning.

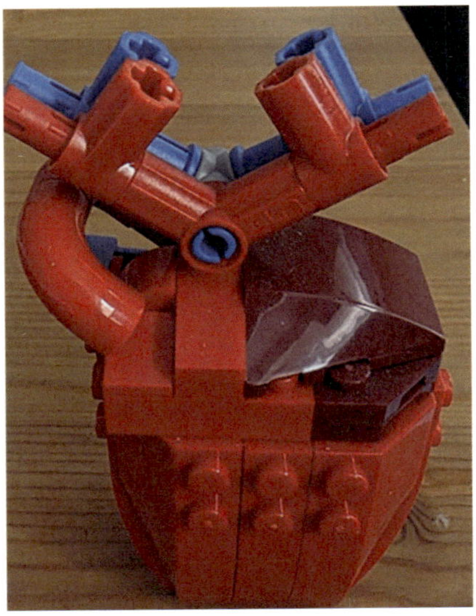

Figure 3.
Miniature Lego heart.

It can be hard to move children accustomed to subject-based teaching into spaces where they are able to think in interdisciplinary ways. One method to enhance this thinking is to provide pupils with a picture of an object such as a bicycle and ask them to identify the different 'subjects' implicit in this. Examples could include physics in the gears/friction tyres, wind resistance, bodies and machine interaction, circle shapes of wheels, and angles of spokes. This simple task unfolds the mind to many different aspects and concepts and is a great way to enhance trans-interdisciplinary thinking.

These examples illustrate how we can deepen learning and engagement and allow pupils to make subject connections. The embedding of building and artwork enhances the social, team-building aspect of learning and supports emotional and mental regulation and expression which can support behavioural and learning subject and emotional barriers for children with special educational needs.

3. Thinking, otherwise

In the challenging, complex landscape of life and in particular education, the idea of freeing the mind, and dwelling in the conscious/subconscious becomes not only appealing but almost a necessity in order to enhance curiosity, innovation and critical thinking. Additionally, it can enable the breakdown of barriers (such as those between subjects and the false binaries between cognition and emotion) and support the development of trans-subject, interdisciplinary connections. In this section we will discuss the role and importance of philosophy in implementing strategies towards the enhancement of these outcomes. Further we will examine the mystique and cyclical nature of time (via texts such as TS Eliot's poem, 'Burnt Norton') and consider how disrupting linear understandings of temporality can influence the ability to think deeply and process information, thereby resulting in novel ideas and discussions.

3.1 Time and the mind

Most would agree that life is run or even "controlled" by time; basic routines are all very much run against the clock. Our modern systems are centred around chronos (linear time) rather than aion (an understanding of time as cyclical, dynamic, insurgent, and connected to nature [10]). As Bridle [11] reminds us, '...the idea of global time - a single, universal standard, divided into hours and minutes, and shifting across time zones - is a recent one.' To imagine time working differently in education (whether it be primary, secondary, college and/or University) feels impossible; teaching sessions all have a set time frame and work to institutional clock rhythms. However, what if we could pause or interrupt time such that we lose that barrier? And how does that impact, not only on mental wellbeing but from an educational standpoint?

The issue of time is discussed eloquently in the late Kate Gross's podcast Private Passion [12]. While she lay gravely ill, Kate talked openly about her struggles with time and how she wanted to pause each moment to enjoy what she called 'the intensity of light and colour.' This has similarities both with the nature writings of authors such as Nan Shepherd and Tove Jansson, both of whom had a deep love of nature and were able to explore as independent spirits with the freedom of no time constraints; Shepherd in the Cairngorms, Scotland, and Jansson in a remote island off the Gulf of Finland. As with Gross, both, in Shepherd's words, 'immerse themselves' in their environments mind, body, soul and senses.

The common theme within the similar yet different scenarios is the desire to fully entangle body and senses with the environment in order to disrupt clock time. However, while Shepherd and Jansson had seemingly no time restrictions, Gross was terminally ill; a 'ticking time bomb'. It is interesting and almost ironic that as soon as the impending knowledge of death lies ahead, we value and want to pause or stop time. However, in life we feel the need to battle with time, unable to appreciate how the sheer enormity of pausing and soaking in the moment allows us to reflect, process and restore. It is as Gross states in her podcast, the need to 'look after our hinterland', not just in terminal illness but life itself.

Taking what Jasmine Ulmer [13] refers to as 'slow ontology' requires us to attune to time in different ways, working with artistic moments of flow, paying attention to the rhythms of nature, and disrupting linear processes which are restrictive and time bound. From an educational perspective this may involve taking learning outside - from closed classroom systems to the open flows of the natural environment - or paying attention to the ways in which learning extends outside school clock time to corridors, playgrounds and homes. These disruptions can be difficult however, particularly in the era of Coronavirus in which the language of speed, acceleration and catch-up is prevalent. In the classroom there is often a constant struggle to get through a dense curriculum or obtain a quick answer to a question so that learning can be evidenced, and the class can move on. The idea of even pausing, giving pupils time to absorb to process information, or reflect individually or with others can feel uncomfortable or risky, yet there is power in that pause regarding deep learning as well as wider human growth concerning being, becoming and thriving.

It is perhaps important at this point to clarify and define the three commonly known states of mind as explained via the Chambers Dictionary [14].

"Consciousness defines our thoughts and awareness, subconscious is defined as memories, motives, intentions and thoughts which the individual is only dimly aware, but which exert and influence on his/her behaviour and unconsciousness is the deepest inaccessible level of the psyche in which repressed impulses and memories are present in a dynamic state".

The concept of time and its relationship with the state of mind, is eloquently expressed in T S Eliot's [9]; the first of the four quartets. At the start Eliot states: 'Time present and time past are both perhaps present in time future'. Alluding to the way in which our understandings and experiences echo the past, as well as recognising their connection to the future, Eliot emphasises the need to value time and take account of the present moment. He goes on to describe the moment of fixity of time when he states, 'Allow but a little consciousness, to be conscious is not to be in time' and it is within this state of mind, we can pause to appreciate to immerse process and deepen our understanding, that is so essential to building transdisciplinary subject knowledge and emotional creative human growth. Such rich experiences we have within the present/past are important not only in our human growth and ability to understand ourselves but to help others understand themselves also. Interestingly, Pullman in The Amber Spyglass ([15], p. 97) states: 'In the landscape of your mind' that 'conscious beings make dust'; emphasising the way in which thought, sentience and recognition can allow us to materialise knowledge (dust here being a physical manifestation of creative and sentient acts). This knowledge he describes as not a 'fixed quantity' but something that gets renewed all the time by thinking, reflecting and by gaining wisdom. Further to this, Pullman describes how through these actions of pause and reflection you can pass on wisdom and help everyone else to learn about themselves, thereby enabling them to 'keep their minds open and free and curious.' This understanding of learning as relational and distributed troubles the individualised nature of education-as-usual, in which the 'dust' of knowledge is owned and used as capital.

This social transfer of knowledge supports Kalanithi's ([16], p. 97) view that 'Human knowledge is never contained in one person; it grows from the relationships we create between each other and the world and is still never complete.' This highlights that the human mind cannot necessarily thrive on constant bombardment of learning content by rote subject knowledge alone, and that there is a need for a humanistic aspect, such that the desire for knowledge incorporates a social and emotional aspect as well as subject authority.

3.2 Philosophical practices to disrupt clock time

Deleuze and Guattari [17] suggest that philosophy should be a practice of creativity and development of new ideas, with education thus being not about repetition but creation: 'Philosophy is the art of forming, inventing, and fabricating concepts' (p. 2). Philosophical concepts can be activators, used to reimagine education, or to defamiliarize ourselves from schooling-as-usual. Bringing art and philosophy together with science thus allows for affective understandings and different modes of thought, or lines of flight, which allow newness in and new understandings of the world to be formulated. One such creative practice, which also subverts linear, individualised and time-bound practices of teaching, is community philosophy, or Philosophy for Children (P4C) [18]. Embedding philosophical inquiry into Science education can allow for slow thinking, deep exploration and connection of science to wider social, environmental and cultural issues. This methodology facilitates active listening, mitigates binary oppositional thinking [1], and empowers young people to reflect critically in a democratic way [19].

In P4C a facilitator introduces a stimulus related to a topic (e.g., photograph, video, news clip); students discuss it in small groups and then identify the related concepts (e.g., justice, health, community). Groups then develop a question (via a democratic voting process), and the facilitator prompts wider discussion. Afterwards, participants engage in active reflection to critically examine how the process unfolded in terms of individual contributions, group dynamics and the creation or enactment of new concepts and perspectives. Such a practice has the potential to deepen understanding of topics such as ethics in science, as well as opening space to deal with complexity, appreciate different contexts and locations, and reimagine new ways to enact ethics in practice.

In a similar example of practice, Science teacher Tait Coles explains in his book, Punk Learning [20] the use of 'Question Formulation Technique' [21] and how it can be employed with pupils to disrupt teacher authority and provoke creative thinking. After introducing a stimulus (for example, an image of the digestive system), pupils are encouraged to create a list of questions about it, no matter how unusual, inappropriate or odd. Working with divergent thinking, rather than replicating the input of the teacher, is shown to foster creative and engaged responses as well as enabling students to work further on the crafting of incisive and thoughtful questions to provoke deeper engagement with scientific concepts. In this way, Socratic questioning techniques, with all their unpredictability, can disrupt 'teaching-as-usual' and open space for exciting perspectives which may well extend beyond the lesson time and walls of the classroom.

4. Creative practice and STEM

In Eliot [9] stated that 'Art required a form which could impose order and meaning on experience.' He goes on to say that '... all art emulates the condition of ritual'; however, this understanding of art as form and cohesion is challenged by outsider practices such as Art Brut. In Peter Shukie's pedagogical practice as a teacher educator [22]. Peter removes any kind of order in his classroom to allow for 'artistic expression' using raw art. Shukie describes art practice as enacting an 'alternate psychic reality' and questions 'what happens when the foundations of normality are removed' (n.p.). It is interesting to consider this as another state of mind or a process of flow [23], where the loss of inhibitions is neither conscious nor sub/unconscious. Within this state can lie a wealth of creativity, innovation and a sheer sense of being, becoming, and thriving. Some of the great artists such as Van Gogh had mental health issues yet produced work of great artistic talent. Keats in his poem 'Ode to Psyche' describes this as an 'untrodden region of my mind, where branched thoughts, new grown with pleasant pain...And in the midst of this quietness... A rosy sanctuary I will dress.' It is within this level of mind where we lose all barriers and restrictions that we experience total freedom of thought and senses and 'face our own intuitive sense.' Deleuze and Guattari's concept of the 'cosmic artisan' reflects this experience of becoming 'otherworldly,' via informal, creative acts which are undertaken in community. They suggest that this melding together of art, philosophy and science can enable new modes of being in the world: 'In this submersion it seems that there is extracted from chaos the shadow of the 'people to come' in the form that art, but also philosophy and science, summon forth...' ([17], p. 218).

The importance of engaging the 'landscape of the mind' and realising its holistic benefits is eloquently and beautifully illustrated in 'The Queen's Gambit'; a fictional but powerful drama about a girl called Beth growing up in a 1950s orphanage. Whilst dealing with childhood trauma and battling addiction, Beth is determined to learn chess and ultimately become a World Chess Champion. The social scenes in the film in which the janitor is teaching Beth chess depicts the distributed and relational aspect of learning within the conscious mind. However, Beth also has an addiction to sleeping drugs whereby her delusional mind allows her to visualise the chess board and create moves, allowing winning formulations and connections to be made. The technique used to demonstrate this in the film, where the chess board appears on the ceiling and moves take place, is an excellent way to show what Keats describes as the 'untrodden region of my mind, where branched thoughts, new grown with pleasant pain'; that is, the delusional yet creative subconscious. Additionally, it supports the idea that in this uninhibited mind space we can be ultimately creative and find our true selves. Seeking out such spaces can be difficult within time-bound learning environments, but nevertheless moments of flow; that is, complete absorption in joyful activity, should continue to be encouraged and elevated once experienced.

5. Bodies in relation

Taking a minor turn [24] means seeing difference (be it as appearance, identity, beliefs, culture etc.) as generative and productive. Difference from normative ways of being in education is rarely appreciated in and for itself, but instead viewed as deficit, with 'inclusion' meaning a layering of additional processes rather than a shift in or redesign of the education system. A holistic, affirmative curriculum practice avoids these universal, normative modes of teaching and moves towards process-led activity, which is grounded in care of individuals and elevation of different modes of being and generating knowledge. There is a central role for neurodiversity, not only for students but for teachers too.

Within Science education, Strom and Kayumova [25] suggest that dialogue is vital for building shared knowledges while recognising such different perspectives, identities and locations. Creating communities and turning away from neo-liberal framings of competition and individualism (many of which are engrained in liberal understandings of science itself) thus form part of an engaged pedagogical approach. Science continues to be based around Western epistemologies, and this in turn creates certain 'science' subjects; young people acculturated to ways of enacting scientific endeavour and practice. The authors suggest instead that we move to 'pluriversal praxis' which '...requires embracing an onto-epistemological shift based on relationality, interdependence, embodiment, ethics, and care towards youth, diverse communities, and more-than-human collectives.' ([25], p. 2).

Further to this, Kayumova and Dou [26] discuss the need to enact 'ethical and sociopolitical accountability' in science. This not only enhances a sense of belonging and engagement, but more importantly '...it helps perhaps even to design worlds in which articulations of nature and culture are intertwined and not separate in ways that we can defend or recreate worlds that relate important and communal dimensions.' Such strategies are necessary for providing an inclusive learning environment such that diverse learners from different cultural backgrounds can experience not only subject knowledge but do this 'outside human experiences' that build new relations with the natural world and new understandings of human/non-human relationality and reciprocity.

Strom and Kayumova also urge us to take account of the body in the classroom, stating that '...embodied onto epistemologies, social and cultural histories, experiences, and performances of different physical bodies (including human bodies) are central to the nature of social relations, including learning interactions.' Taking a material turn in this way means paying attention to the way bodies move in space and the role (or not) of bodily autonomy, as well as the structure of learning environments, the bodily effects of student-teacher relations, the impact of school rules on the body, and so on.

Francis Bacon was an advocate of the notion that the 'destiny of science was not only to enlarge human beings' knowledge but to improve human beings' life on earth' ([7], p. 28). He also believed that the construction of a better world was achieved through ascertaining the truth about natures' workings. The emphasis on knowledge retention currently seen in schooling frequently overlooks the ethical component of science learning, along with the understanding that learning is a relational and distributed activity. Shifting attention to the body and thereby appreciating inequalities, located and political perspectives and oppressive practices is a vital way of addressing this call for improvement, not only for humans but our more-than-human counterparts.

6. Conclusion

In this chapter we have drawn together philosophical ideas and theory with historical writings and contemporary works to demonstrate how learning-as-usual can be disrupted when we think across disciplinary boundaries and understand the mind/body binary as a false and unhelpful distinction that is engrained in educational systems. By drawing together science education with artistic, social, and philosophical practices, we have suggested here that new connections and ideas can be formulated which are valuable in our contemporary predicament. As Dewey urged us in [27], we must resist the compartmentalisation of subjects as this practice does not reflect the real world as experienced by children both externally and internally:

> *... [the] universe is fluid and fluent; its contents dissolve and re-form with amazing rapidity. But, after all, it is the child's own world. It has the unity and completeness of his own life. He goes to school, and various studies divide and fractionize the world for him. Geography selects, it abstracts and analyses one set of facts, and from one point of view. Arithmetic is another division, grammar another department, and so on indefinitely. (p. 5)*

Although we are necessarily part of systems that uphold disciplinary boundaries, we have attempted to demonstrate here that there are always methods by which we can integrate subjects and reterritorialize fixed ways of teaching and learning. Braidotti [1] urges us to undergo processes of de-familiarisation and dis-identification in order to give us critical distance from the systems that constrain us. Diffracting texts and media of different kinds through our teaching and thinking outside of disciplinary silos are examples of ways in which we can find ways of teaching and learning otherwise. Oppressive practices have become so normalised that we cannot imagine other ways of doing things. Yet when we examine structures, organisations and political systems from a critical distance, or with different understandings of what it means to be human, we begin to see the cracks and even feel a sense of the ridiculous. Owning 'radical space' alongside students is possible, both via the individual process of working ourselves from pain to knowledge, and collective thought and action. We close with bell hooks' important reminder and call to action: 'The classroom remains the most radical space of possibility in the academy' ((1994), p. 12).

Acknowledgements

I would like to express my gratitude to Professor Laura Stroud, Pro vice Chancellor and Dean of Health Sciences at Sunderland University, for her generous contribution to publishing costs. Also, to Mrs Sue Brent (Head) and Mrs Karen Giles (Associate Head) of the school of Nursing and Health Sciences at Sunderland University for their continued support in my professional development.

Author details

Christine Challen[1*†] and Kay Sidebottom[2*†]

1 Sunderland University, Sunderland, United Kingdom

2 University of Stirling, Stirling, Scotland, UK

*Address all correspondence to: christine.challen@sunderland.ac.uk and kay.sidebottom@stir.ac.uk

† The authors are equal contributing.

References

[1] Braidotti R. Transversal posthumanities. Philosophy Today. 2019;**63**(4):1181-1195

[2] Braidotti R, Hlavajova M. Posthuman Glossary. London: Bloomsbury Academic; 2018

[3] Rich A. Blood, Bread, and Poetry: Selected Prose 1979-1985. New York: W. W. Norton & Company; 1986

[4] Ball S. Neoliberal education? Confronting the slouching beast. Policy Futures in Education. 2016;**14**(8):1046-1059

[5] Woolf V. Three Guineas. London: Blackwell; 1938

[6] Lykke N. Postdisciplinarity. In: Braidotti R, Hlavajova M, editors. Posthuman Glossary. Cambridge: Bloomsbury; 2018. pp. 332-335

[7] Sidebottom K, Carlyle D. From reflection to diffraction: What Toto teaches us about "thinking-with" multispecies companions in education. Cultural and Pedagogical Inquiry. 2023;**14**(2):8-21

[8] Deleuze G, Guattari F. In: Massumi B, editor. A Thousand Plateaus: Capitalism and Schizophrenia. London: Bloomsbury Academic; 1987

[9] Eliot TS. The Complete Poems and Plays. London: Faber and Faber Publishers; 1934

[10] Braidotti R. A theoretical framework for the critical posthumanities. Theory, Culture and Society. 2018;**36**(6):31-61

[11] Bridle J. Ways of Being: Animals, Plants, Machines: The Search for a Planetary Intelligence. London: Penguin; 2022

[12] Gross K. Private passion. 2015. Available from: https://www.bbc.co.uk/programmes/b04wmg26 [Accessed: 26 July 2023]

[13] Ulmer J. Slow ontology. Qualitative Inquiry. 2017;**23**(3):201-211

[14] The Chambers Dictionary. 13th ed. London: Chambers Harrap Publishers Ltd.; 2014

[15] Pullman P. The Amber Spyglass. New York: Random Books; 2000

[16] Kalanithi P. When Breath Becomes Air. London: Random House; 2016

[17] Deleuze G, Guattari F. In: Burchell G, Tomlinson H, editors. What Is Philosophy? London: Verso; 1994

[18] Society for the Advancement of Philosophical Enquiry and Reflection in Education (SAPERE). Community philosophy. n.d. Available from: http://www.sapere.org.uk/MembersSchoolsPartners/CommunityPhilosophy.aspx [Accessed: 21 June 2023]

[19] Sidebottom K. From prevent to enable: Reclaiming radical thinking spaces through democratic education. Feminist Dissent. 2019;**4**:246-273. DOI: 10.31273/fd.n4.2019.189

[20] Coles T. Never Mind the Inspectors: Here's Punk Learning. Carmarthen: Independent Thinking Press; 2014

[21] Right Question Institute. What is the QFT?. n.d. [Online]. Available from: https://rightquestion.org/

what-is-the-qft/#:~:text=Developed%20by%20the%20Right%20Question,%2C%20accessible%2C%20and%20reproducible%20technique [Accessed: 19 October 2023]

[22] Sidebottom K, Shukie P. Dwelling in the Dissolve..... Towards Transdisciplinary Posthuman Pedagogies for Complex Times: Pedagogies/Posthumanities/Research Convergences. Edinburgh: Edinburgh University Press; forthcoming, 2023

[23] Csikszentmihalyi M. Flow: The Psychology of Optimal Experience. London: Harper Collins; 2008

[24] Manning E. The Minor Gesture. Durham NC: Duke University Press; 2016

[25] Strom K, Kayumova S. Ontology, Epistemology, and Critical Theory in STEM Education. Oxford: Oxford Research Encyclopedia of Education [Online]. 2023. Available from: https://oxfordre.com/education/view/10.1093/acrefore/9780190264093.001.0001/acrefore-9780190264093-e-1508 [Accessed: 29 August 2023]

[26] Kayumova S, Dou R. Equity and justice in science education: Toward a pluriverse of multiple identities and onto-epistemologies. Science Education. 2022;**106**:1097-1117. DOI: 10.1002/sce.21750

[27] Dewey J. The Child and the Curriculum. Chicago: University of Chicago Press; 1902

Chapter 7

Evaluating STEM Lecturers' Experiences with Digital Assessments and Continuity of Digital Transformation in Higher Education

Innocent Zitha, Georgina Mokgaetji Mokganya and Orifha Sinthumule

Abstract

There are clear signs that the COVID-19 pandemic has had a straining impact on the higher education sector in South Africa and the far reaches of the world. This chapter examines (Science, Technology, Engineering, Mathematics) STEM lecturers' experiences with digital assessments and perspectives on the continuity of digital transformation in the higher education sector in the post-COVID-19 era at one of South Africa's rural universities. In 2022, a survey of ten (10) lecturers was conducted and the reports were also consulted by four (4) e-learning practitioners to obtain their views on the use of online learning and contact courses in the post-pandemic period. A qualitative research approach was used to understand the puzzling issues faced by STEM lecturers at a previously disadvantaged institution in Limpopo Province. The results of this study show that there is a puzzling conundrum that hinders the integration of effective online learning. Furthermore, online education as the new normal in the post-COVID-19 era seems unattainable due to extremely complex and difficult-to-understand mathematical symbols. Software and tools proved to be limitations in terms of establishing an interactive learning environment for students and lecturers on online learning platforms. Lecturers' use of learning and teaching platforms deteriorated due to numerous unknown challenges, but also the tightening of COVID-19 restrictions. Furthermore, the results showed that science and mathematics lecturers believe that contact learning remains irreplaceable across generations and eras as some are not technology enthusiasts. This paper recommends that the lack of technological systems as well as technological resistance and reluctance to adopt transformative pedagogy and innovative practices should not be viewed as a thorny issue to end the dominance of online education in STEM courses.

Keywords: connectivism, Learning Management Systems, affordances, technological resistance, Moodle, transformative pedagogy

1. Introduction

In these uncertain times of the Fourth Industrial Revolution (4IR), faced with advanced technological changes, globalization, digitalization, and decolonization of curricular approaches in higher education, the question of what comes next in the higher education sector is likely to require calculated speculation and some risk-taking initiatives. It is important to note that the COVID-19 pandemic has not only created significant financial and operational challenges for higher education institutions in the short term but has also magnified the impact of demographic, financial, technological, and political structures that have been long-standing challenges in higher education sector. While lecturers remain committed to their institution's core mission in the face of these challenges, the lessons of COVID-19 have left a legacy that every institution must rethink the core values of the teaching and learning community. This paper has prepared this report to help institutions navigate the uncertainties ahead.

The new urgency for online or blended learning in higher education institutions worldwide and in South Africa caused by the COVID-19 pandemic has created an opportunity to rethink and reexamine the higher education sector. Globally, this has laid the foundation for institutions such as higher education structures, university councils, and public and private education-related professional associations to design and adopt policies to accelerate blended learning practices to support teaching and learning post-pandemic. With some academic programs in higher education being delivered entirely online in 2020 due to the COVID-19 pandemic, major changes have been required in the academic staff who design and present the programs, as well as the students who deliver them. The change and adaptability to change have been somewhat challenging for both the teacher and the learner, with less training on some Learning Management Systems (LMS) offerings during the pandemic. In a sense, both lecturers and learners learned how to navigate the online platforms to initiate learning and teaching activities.

Traditional on-campus, face-to-face instruction will most likely return as the pandemic subsides, providing the sociocultural elements of the instructor, learning environment, technology, learning activities, and peers that support effective student engagement in undergraduate education. Crosling et al. [1] the development of students' higher-level skills such as problem-solving skills and technical skills appear to be the focal point of effective and interactive learning space.

It is interesting to note that during the pandemic, students' preferences for online compared with traditional face-to-face learning and teaching activities shifted worldwide. This might have been due to some flexibilities toward their studies rather than having a fixed and predetermined teaching and learning pace. Kandri et al. [2] argued that about 78, 83, and 78% of tertiary students in Malaysia, Canada, and China, respectively, indicated a preference for online learning dependent on the program tuition fees if were to be reduced accordingly. Generally, it is expected that an online program should cost less than a blended or full contact program. Evident to Kandri et al. [2] who highlighted that globally institutions of higher learning saw online learning as the solution to advance teaching and learning during the pandemic. Alammary et al. [3] attested that the Singaporean Government deemed it a general requirement of the education sector to conduct their teaching and learning online. Alammary [3] and Zitha et al. [4] argued that it is expected that post-COVID-19 there will be new educational models, policies, and learning methods to emerge or be adopted by institutions of higher learning. The likely teaching and learning method to be adopted is either a blended learning method, which melds face-to-face teaching with online tools, or institutions going fully online for their teaching and learning activities. Zitha et al. [4] argued that blended learning has become immensely integrated in contemporary times owing to the emergence of COVID-19. Furthermore, they stressed that blended learning is an effective approach to meet the accelerated demands of the diverse student population in universities and colleges.

Emerging from an in-depth study of the uses and perspectives of academic staff with Blended Learning in a Singaporean HEI [5], the science, technology, engineering, and mathematics (STEM) lecturers seemed to experience challenges in online learning platforms as learning and teaching activities and all forms of assessments were to be given and graded on the Moodle LMS. Technically, the institution-prescribed LMS was not intensively explored on its affordances and used before the COVID-19 pandemic. Learning and teaching activities by then were conducted on traditional face-to-face and paper-based learning. Henceforth, the instant shift from minimal use of the Moodle LMS to being the core tool used to enhance learning and teaching within the institutions brought about challenges to both lecturers and students.

Moreover, the academics at one of the South African rural-based and historically disadvantaged universities constantly raise concerns regarding the Moodle platform in mathematics and science-related modules due to unduly accommodation in the assessments uploaded on the platform. The institution encourages academics to regularly use Moodle over other platforms to integrate online classes and assessments. Moreover, the assessments are mostly not easy to mark on the platform. The LMS affordances to STEM subjects seem lacking, which had a negative impact and vast challenges to teaching practitioners as well as to students when they attempt some of their assessment and module core-competencies tasks.

2. Purpose of the paper

The aim of this study is to evaluate STEM lecturers' experiences with digital assessments and the continuity of digital transformation in the higher education sector in the post-pandemic period at one of South Africa's rural universities.

3. Research questions

- What experiences have STEM instructors at a rural university had with digital assessments in the COVID-19 era?

- How do STEM lecturers perceive the continuity of online learning as the new norm in higher education?

4. Theoretical framework

Connectivism is a well-known theory that became popular in published articles in 2005, for which there were two publications: "Siemens Connectivism: Learning as Network Creation" [5] and "Downes An Introduction to Connective Knowledge" [6]. The current digital and e-learning space focuses on teaching and learning activities and cognitive theoretical foundations based on connectivism, which is a departure from constructivism. Considering connectivism and its technological implications, modern teaching and learning activities are based on this learning theory, which is based on a change in technology and the mandated LMSs by higher education institutions. Furthermore, this theoretical framework is used in this article as it focuses on understanding education in a period of digital transformation. It then emphasizes how Internet technologies such as web browsers, search engines, wikis, online discussion forums, and social networks relate to current ways of learning. The affordability of technology has enabled academics to acquire skills and share knowledge across the World Wide Web and with each other in ways that were not possible before the transition to digital transformation.

Learning does not simply happen in an individual, nevertheless, it is across the networks. Connectivism comprehends knowledge as a network and learning as a process of pattern recognition [7]. The phrase "a learning theory for the digital age" indicates the emphasis that connectivism provides to technology's effect on how people live, communicate, and learn. In relation to this paper, connectivism becomes an integral part of establishing the interconnectedness between the lecturer and students with transformative principle-based pedagogy where the emphasis is on the joint construction of knowledge in a community of students through the aid of e-learning education systems and platforms.

5. Material and methods

In this study, the researchers used a qualitative research method in which the survey questionnaire was later quantified to determine the degree of the problem and the prevailing problems and challenges from the researchers' responses. Miles et al. [8] and Kuper [9] demonstrate the importance of using a qualitative data collection approach in the social sciences extremely effectively for the study of areas such as education, sociology, and anthropology. Most significantly, the use of this approach has increased in the health professions and education sectors. This method was used to obtain descriptive and empirical data from participants [10]. This included frequency counts and percentages to identify the most common perspectives and challenges related to the questions. The researchers used a purposive sampling technique as the participants were interested in interest of the researchers' criteria as they used technological devices and navigated the online space during the hard lockdown period. The sample included a total of ten (10) STEM lecturers from different departments of the Faculty of Science, Engineering, and Agriculture and four (4) e-learning practitioners from the rural university in Limpopo Province, all of whom were scheduled to participate in the study, by first signing the ethical review consent form and later completing the questionnaire and narratives.

However, only ten (10) academics and four (4) e-learning practitioners from the entire population of science lecturers responded to the questionnaire. This survey aimed to assess lecturers' experiences with digital assessments while enabling online lectures on various platforms. The data collected for this study were analyzed using thematic coding and descriptive statistics, with researchers carefully going through responses to classify the data by themes and academics' perspectives on the use of technology beyond the COVID-19 era in higher education institutions among the students from rural areas to better understand the challenges encountered with respect to online education. The researchers analyzed publicly available data, and surveys and interviewed lecturers and e-learning practitioners from rural universities in the South African higher education sector to large public flagship universities.

6. Results and discussion

6.1 Sociodemographic data of the participants

The population comprised six male academics and four female academics who are lecturing the science modules shown in **Figure 1**, wherein there is an integration of Moodle LMS. Subsequently, it was quite difficult to mark digital assessments for STEM courses and provide timely feedback to the students. The four male e-learning practitioners were interviewed to report on their observations and challenges encountered by the academics in the different faculties within this institution.

The responsibilities of the three e-learning practitioners focused mainly on supporting staff academics with mitigation of conundrums faced during the digital engagements and grading assessments for STEM courses (**Figure 2**). On the other hand, challenges around digital learning and teaching are not encountered by academics only, however, students are affected and therefore one of the participants was responsible for the provisioning of academic support to students (**Figure 2**).

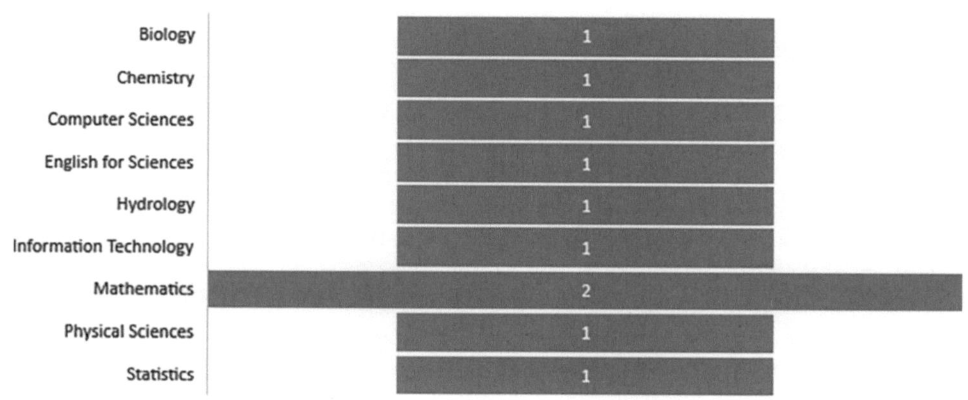

Figure 1.
Science modules taught by the ten academic participants.

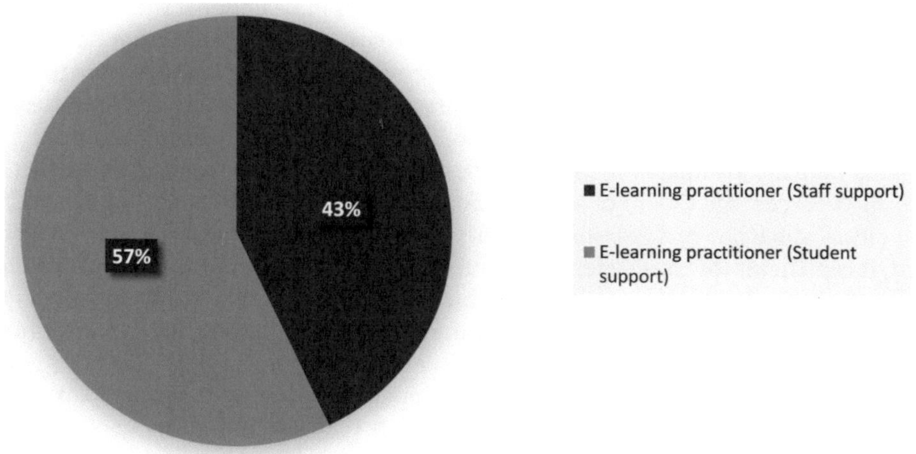

Figure 2.
Accountabilities of e-learning practitioners.

Moreover, the data were collected through interviews among the lecturers in the science-related courses. Subsequently, the consultation of the report was facilitated by the e-learning practitioners to obtain the data for comprehension of the challenges observed through the teaching and learning activities. The students whom the lecturers taught were at the first-year level in the Science Foundation, Earth Sciences, and Mathematical and Computational Sciences departments at one of the rural-based universities in South Africa.

The findings of this study exhibit that there are major issues and challenges that many academics and e-learning practitioners are encountering regarding the integration of blended learning as the new norm even in the post-COVID-19 era in higher education. Technological continuity as a new normal in the post-COVID-19 era seems to be unattainable.

6.2 Moodle LMS adoption

The table titled "Moodle Adoption Report" shows the active and inactive academics in line with the minimum online presence policy within the four faculties namely, the Faculty of Science, Engineering, and Agriculture, the Faculty of Health Sciences, the Faculty of Management Commerce and Law, and the Faculty of Humanities, Social Sciences, and Education. Moreover, the report further indicated the percentage of adoption within the four faculties in the university. The Faculty of Science, Engineering, and Agriculture recorded 29, 57%, the Faculty of Management, Commerce, and Law recorded 29, 09%, the Faculty of Health Sciences recorded 10, 62%, and the Faculty of Humanities and Social Sciences reported 26, 76% in relation to the Moodle LMS usage. Nevertheless, this study focused on participants sampled from the three departments in the Faculty of Science, Engineering, and Agriculture (**Table 1**).

Interestingly, the results of the report clearly illustrated a total of 16 online active academics out of a total of all teaching staff members of the three participating departments. Frankly, this notion resulted in a huge number of 40 online inactive academics within these departments.

The usage of the Moodle LMS has deteriorated and the adoption is quite cumbersome in terms of the percentage of academics who are actively integrating blended learning. Moreover, the continuity of digital education seems to be questionable based on the current realities of usage. As such, some of the lecturers indicated that the assessments will forever need a physical setting, not an online space, particularly in rural-based and historically disadvantaged institutions where connectivity infrastructures seem to be a major conundrum for students from remote areas. Some honestly indicated that they are reluctant to undergo proper training for Moodle and other online platforms and their students were mostly unable to engage in the ongoing discussion during the session.

A high percentage (70%) of some academics narrated that face-to-face is better than online since one can monitor students who attend the classes while on the platform, it is difficult for the academics to trace the students who are actively engaging in the content and those who are only available for compliance (**Figure 3**). Hwakoh [11] concurs with the difficulties of student online engagement when students show no interaction during the online lecture.

On the other hand, 30% of academic participants thrived with learning the usage of the Moodle platform (**Figure 3**). These lecturers realized the significance of LMS usage as it brings up an improvement in terms of the quality of teaching [12]. Over and above, the online preference of some academics resulted from the LMS benefits such as interacting with students through assigning essay assignments, testing *via*

Faculty	**Online active staff**	**Online inactive staff**	**Staff total**	**Percentage adoption**
1. SCI_ENG_AGAIC	55	131	186	29.57%
2. FMCL	32	78	110	29.09%
3. HUMAN_SOC_SC	12	101	113	10.62%
4. HEALTH_SC	19	52	71	26.76%
11. Active with no faculty	28	0	28	100.00%
12. TEACH ASSISTANT	80	0	80	100.00%
A. Faculty of science mathematics and engineering				
Dept: Department name	No of active staff	No of inactive staff	Total no	% Adoption
1. Dept: AGRICULTURAL_ECONOMICS_AND_AGRICULTURE	1	2	3	33.33%
2. Dept: ANIMAL_SCIENCE	3	3	6	50.00%
3. Dept: BIOCHEMISTRY_ANL_MICROBIOLOGY	2	12	14	14.29%
4. Dept: BIOLOGICAL_SCIENCES	6	12	18	33.33%
5. Dept: CHEMISTRY	3	13	16	18.75%
6. Dept: EARTH_SCIENCE	5	11	16	31.25%
7. Dept: FOOD_SCIENCE_AND_TECHNOLOGY	2	5	7	28.57%
8. Dept: FORESTRY	1	4	5	20.00%
9. Dept: GEOGRAPHY_AND_ENVIRONMENT_SCIENCE	9	7	16	56.25%
10. Dept: INSTITUTE_FOR_RURAL_DEVELOPMENT	0	5	0	00.00%
11. Dept: MATHEMATICAL_AND_COMPUTATIONAL_SCIENCE	9	26	35	25.71%
12. Dept: PHYSICS	1	12	13	07.69%
13. Dept: PLANT_AND_SOIL_SCIENCE	6	6	12	50.00%
14. Dept: SCIENCE_FOUNDATION	2	3	5	40.00%
15. Dept: URBAN_AND_REGIONAL_PLANNING	3	7	10	30.00%
16. Dept: AGRICULTURAL_AND_RURAL_ENGINEERING	2	3	5	40.00%

Table 1.
Moodle adoption report.

multiple-choice questions, provision of presentation slides prior to lecturers [13], and detecting students' participation through the online participation register.

Furthermore, the results of this study indicate the preferred platforms (WhatsApp, Moodle, Microsoft Teams, and Google Teams) shown in **Figure 4** that

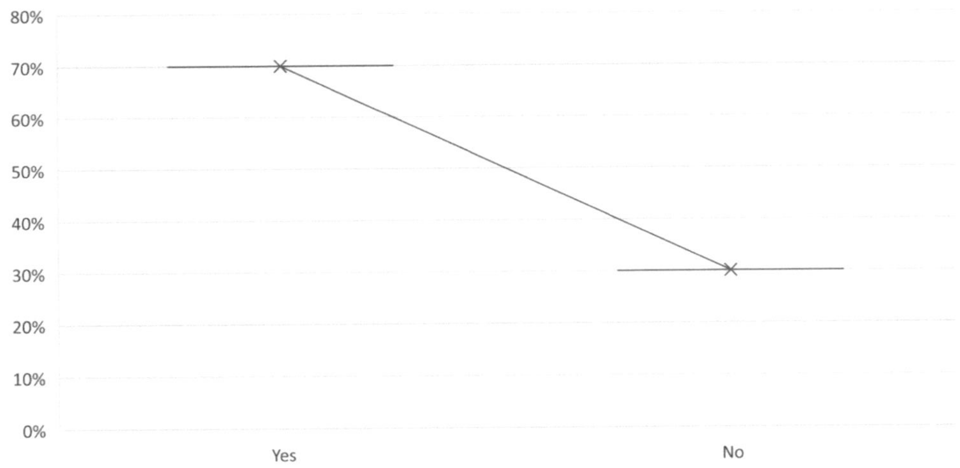

Figure 3.
Lecturers' face-to-face preferences.

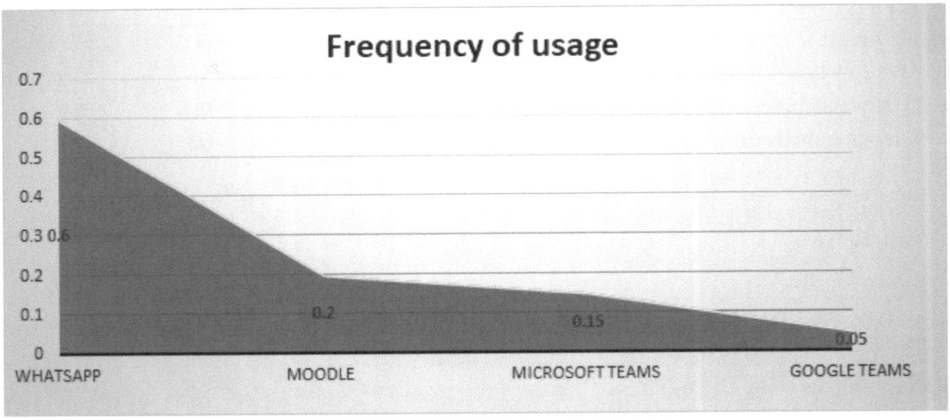

Figure 4.
Frequency of use of online platforms by lecturers.

academics use to supplement their teaching and learning with the students if they are not available on campus for contact sessions.

WhatsApp is more used than the other platforms due to students' data challenges as they are presently not receiving incentives from the institution. Moreover, the computer labs were not accessible to the students, and they had problems with the lack of devices for online classes. Students supported the use of WhatsApp while considering their financial circumstances and lack of technological devices to attend online classes and network connectivity problems in the comfort of their learning space. The lecturers provided feedback on students' assessments using these platforms, while interaction between the lecturers and students was quite unsatisfactory due to the network connectivity.

This resonates with Tay et al. [14] articulated in their study that online learning depends on the technologies used at the time and the curriculum that is being taught. The researchers suggest that the platforms and different approaches should be integrated to achieve maximum learning. Moreover, in congruence with the retrospect of studies conducted in this area, the findings of this study report that the digital competence of the users may be a solution to online learning.

6.3 Usage of the Learning Management System between students and academics

The e-learning practitioners expressed their concerns and major conundrums during their training, observations of minimum online presence, and daily consultation with the academics and students. The students' daily consultation had common problems, which among others encompass network connectivity, struggling to log into the Moodle LMS, and inaccessibility to the assessment activities uploaded by their module lecturers. Furthermore, e-learning practitioners further elucidated that academics have challenges concerning marking assessments and providing feedback on the platform due to inadequate mastery of the platform operation and navigation skills. This study is validated by the findings of Karuppiah et al. [15] who elucidated that it is imperative to recognize the type of student enrolled in the course to design a suitable interaction system that enables the students not only to learn but even to interact among themselves. Therefore, LMS's affordances come in handy in any prescribed learning and teaching e-platform as it determines and prescribes efficacy through use in learning activities within a course.

6.4 Complacency in the online setting

The findings of this paper reported that the biggest problem among others is complacency in the online setting and the academics tend to think that the online platform is conformable to constantly facilitate online classes while many students are mostly not participating and only available for compliance and procedural record keeping. Over and above, online classes have proven not to be solely independent due to the higher number of students who are not capacitated, and this poses an academic trajectory in many institutions. This study concurs with Dhawan [16] who pinpointed that face-to-face teaching has received strong support from students at all levels and has been viewed as more effective than online interaction as the lecturers and students tend to experience challenges. Lai et al. [17] argued that within a learning and teaching setup, its deliverables are on a web-based Learning Management System, results showed that students' frustrations were found in three folds:

(i) *Lack of prompt feedback from the lecturer*
Lack of prompt feedback denotes that STEM students were impeded from receiving proper guidance and timely feedback on learning and teaching progress. This ultimately relates to a lack of guidance, which, in the long run, puts students at risk of not mastering concepts and not passing the module.

(ii) *Ambiguous instructions on the Web*
Learning and teaching instruction must be explicitly stated and directed to students, however, in an online platform, students find that teaching and learning activities with competing deadlines may have ambiguous instruction and it is practically not viable to continue with the task until such time the lecturer may clarify the instructions.

(iii) *Technical problems*
Amid online learning and teaching activities, STEM students raised concerns about technical problems they encounter on and with the LMS platforms. These

challenges cannot be resolved as expeditiously as needed; they may be escalated to e-learning support personnel. This could be frustrating to students as they may be facing an immediate need to use and participate in learning and teaching activities but fail to do so because of technical problems they encounter. Therefore, students get support from the e-learning whenever they are on the verge to submit their projects and this poses learning challenges with respect to the submission deadlines.

6.5 Fixed mindset and expectation

In this study, it was mainly reported that the transformative and flexible strategies and approaches were reluctant considered by the academics in their attempts to adapt to the new normal. Furthermore, the findings revealed that most academics believed that contact learning remains irreplaceable throughout generations and ages as some are technophilic. This study is incongruent with the findings of Lai et al. [17]. The study outlined that face-to-face learning is irreplaceable and a cornerstone of any learning institution, even if the current discourse and technological revolution require the use of e-Learning. Attard [18] further highlighted in comparison to a program that is run online and one that is not, students who are attending their course online face a few barriers to their full participation in coursework units. Henceforth, face-to-face learning is more powerful in terms of student engagement and participation than online learning. Moreover, online learning was and is still an effective approach to meet the accelerated demands of the diverse student population in universities and colleges.

6.6 Resistance to using technology

The e-learning practitioners who were interviewed about their perceptions and perspectives on the continuous usage of technology in the higher education sector in the post-COVID-19 era indicated that the academics deteriorated the use of blended learning in their respective programs. This has been necessitated by the adjustment of COVID-19 regulations imposed on the indoor gathering capacity wherein there is a resurgence in COVID-19 cases. Subsequently, the findings of this study are consistent with Attard [18], which articulated that a major challenge for online learning is the one-size-fits-all approach, which does not do justice to student differences.

This is further validated by the authors in Orlando and Attar [19] elucidated that emerging and experienced academics are reluctant to adapt to the digital transformation in their teaching practices and approaches. The researchers suggest that online learning should be integrated with contact sessions to mitigate decontextualized teaching classrooms while technical support ought to be provided to the academics who are unable to effectively use blended learning in their teaching.

6.7 Lack of technological skills

Based on the findings of this study, this aspect of technological deficiency has largely been in the responses of both academics and e-learning practitioners. Most lecturers stated that they are not well equipped with the technological devices usually used for online teaching and learning environments. Academic institutions were impelled by the COVID-19 pandemic to integrate technology when facilitating lessons. However, academics made use of social media platforms to compensate for their incompetency with Moodle utilization. Nevertheless, this strategy did not yield positive results as the students were unable to comprehend the feedback on their assessments owing to the unsatisfactory interaction with their lecturers. Significantly, there is a need for academics to be well-trained to use technology and LMS affordances skillfully and excellently. Unfortunately, Orlando and Attar [19] posit that the online environment presents several challenges for both academic staff and students who increasingly require higher levels of technological competency and proficiency on top of their regular academic workload. This should be done to ensure that they have the necessary competencies and proficiency to use the prescribed LMS to aid their academic activities. This study corroborates with Khoa et al. [20] whose findings exhibited that lecturers and students are mainly not thoroughly trained to navigate the Moodle platform affordances. Moreover, it attests to Khoa et al. [20] espoused that technology is expected to improve access to education, reduce costs, improve the cost-effectiveness of education, and maintain the competitive advantage in recruiting students in higher education.

7. Conclusion

In keeping with the purpose of this study, it is crystal clear that the continuity of technology and digital learning still appears to be a difficult puzzle, as the STEM academics who participated in this study were completely unwilling to adopt transformative pedagogies and adapt to the online learning environment. Furthermore, the full realization of technology as the new norm in the post-COVID-19 era seems virtually unattainable due to the challenges and issues expressed by STEM academics and e-learning specialists. This is a prevalent area of concern as the COVID-19 pandemic state of emergency has exacerbated teaching and learning challenges due to distanced interaction. Conversely, it is important to note that the online space has been viewed as a temporary learning environment as most lecturers do not seem to welcome blended learning beyond the strict COVID-19 restrictions. Even though there is a paradigm shift in teaching and learning approaches. Most academics believe that contact learning remains irreplaceable across generations and ages as some are technology enthusiasts. Most academics used social media platforms to compensate for their inability to use Moodle.

8. Recommendations and implications for practice

To increase the acceptance of the use of the Moodle LMS by STEM lecturers in lectures, some business aspects should be considered, for example, the establishment of some guidelines. The policy needs to be implemented for the minimal online presence of academics and students to improve the use of the platform. Moodle training should be mandatory for both students and academics to provide them with the basic skills to use it. In addition to the training sessions, user-friendly manuals for STEM students should be developed to guide LMS users. Additionally, it is important to promote the utility of the platform by recognizing and rewarding instructors who use the platform effectively and imaginatively in the teaching process. Recognition and reward can be determined by clear criteria for assessing the quality of students' pass rate or outcome after acceptance by lecturers. Another useful indicator would be to compare student scores before and after using the LMS, as well as student satisfaction reports before and after implementing the LMS. In addition, the formative assessments should be submitted in each module on the online platform every semester. The Moodle platform should be used to track students' active participation on the platform and impose penalties on those who do not actively use the assessment and learning tools. Lack of access to devices should not be used as a barrier to consider e-learning as a viable option to discontinue learning. The connectivity challenge in rural areas should not be used as a reason to ignore the significant potential of e-learning as an enabler to access and advance education. The digital learning challenge should be prioritized to encourage the lecturers to use it always, while all academics should be advised to conduct at least 30% of their assessments on the online system to increase the usage of blended learning.

Author details

Innocent Zitha[1*], Georgina Mokgaetji Mokganya[1] and Orifha Sinthumule[2]

1 Faculty of Science, Department of Science Foundation, Engineering and Agriculture, South Africa

2 Faculty of Management, Extended Curriculum Programme, Commerce and Law, South Africa

*Address all correspondence to: innocent.zitha@univen.ac.za

References

[1] Crosling G, Edwards R, Schroder B. Internationalising the curriculum: The implementation experience in a faculty of business and economics. Journal of Higher Education Policy and Management. 2008;**30**(2):107-121

[2] Kandri JJ, Marić IG-A. Implications of the COVID-19 pandemic in education and vaccine hesitancy among students: A cross-sectional analysis from France. International Journal of Logistics Research and Applications. 2020:**4**(1):09-20

[3] Alammary A, Sheard J, Carbone A. Blended learning in higher education: Three different design approaches. Australasian Journal of Educational Technology. 2014;**30**(4):10-19

[4] Zitha I, Mokganya MG, Manyage T. Integration of blended learning in the advent of COVID-19: Online learning experiences of the science foundation students. Education Sciences. 2023;**13**(7):8-11

[5] Siemens G. Connectivism: Learning as network-creation. ASTD Learning News. 2005;**10**(1):1-28

[6] Downes S. An Introduction to Connective Knowledge. Washington: Routledge Press; 2005:**1**(5):5-19

[7] Ayebi-Arthur K. E-learning, resilience and change in higher education: Helping a university cope after a natural disaster. E-learning and Digital Media. 2017;**14**(5):259-274

[8] Miles E, Crisp RJ, Husnu S. Support for the replicability of imagined contact effects. Social Psychology. 2014;**45**(4):303-304

[9] Kuper A, Lingard L, Levinson W. Critically appraising qualitative research. BMJ. 2008;**33**(7):14-22

[10] Jacobs D. Teaching Responsibility through Physical Activity. Champaign, IL: Human Kinetics; 2013

[11] Hwakoh Y. Internet-based distance learning in higher education. Tech Directions. 2020;**62**(1):28-32

[12] Liu J, Long Y. Critical thinking and the disciplines reconsidered. Higher Education Research and Development. 2014;**32**(4):529-544

[13] Gillett-Swan J. The challenges of online learning supporting and engaging the isolated learner. Journal of Learning Design. Special Issue: Business Management. 2017;**10**(1):20-30

[14] Tay Y, Bahri D, Metzler D, Juan DC, Zhao Z, Zheng C. Synthesizer: Rethinking self-attention for transformer models. In: International Conference on Machine Learning. New York: PMLR; 2021. pp. 10183-10192

[15] Karuppiah K, Sankaranarayanan B, Ali SM, Paul SK. Key challenges to sustainable humanitarian supply chains: Lessons from the COVID-19 pandemic. Sustainability. 2021;**13**(11):5850

[16] Dhawan S. Online learning: A panacea in the time of COVID-19 crisis. Journal of Educational Technology Systems. 2020;**49**(1):5-22

[17] Lai J, Widmar NO. Revisiting the digital divide in the COVID-19 era. Applied Economic Perspectives and Policy. 2021;**43**(1):458-464

[18] Attard M, McArthur A, Riitano D, Aromataris E, Bollen C, Pearson A. Improving communication between healthcare professionals and patients with limited English proficiency

in the general practice setting. Australian Journal of Primary Health. 2015;**21**(1):96-101

[19] Orlando D, Attar E. Digital natives come of age: The reality of today's early career teachers using mobile devices to teach mathematics. Mathematics Education Research Journal. 2015;**28**:107-121

[20] Khoa BT, Ha NM, Nguyen TVH, Bich NH. Lecturers' adoption to use the online Learning Management System (LMS): Empirical evidence from TAM2 model for Vietnam. Ho Chi Minh City Open University Journal of Science-Economics and Business Administration. 2020;**10**(1):3-17

Chapter 8

STEM Integration in Resource Constrained Environments

Majid Nawaz, Irfan Ahmed Khan and Muhammad Imran Ahmad

Abstract

The revolutionary pace of technological advancements of present day offer a silver lining for bringing about much needed and desired change in education systems in developing countries to bring them at par with the requirements of the 21st century. STEM education promises the necessary impetus to catapult the orthodox education systems to modernity in an accelerated manner. In this chapter, we discuss the challenges and opportunities of STEM integration in the context of resource-constrained environments, offer our insights and present novel approaches developed based on our experiences from STEM implementation in urban and rural areas of Pakistan. From setting up purpose-built STEM labs to conducting STEM sessions in tight spaces or open areas, development of STEM curriculum based on relevance, authenticity and affordability, from robotics to do-it-yourself (DIY) projects with low cost electronics or recyclable and reusable materials, a range of possibilities are extensively covered. Teacher training and program sustainability are also discussed for STEM integration in resource-constrained environments of developing countries.

Keywords: STEM implementation, STEM methodology, STEM education, framework, effectiveness of STEM

1. Introduction

The 21st century, with all its marvels and awe-inspiring innovations, poses a very apt and pertinent question to parents and educators around the world; in this age of technology and massive information inflow, how do we engage our children in creative activities? The introduction of STEM methodology in the last couple of decades in education systems around the world has attempted to address this challenge. STEM methodology, based on Engineering Design Process, invites students to collaborative problem solving, engages them through relevant and authentic topics, motivates them to come up with creative solutions and in doing so prepares them for the challenges of today and tomorrow.

The constructivist approach to learning and the concept of learning through experience has been around for over a century. Jean Piaget's constructivist theory focused on knowledge development based on experiences, in four stages of cognitive development [1]. Lev Vygotsky introduced Zone of Proximal Development (ZPD) in his social-cultural development theory emphasizing the process of learning by doing [2]. Jerome Bruner supported guided discovery learning [3].

However, STEM methodology introduces a systematic approach to problem solving through interdisciplinary, crosscutting and hands-on experiential learning. In their journey from an idea to a solution, students learn and apply integrated curriculum concepts, use technology to create a model and apply the principles of Engineering Design Process iteratively to achieve the desired results. Failures are seen as opportunities to learn and course correct.

STEM methodology has gained significant popularity in the last couple of decades, and for good reasons. Many countries around the world have either experimented with or successfully incorporated STEM methodology in their school systems. The improvement in Science and Math learning, nurturing of core competencies such as Creativity, Communication, Collaboration, and Critical Thinking also called 4Cs, and development of hands-on technological skills are the common denominations that drive the uptake of this method in education systems around the world.

In the following pages we would like to share our experiences, as STEM service providers, of implementing STEM methodology in urban and rural areas of Pakistan. The unique challenges faced, opportunities and the lessons learned during our journey are presented in the context of resource constrained environments.

1.1 STEM as a great equalizer?

The 21st century has ushered in technological advancements at a mind boggling pace. The fourth industrial revolution led by Artificial Intelligence, Blockchain and nano technology is threatening to shake the very foundation of our societal values, our way of thinking, way of working and way of receiving education. The education systems of developed countries are striving to keep up with the pace, and those of the developing countries are spinning their wheels to come out of the 20th century. The gap is increasing at an alarming rate. Beyond any doubt, it is the integration of Science, Technology, Engineering and Math that has enabled significant developments and innovations in the past few decades and it is this integration of these disciplines that holds the panacea for education systems to prepare our children for the world as it exists today and for what it is about to become in the future.

The requirements of modern education systems include not only foundational literacies such as reading, writing, math, Internet and Communication Technologies, sciences but also skills, competencies and attitudes to become a creative and collaborative problem solver and therefore a contributing citizen. STEM education holds the promise to deliver this and more. Adoption of STEM methodology in developing countries can not only address some fundamental issues such as quality of science and math education, acquisition of essential 21st century literacies, skills and competencies, but also help improve enrollment and retention rates, gender parity in STEM fields, and provide an accelerated path to catch up with the rest of the world.

1.2 Challenges of STEM integration in developing countries

There are numerous challenges in introducing STEM methodology in resource-constrained environments. The schools are over-crowded, infrastructure is lacking, teachers are fewer in numbers, not adequately qualified and trained, over-burdened and less motivated, if not completely averse, to bringing any change or improvement in the status quo. The curriculum is often antiquated, standards are missing or obsolete, and the textbooks void of interesting content.

The school leadership despite having the desire for improvement and innovation find themselves in a quagmire to either focus their energies and resources on running day-to-day operations and meeting the bare minimum requirements or risk disturbing the apple cart for experimentation with new methods, tools and pedagogies. The former takes precedence in almost all cases.

At the government level, things become even more complicated. Even if the policy makers are proactive in adopting measures to bring about improvement in the system, the sheer inertia of the behemoth system coupled with inherent inefficiencies create significant hurdles and drag in implementation of new initiatives seeking improvement in quality of education resulting in either failure of the interventions to take off or bring about desired changes at all.

The cost of STEM implementation is a major concern for all stakeholders. For most people, one of the first exposures to STEM comes from Robotics. Robotic kits are expensive in general and in developing countries they are simply cost prohibitive. This often creates misunderstanding that STEM being Robotics only is simply not affordable. In most, if not all, developing countries, schools are over-crowded and space comes at a premium. This is particularly true for government owned and run schools. It is extremely difficult to conduct STEM sessions in classrooms effectively.

Turning these challenges into opportunities require some very well thought through and innovative solutions. Therefore, we offer our perspective of integrating STEM education in resource constrained schools in the following sections.

2. Integration of STEM education in a school: A phased approach

Introduction of STEM education requires a phased approach to give time to the school administration and faculty to fully understand and prepare for a full integration of STEM methodology in their regular classroom sessions. Starting with a STEM club, then introduction in a couple of classes and finally expanding gradually to cover more classes allows the system to ease in into the integration of STEM throughout the school.

2.1 STEM lab design and setup

STEM lab provides an environment for students to foster learning in a fun and exciting way. It provides space, tools and means for students to practice their theoretical knowledge by creating solutions for real-world problems that are both authentic and relevant to them. In a STEM lab, students not only engage in learning of core-curriculum concepts and nurture core competencies but also acquire new psychomotor skills.

A STEM Lab must be designed to facilitate hands-on experiential learning in an effective manner utilizing the space optimally and efficiently within the confines of specified constraints and limitations. This is a balancing act. The lab design process begins by understanding the purpose and basic requirements of the lab. These requirements emanate from the planned activities and the number of students that will be accommodated for these activities at any given time. Unlike a typical classroom where the setup is static, a STEM lab is designed on the principles of freedom of movement and freedom of choice. Dynamic spaces are created through modular and mobile furniture, which offers the flexibility for the students to move around and work in groups. The furniture is designed to promote active lean-forward learning instead of passive lean-back information transfer.

Modular and mobile furniture provides opportunities to create open spaces when needed so that the students can work on the floor or gather around for some activities. Sufficient storage spaces are provided in the form of shelves and cabinets to stow away tools, equipment and materials used in STEM activities and also showcase student projects. The material used in furniture needs to be lightweight, durable and water proof.

Continuous electricity supply, especially in the rural areas is one of the many challenges in developing countries. Backup power becomes a necessity. Similarly, Internet connection in remote areas is mostly available through wireless carriers and as such needs to be arranged for the STEM lab. A view of such a model STEM lab is shown in **Figure 1**.

2.2 STEM in classrooms in rural schools

Most schools in rural areas, whether government, private or community owned are extremely resource constrained. There are barely enough rooms to accommodate all students and in most cases there aren't any computer labs or libraries. Constructing a new facility for STEM activities is out of question. So what to do in these cases? For situations like these, the lesson plans are adapted to make them locally relevant and the associated STEM activities are designed such that these can be conducted either in the classrooms or in an open area usually used as playground or assembly area.

Figure 1.
STEM Lab: A project of Durrani STEMs @ Service (DS@S) at Qadir Nagar High School, Buner, Pakistan.

Here it will be pertinent to mention that STEM is not so much about the equipment, materials and tools used to perform activities but it is the iterative process of brainstorming ideas, creating plans, crafting solutions and testing and reflecting upon this process. If we carefully examine this process, STEM sessions can be effectively carried out with very limited and meager resources. The ideation, planning and reflection phases of the process simply require a piece of paper, a pencil and some peers to discuss, record and communicate whereas the creation and testing part can be carried out using mostly recyclable and reusable materials.

2.3 STEM on wheels

In one of our visits to a remote area in the North of Pakistan, where we were starting to implement STEM in community owned schools with the help of an organization that was actively involved in improving the infrastructure and quality of education in these schools, we came across one of their initiatives in which a pickup van was converted into a mobile library. The van was filled with colorful children's books of all sorts. The van visited around 30 schools a month and every school would receive one visit a month. The coordinator of program mentioned that since these schools didn't have libraries, the students waited eagerly for the Mobile library van to show up. Quite remarkably, the coordinator had fitted in a couple of boxes of sports goods and gardening tools in the van as well to give children opportunities to play and also learn the basics of gardening.

This gave us an idea. If we can't have a dedicated STEM lab in each of these schools for the students to experience the wonders of science, why not use the Mobile library as a vehicle to reach out to these kids by including some very basic STEM kits and activities that will pique their interest and nurture their curiosity. The van will visit each school in the area and the students will not only have the option to read books but also experience the wonders of science with their peers.

2.4 Community STEM labs

Another effective way of providing dedicated spaces for science exploration is setting up of community STEM labs. Just like public libraries, each locality can have a STEM lab either as a separate facility or even co-located with the library providing children opportunities to nurture their curiosity by experiencing the wonders of science. This will not only promote scientific literacy but also encourage children to read.

3. STEM curriculum

STEM curriculum may comprise of a set of lesson plans for a few selected topics that are designed to augment the understanding and practice of the core curriculum concepts and develop core competencies, i.e. Critical Thinking, Collaborative Problem Solving, Communication as well as digital literacies, psychomotor skills and desired behavioral attitudes.

The selection of topics for STEM sessions must take into account the local curriculum requirements, standards, constraints and limitations such as culture and available resources. The lesson plans have to be tailored to the target student, teacher and the school system. Since the STEM program revolves around this curriculum and its outcomes, this is an exercise that must be carried out meticulously and methodically.

3.1 STEM curriculum development methodology

STEM lesson plans help students make connections between their theoretical knowledge and the real world through collaborative problem solving. One of the main objectives of STEM curriculum is to prepare our children for the challenges of today and tomorrow. These challenges, whether local, national or global, have to be identified and tied with the core curriculum concepts. STEM lesson plans also provide an opportunity to address gaps in the curriculum.

It is needless to say that not all concepts can be covered through STEM during an academic year given time and resource constraints. A carefully designed selection process covering ten to twelve crosscutting topics in an academic year through well-designed comprehensive lesson plans can achieve the desired learning outcomes.

The process of topic selection is influenced by many factors. These mainly include the national curriculum standards, problems affecting the population, desired literacies and skills, students' interest, and topics that lend well to hands-on activities. The topics identified and selected through this process may be called Key Curriculum Concepts (KCCs) as shown in **Figure 2**.

One of the first challenges that a STEM lesson plan has to address is to invite and engage students in a manner that piques their interest and gets them sufficiently motivated to get their creative juices flowing. The relevance and authenticity of the topic, the method of introduction to the topic, whether in the form of a question, a story or a presentation, verbal or with the help of audio-visuals, sets up the stage for the level of engagement and immersive experience that the children are going to have.

Figure 2.
Selection of key curriculum concepts for STEM lesson plans.

Similarly, the hands-on activities for the lesson plans are designed in the context of learners' needs and abilities making sure that they are in the Zone of Proximal Development, i.e. neither too easy nor too difficult but just the right amount of difficulty level to enable them to construct new knowledge autonomously through self-discovery with minimum scaffolding from teachers. This helps students develop industry and agency.

The lesson plans may typically be accompanied with worksheets that provide students the opportunity to practice and demonstrate their learning. Instead of simple question-and-answer (Q&A), the worksheets incorporate a variety of methods to assess whether the outcomes of lesson plans have been effectively achieved and identify areas that need further attention.

Finally, following the principles of assessment for learning, rubrics are created for formative assessments. This includes both teacher review and peer review by students. Since students are provided with multiple opportunities to express themselves during the course of a STEM lesson plan, starting from the introduction to the topic to the final presentation by students of their projects, it provides teachers and students ample opportunities to observe, capture and review their performances. The formative assessment also creates opportunities for teachers to apply differentiated learning to ensure an equitable learning environment.

3.2 Essentials of STEM lesson plans

The STEM Lesson Plans are mapped to the core curriculum concepts covering the instruction, practice and assessment of curriculum and also include supplementary concepts, skills, and attitudes.

Some of the key components of a STEM Lesson Plan are as follows:

1. Engineering Design Process (Ask, Imagine, Plan, Create, Test, Improve).

2. Interdisciplinary or Cross-cutting topics, i.e. concepts from different subjects are combined to solve problems making it easier for students to understand the big picture.

3. Collaboration – Team Work, i.e. students work in small groups to solve problems and thus learn to work with other people, manage expectations, resolution of conflicts, effectively communicate their point of view, respectfully listen to others, make decisions objectively with empathy.

4. Real-world connections and/or problems making them authentic and relevant to students thus resulting in enhanced student engagement and interest.

5. Use and/or creation of technology, i.e. students use technology to create solutions.

6. Activity-based, student-centered, hands-on experiential, making learning fun, exciting and an active lean-forward activity.

7. Failure is seen as a learning opportunity, i.e. Engineering Design Process is an iterative process in which every iteration is an opportunity to learn whether desired results are achieved or not.

4. STEM: Robotics and DIY

Robotics has been successfully employed as a very effective STEM tool to inculcate not only useful digital skills such as programming and automation that form the basis of more advance and crucial technologies of the modern era such as Artificial Intelligence but also facilitate development of collaborative problem solving competencies. It has served as a very good advertisement for STEM. However, counter to popular belief, STEM is not just robotics. As a matter of fact, STEM methodology is not dependent on any specific hardware or software. It is the process of understanding a problem and then going through the Engineering Design Process steps to find a solution that forms the basis of STEM methodology.

With the advent of STEM methodology, many commercial STEM products and solution providers sprang into action and introduced very high quality STEM kits. These kits for the most part were Robotics kits, which were developed with the help of teams of engineers, educationists and neuroscientists. The Robotics kits came with a curriculum and a complete set of lesson plans and proved to be highly effective in achieving desired outcomes. However, they carried a very steep price tag that made it nearly impossible to employ them at mass scale, particularly in developing countries.

Fortunately, by virtue of explosion in manufacturing of low cost electronic components in the last couple of decades and availability of freely available open-source educational softwares, robotics and automation that lay the foundation for more advance technologies such as Artificial Intelligence have become more and more accessible to the less and less privileged. For example, as a substitute to the high-end Robotics kits, a locally available robot chassis, Arduino microcontroller, low-cost sensors and freely available open-source softwares including block-coding applications provide a reasonable and affordable alternate for STEM activities.

Although these open-source hardware and software provide a lot more versatility and flexibility in terms of adding new functionality, they do come with a catch. Like all open-source solutions, they require a fair amount of technical understanding and ability to do system integration. In addition, these require meticulously created lesson plans that make it easy for the facilitators to conduct the sessions and the students to create projects without any technical hiccups arising from the hardware or the software. In most urban schools where computers are available, the locally developed kits are an affordable and effective substitute for expensive STEM kits.

However, in rural areas where computers are not available, STEM lesson plans require careful integration of hands-on activities that not only facilitate achieving desired objectives but are also feasible and affordable in the local context.

In this context, DIY projects using low cost daily life materials or recyclable and reusable materials prove extremely effective. The efficacy of these activities has been proven in many urban and rural setups in extremely resource-constrained environments. Discarded plastic bottles, cups, plates, cardboard boxes, rubber bands, balloons, straws, etc. all are valuable resources for STEM activities. Lack of resources is not necessarily a handicap; it is often an opportunity to be as creative as possible with whatever means available.

5. Teacher's role in STEM education

The role of teacher in a STEM classroom is different than that in a conventional classroom. The teacher in a STEM classroom is more of a facilitator, a guide and a mentor who poses the right questions at the right time, nudges students in the right direction, encourages students to collaborate and co-create, facilitates construction of knowledge and creates opportunities for students to express themselves without any apprehensions and inhibitions. This role requires an intentional effort to keep the teacher's interference to a minimum and let the children's creativity guide the process of self-discovery.

In this regard, training teachers for this role becomes a very important aspect of any STEM program. No amount of technology, hardware and software can make a program successful unless fully trained and motivated teachers are there to support and own the program with full enthusiasm. Some of the key objectives of a Teacher Training Program are to make teachers understand the full potential of STEM methodology, equip teachers with required knowledge, skills and attitudes that will enable them to be successful in their role. Particular emphasis is given on the Engineering Design Process (EDP), interdisciplinary approach, key elements of STEM lesson plans and STEM sessions, tools and strategies for classroom management during STEM sessions, assessment for learning and differentiated learning.

Continuous Professional Development (CPD) courses, whether online or in-person provide an effective means to stay abreast with new developments, learn new skills particularly related to technology, and address challenges and shortcomings encountered during prior STEM sessions. These also serve as a platform to share experiences and promote cross-fertilization of ideas for continuous quality improvement.

6. A framework for assessing effectiveness of STEM learning

A framework for reflecting upon effectiveness of our STEM lessons and activities, adopted and modified from Larmer et al. [4] is presented in **Figure 3**:

Key Curriculum Concepts or learning outcomes essentially relate to change in knowledge, skills, and attitude encompassing success skills for 21st century. In view of the key curriculum concepts a challenging problem or questions serve as driving force that on one hand help teachers plan learning experiences and on the other hand make learning meaningful for students by focusing their attention towards relevant information, previous knowledge, and enable them to transfer their learning in future. When students find answers, more questions arise in the form of a spiral, i.e. an iterative process resulting in deeper learning which is John Dewey's philosophy of learning stressing on inquiry forming the basis of project activities [5]. STEM activities could be made 'real' for students if the nature of problem or questions relates to our society, or knowing that the project may have positive implications for our society. Thus authenticity demands that our students may also use the tools that people would normally do in the real world for completing tasks as part of their STEM activities. For example, comparing alternatives, conducting surveys or communicating with experts, organizing exhibitions require completion of authentic tasks.

John Dewey suggested student voice and choice as a pre-requisite to critical thinking and problem solving [5, 6]. The degree of voice and choice given to students is a decision in view of their readiness, the scaffolds and coaching a teacher is ready to offer. Typically, scaffolding may include direct instruction, handouts and readings as well as other tools and processes that may help students in STEM activities.

Obstacles may be treated as opportunities for students as well as teachers to reflect on how to overcome these challenges as well as to reflect upon effectiveness of their inquiry leading to STEM activities.

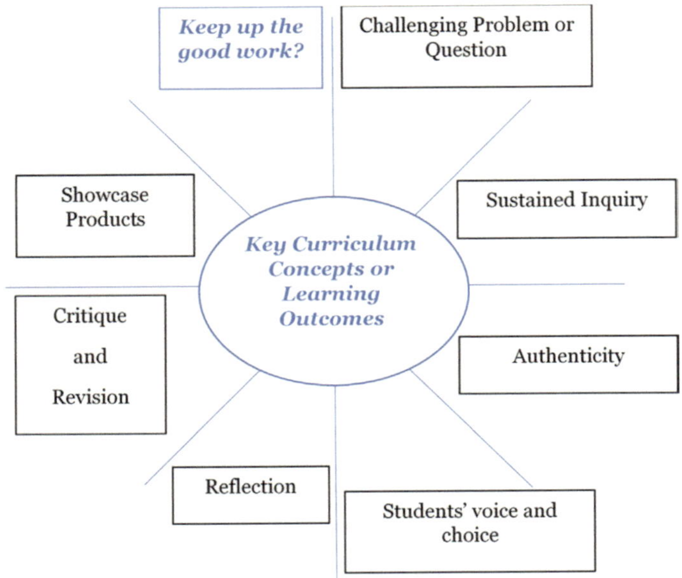

Figure 3.
A framework for assessing effectiveness of STEM learning, adopted and modified from Larmer et al. [4].

Assessment of STEM activities may comprise of both formative assessments to gauge progress in collaborative tasks, and summative assessments about the degree of achievement in learning outcomes by individual students. Traditional assessment tools such as quizzes, assignments, and tests may be more suitable for assessing content knowledge (cognitive domain), however, success skills such as critical thinking, collaboration, and communication may be evaluated through rubrics.

Showcasing the products developed by students, to a wider audience through exhibitions, poster competitions serve as opportunities to celebrate the efforts of our students, encourage them, and share constructive feedback on how their learning may be extended while teachers may benefit from ideas and discussions on how to keep up the good work.

7. Program sustainability

In Pakistan, introduction of STEM methodology is still in its infancy and a long way to go before it becomes widespread. Based on our experience, we have come up with a set of recommendations that we believe can make the STEM initiatives sustainable in the long run:

1. When introducing STEM education in an area, form school clusters for schools in close proximity of each other to foster collaboration between teachers. Periodic communications between teachers involved with STEM implementation will help them learn from each other's experiences.

2. Incentivize STEM implementation program with monetary and non-monetary rewards and recognition.

3. Periodic STEM events for students to exhibit their talents and creations.

4. Sponsorships for STEM events by local industry partnerships.

5. Incubation of Ed-Tech startups to come up with locally contextualized, affordable and scalable STEM solutions.

6. STEM Training programs and Continuous Professional Development (CPD) for teachers with certifications and points for achievements.

7. Involvement of School leadership in planning and implementation of STEM integration.

8. Ongoing monitoring and evaluation of Key Performance Indicators (KPIs) through school visits, interactions with teachers and students, review of progress reports and student assessments.

Some KPIs at the policy or regulator level may include:

1. Number of STEM labs implementing STEM methodology.

2. Time allocated to STEM activities.

3. Number of STEM events at each school where a STEM lab is setup, i.e. science exhibitions, tournaments, competitions.

4. Teachers' professional development for STEM, i.e. STEM education trainings attended, number of training hours, content covered.

5. Performance improvement in Math & Science compared to baseline (3rd party evaluations).

6. Decrease in absenteeism leading to improvement in student retention rate.

7. Improvement in gender parity from exposure to science education.

8. Role of STEM service providers

At the very heart of STEM eco-systems, in resource constrained environments, are STEM service providers serving as centers of excellence, hubs of learning innovation in STEM education, and fostering communities of learning in schools. STEM service providers typically maintain a handsome inventory of STEM kits, gadgets, tools, digital accessories, and other resources including softwares. Most importantly STEM service providers have experts passionate in STEM learning and capable of launching STEM programs for schools, train teachers, and run STEM sessions at schools or at their own STEM labs.

STEM service providers play a key role in making STEM initiatives sustainable by enabling school teachers implement STEM methodology through trainings, fostering collaboration and communication between teachers engaged in STEM activities typically via online groups following Continuous Professional Development (CPD) sessions.

School management and Academic leaders seek help of STEM service providers in planning and implementation of STEM integration as well as STEM Training programs and Continuous Professional Development (CPD) for teachers. In Pakistan a slow but steady change in Policy makers' and regulators' perspective is observable, resulting into pilot projects of STEM education, primarily because STEM service providers engaging with decision makers in Government bodies by virtue of their strategic goals tied to growth in STEM education in schools.

Finally, STEM service providers organize regular events and competitions wherein students from various schools are able to participate and showcase their products. Such events generate a structural tension that forces teachers and student teams out of their comfort zone in conventional educational settings. Some STEM service providers venture as Ed-Tech startups to come up with locally contextualized, affordable and scalable STEM solutions.

9. Final thoughts

A purpose of any education system ought to be to create an enabling environment for individuals to realize their true natural potential. It should develop them into independent thinkers and autonomous agents who are socially responsible and contributing citizens of this world. It should cater to the innate curiosity of mankind to observe, experiment and make sense of the world around them and understand their place in the universe. Above all, it needs to be dynamic, anticipating the changes and challenges of future and ever-evolving accordingly. The massive inertia of the conventional education system however makes this a daunting task and an uphill battle albeit one, which is an absolute necessity, for the failure to keep up has consequences beyond imagination.

Acknowledgements

Authors would like to acknowledge the contributions of DS@S team including Engr. Ali Ahmed Durrani (CEO, DS@S), Engr. Mian Afzaal Zahid, and Engr. Muhammad Salman Chaudhry throughout our journey of STEM implementation in urban and rural areas of Pakistan since 2017.

Conflict of interest

The authors declare no conflict of interest.

Author details

Majid Nawaz[1], Irfan Ahmed Khan[2] and Muhammad Imran Ahmad[3,4*]

1 Durrani STEMs @ Service (DS@S), Peshawar, Pakistan

2 Qadir Nagar Trust, Buner, Pakistan

3 Department of Chemical Engineering, University of Engineering and Technology, Peshawar, Pakistan

4 Qadir Nagar High School, Buner, Pakistan

*Address all correspondence to: imran.ahmad@uetpeshawar.edu.pk

References

[1] Slavin R. Educational Psychology: Theory and Practice. London: Pearson; 2012

[2] Gredler ME. Learning and Instruction: Theory into Practice. New Jersey: Prentice Hall; 1997

[3] Driscoll MP. Psychology of Learning for Instruction. Boston: Allyn & Bacon; 1994

[4] Larmer J, Mergendoller J, Boss S. Setting The Standard for Project Based Learning. USA: ASCD; 2015

[5] Dewey J. The School and Society and the Child and the Curriculum. Chicago: University of Chicago Press; 1915

[6] Dewey J. Democracy and Education. New York: The Free Press; 1944

Chapter 9

Developing Collaborative Skills through STEM Approach

Chairil Faif Pasani and Rizky Amelia

Abstract

Collaborative learning is a practice that dates back centuries. In Confucius' classic text, Li Ji, there is a famous saying that one becomes narrow-minded when learning without friends. Therefore, collaborative skills not only allow students to interact with others but also enhance their opportunity to learn, which differs significantly from individual learning. Collaborative problem-solving is crucial in science, technology, engineering, and math (STEM), which are fraught with complex challenges like climate change, overpopulation, welfare, resource management, health, and biodiversity. The Systematic Literature Review (SLR) approach was used to find, assess, evaluate, and interpret all of the research that was accessible in order to create this book chapter. Preparing the youths with a collaborative mindset is crucial for addressing the issues and difficulties that arise in real life. This review specifically focuses on developing technology-based collaborative skills through STEM approach to reflect the trend of integrating technology into education.

Keywords: collaborating skills, STEM approach, technology, education, Systematic Literature Review

1. Introduction

The abilities in question are theoretically grounded on the principles of social constructivism, wherein learning takes place through social interactions [1]. This section adopts Dillenbourg's broad definition of collaborative skills, which refers to a situation where two or more people learn or attempt to learn something together [2]. There is also an increasing demand for new collaborative forms due to the widespread application of modern digital tools and social networking platforms [3, 4]. Common scenarios today in both in-school and out-of-school environments may include (a) a community of gamers discussing mechanisms to solve emerging technical bugs; (b) a small team working in an online chat room to solve a math problem; and (c) a class using discussion boards asynchronously to brainstorm ideas on a science topic [5]. Meanwhile, in practice, the key to defining specific forms of collaborative skills is to determine their various pragmatic parameters:

1. The scale of collaborative skills can range from as small as a couple to a large community. The timing can vary, such as in a year-long course or a one-time activity (for example, a team-building activity in a company).

2. The setting can change from physical collaborative skills to virtual learning.

3. The medium can include various technology platforms and others [6].

Besides these pragmatic parameters, the dynamics of collaborative skills are influenced by various forces, such as the preference for individualism or collectivism in relation to learning objectives. When learning goals are established for individuals, collaborative skills prioritize improving individual knowledge or skills. In the Programme for International Student Assessment (PISA) test, the Organization for Economic Co-operation and Development (OECD) defines students' collaborative problem-solving competence as the capacity of individuals to effectively engage in a process, where two or more agents attempt to solve a problem [7]. This entails the collaborative sharing of knowledge, skills, and efforts in order to collectively work toward the attainment of a solution.

In summary, when education is focused on communal objectives, the emphasis is placed on fostering community cohesion and collaborative knowledge development. The pedagogical approach known as 'team-based learning' centers on fostering small group dynamics within a classroom environment, wherein tasks are designed to facilitate both individual learning and team growth [8]. Similarly, advocates of 'communities of practice' [9] or knowledge building [10] emphasize the importance of building communities among students. Communities of practice are groups of people who share a common concern or passion for what they do, and regularly interact to learn how to do it better [9]. Knowledge building emphasizes collective knowledge creation and innovation as well as prioritizes the advancement of community knowledge over individual achievement [11].

Another force that influences collaborative skills is the division of work. In many learning scenarios, the teacher organizes the work division, where tasks are divided into independent sub-tasks, and the group assembles the different parts together [12]. Collaborative learning is often utilized in these cases. On the other hand, in other collaborative skills opportunities, work is not divided but rather negotiated and completed by the individuals involved [13].

All group members contribute to the same learning task in a collaborative learning environment. In contrast, inside a competitive academic setting, pupils tend to operate autonomously and exert efforts to surpass their peers. Nevertheless, it is important to note that collaboration and competitiveness are not inherently contradictory concepts. Contradictions can play a pivotal role in propelling the development of an activity system [14]. Similarly, competition can be a positive element for collaborative skills [15]. This often occurs alongside collaboration within or across group learning settings [16]. A competitive mindset can be a double-edged sword and can facilitate collaborative problem-solving processes when carefully crafted and well-utilized.

Many curricula, instructional guidelines, and instructional materials connected to STEM have emerged in recent years. Even though the majority of the early attempts at STEM education focused on one or more of the STEM topics alone, proponents for stressing links between or among the subjects in STEM education are growing [17]. Due to the earlier lack of integration, STEM-related research frequently has a wide variety of foci and contexts. While other more thorough studies may address additional and deeper relationships among the STEM disciplines, the study contexts of some STEM-related studies may only involve one of the STEM fields [18]. The potential benefits of using educational technology to enhance STEM learning outcomes are being emphasized by educators and academics more and more as a result of the rapid growth of information and communication technology [5, 19]. This study provides compelling evidence that collaborative skills effectively address critical issues in STEM education.

The Systematic Literature Review (SLR) approach was used to find, assess, evaluate, and interpret all of the research that was accessible in order to create this book chapter. With this approach, the author conducts systematic journal reviews and identifications, going through each stage in a set order. Researchers gathered journal articles from Google Scholar, Research Gate, SINTA, DOAJ, and Scopus to finish this manuscript. STEM education and collaborative education are key terms.

2. Benefits of collaborative skills in STEM education

Collaborative skills have been well-embraced in STEM education [20, 21]. It is advisable to establish student learning communities as a means to enhance student engagement and perseverance in STEM programs inside educational institutions [22]. Multiple meta-analysis studies have demonstrated that the acquisition of collaborative abilities has a positive impact on overall student learning [4, 23]. Moreover, a more recent meta-analysis revealed that collaborative skills supported by computer technology have demonstrated notable effectiveness in STEM education, specifically in process, knowledge, and affective outcomes [24].

The study presents substantial evidence that collaborative skills are an effective means of addressing crucial challenges in STEM education [25]. An issue that can be identified is the phenomenon of low enrolment coupled with great interest [22, 26–28]. The cultivation of collaborative skills possesses the capacity to enhance students' self-esteem and sense of accomplishment through the facilitation of their assistance to others and engagement in the co-creation of classroom activities. Students can develop HOTS (Higher Other Thinking Skills) and learn better in STEM content through collaborative skills than individual learning [29]. In addition, collaborative skills can foster students' communication skills to resolve conflicts [30].

Another issue is equity in STEM education. Female students generally have lower enrollment rates but higher dropout rates in STEM majors compared to male students. Similarly, students from low-income families show similar patterns compared to high-income families [26, 27]. Previous studies have shown that students' satisfaction with the positive experience of collaborative skills makes them more intrinsically interested in learning. This has consequently made them more willing to attend school and persist in STEM learning. In addition, the cultivation of collaborative abilities facilitates effective communication and interdependence among individuals within a group, irrespective of their racial, gender, or academic backgrounds. This, in turn, contributes to the establishment of a learning environment that is characterized by fairness and equality [31, 32].

3. Collaborative skills and technology in STEM education

The utilization of technology is of paramount importance in enabling the development and enhancement of collaborative abilities. The utilization of visual representations in educational contexts can facilitate the comprehension and completion of learning activities, facilitate collaborative processes, and function as a supportive framework for the construction of shared knowledge [33]. This study centers its attention on instances of technology that provide direct support for collaboration within the learning process. While several technologies have the potential to support collaborative abilities, their impact is not considered important. This assertion is substantiated by the findings of Lee et al. (2016) in their report, wherein students engaged in the collection of their activity data, such as step counts over a specific duration, through the utilization of sports watches. Subsequently, these students acquired statistical knowledge by collaboratively examining and analyzing the collected data within group settings [34]. The sport watches were solely used for data collection, and not to facilitate collaboration. This type of study is beyond the scope of this chapter.

4. Developing collaborative skills through STEM approach

4.1 The environment of remote collaborative skills

Audio and video conferencing platforms, such as Zoom and Skype, are extensively utilized in contemporary synchronous remote learning environments. The aforementioned tools facilitate the connection of students in different locations [35]. They frequently incorporate distinct functionalities that enhance collaboration abilities, such as screen sharing, chat rooms, and annotations. Babaian and Schiano presented a set of practical recommendations aimed at enhancing the effectiveness of small-group collaboration within the context of online learning [36]. The aforementioned skills encompass the ability to facilitate teachers and students in becoming acquainted with functionalities that facilitate small-group collaboration. This entails effectively managing display names and altering screen names to reflect group identification or names, seamlessly transitioning between group workspaces and the primary space, communicating messages to all groups or specific groups within breakout spaces, and possessing a comprehensive understanding of sharing and annotation tools.

Furthermore, research has been performed to investigate these characteristics. Singhal (year) demonstrated the utilization of breakout rooms within the Zoom platform for the purpose of conducting a virtual pharmacotherapy course throughout the COVID-19 pandemic [37]. In class, every five to six students were grouped and placed in a breakout room, hence the teacher could move around the room to facilitate discussions. The study emphasized that the use of breakout rooms can assist students' small team tasks in promoting active learning, consequently leading to more participation.

4.2 Multi-user virtual environments (MUVEs)

MUVEs have been used to facilitate collaborative skills. In these environments, students typically use avatars to interact with each other in a virtual world, similar to multiplayer online role-playing games [38]. Additionally, these environments have the potential to provide educational advantages for students diagnosed with autism spectrum diseases [39]. River City is an exemplary instance in the field of science education, as it serves as a pioneering model for a Multi-User Virtual Environment (MUVEs) learning platform. This innovative educational tool has been specifically developed to effectively involve students in middle grades in the application of scientific inquiry techniques that align with established criteria (https://muve.gse.harvard.edu/) [40].

The curriculum was enhanced through several development versions [41]. For example, the first unit focuses on biology and ecology. Students can connect through avatars, and investigate an authentic problem (e.g., disease outbreak in a city) in a virtual city with a river running through the environment. In order to solve a problem, students can collaborate by sharing information and communicating with team members about the collected data through a chat text tool. In retrospect, [42] concluded that MUVEs have been identified as effective instruments for facilitating student-centered collaborative inquiry learning, offering a viable alternative to traditional inquiry-based science instruction methods.

In this study, Ibanez et al. introduced a novel 3D virtual collaboration skills model designed to facilitate and direct students' collaborative activities within Multi-User Virtual Environments (MUVEs) [43]. The model establishes synchronization learning points to achieve individual or group desired learning goals, as well as work division to promote positive interdependence. Based on the model, students conducted a case study in which participants (through avatars) navigated a 3D environment and collected information related to available theater performances in order to purchase tickets for the shows they wanted to attend. They also performed several collaborative activities, such as information sharing and group assessment. The factors affecting interactive collaboration were identified in the first iteration and the design was refined. This resulted in improved collaboration in the second iteration [43].

Unlike the examples mentioned earlier that gave a game-like setting, [42] described a geometry unit for fifth graders, wherein students engaged in various activities within a virtual classroom. In this environment, they were able to manipulate virtual 3D geometric objects, observe their classmates' problem-solving processes, as well as provide feedback and critique to each other. According to their findings, the experimental group outperformed the control group in terms of geometry learning. Moreover, the most effective way for students to acquire abstract geometry ideas is to collaborate with their classmates and practice manipulating virtual objects (e.g., by viewing and critiquing each other's answers) [44].

It is noteworthy that manipulating virtual items can enhance peer-to-peer interactions in real life. Jackson et al. designed a math game activity on a table to augment and reinforce what students had learned in their fourth-grade math class [27]. The students were expected to use components from their resource pool to fill in the blanks on a math problem displayed in the center of the table screen (allowing multiple users to edit the content simultaneously). Four students work together in a group, with each possessing resources that could potentially contain important components of the problem-solving solution. They were required to collaborate in order to complete the task. The students responded positively to the collaborative game-based learning exercise. Upon completing the program, male students' arithmetic scores showed substantial improvement.

In conclusion, MUVEs and audio/video conferencing systems are examples of mechanisms that connect students online. Although these technologies are now widely used in everyday life, there is still a need for further empirical studies to determine the exact elements of these tools that can most effectively enable collaborative learning [45]. Meanwhile, MUVEs not only provide alternative and shared areas for students to learn, but they also simulate various work environments [46]. Students are likely to feel more engaged in the problem-solving process when they can experience it authentically by exploring virtual environments and interacting with virtual objects.

4.3 Facilitate written discourse in groups

Tools for audio or video conferencing that make remote communication more convenient and effective, require high-speed Internet connection and gadgets equipped with cameras and microphones. In contrast, written discourse can fulfill the needs of collaborative skills without such constraints.

a. Synchronized discussion forum

In synchronous discussion forums, students can participate in real-time using written text or symbols [47, 48]. This type of forum is also known as online chat. Some advanced forms of this forum, designed for educational purposes, incorporate simulated intelligent agents that can offer students real-time feedback on the discussed topics and facilitate communication between participants. According to Wang, Rosé, and Chang, high school students collaborated in pairs in several geology-related brainstorming activities [49]. They used a synchronous chat program called VIBRANT, which included an integrated virtual agent capable of providing real-time comments.

Using a similar platform called MentorChat and virtual agents, Tegos et al. demonstrated collaborative learning among 96 university students on human-computer interaction [50]. In both educational settings, virtual agents were seen as additional participants in the team dialog, offering remarks and posing leading questions derived from their discourse content. Both studies demonstrated the beneficial impact of virtual agents on enhancing students' collaboration abilities.

b. Asynchronous discussion forum

An asynchronous discussion forum allows students sufficient time to analyze the posts made by their peers and provide more useful comments to the forum. This tool provides the benefit of meeting students' requests for rapid interaction and feedback. In a study examining the efficacy of collaborative learning in an undergraduate engineering course, participants were instructed to engage in group discussions pertaining to assigned teamwork and afterward submit a collective report derived from their email-based communication [51].

The web, shared by multiple users, such as blogs, can serve as an asynchronous discussion forum to exchange viewpoints. For example, [42] used an educational blog to engage 21 high school students in examining the issue of acid rain. A learning community was established for each student on the blog, where they shared their results and commented on their colleagues' findings. According to the results, students showed increased interest in the activity [52].

Through their activities on the blog, the students have shown social and emotional support for each other, leading to increased construction of new information. An investigation into how students collaborate was conducted with a total participation of 25 pre-service teachers who used an asynchronous online whiteboard to communicate with each other [53]. The students were divided into groups of five and tasked with developing an interdisciplinary education module that included elements of math and science. Each member in the group can upload their design ideas, while others can comment, make suggestions for modifications, and rate them. Despite the absence of an initial allocation of tasks by the teacher, the students assumed distinct responsibilities and assumed leadership positions (such as soliciting feedback, contributing knowledge, and exerting control over the topic), which were evenly distributed among team members. This occurred even though there was no initial work division by the teacher.

Discussion forums, both synchronous and asynchronous, have been around for quite some time. However, they are constantly evolving with the emergence of new technologies. In recent years, multimodal inputs like audio or video were incorporated into online discourse platforms, such as the Flipgrid app. Moreover, machine-based learning algorithms have been embedded to support automated and customized student feedback [48].

4.4 Provide direction for the collaborative processes

The previously mentioned technologies generate more opportunities for students to work together by connecting across different locations and times. Nonetheless, the collaboration may not always be successful due to various reasons, such as inefficient communication, lack of interaction, and inappropriate work division [54]. Affordable collaborative scripts in computer-based learning environments have been investigated as a means to guide collaborative activities. Rummel, Mulins, and Spada used a total of 106 middle school students to investigate scripted collaborative arithmetic learning. They modified a computer-based teaching program called Cognitive Tutor Algebra to include scripts that allowed users to work together [55].

After initially working on separate challenges individually, two students eventually teamed up at a computer to address a challenge presented by a script, integrating their respective problems. During the collaboration, additional scripts emerged to encourage collaborative behavior at specific times (e.g., contributing based on independent problem-solving experience, listening to peers, and asking questions for clarification). The results showed students who followed the scripts demonstrated better collaborative skills and were more successful in finding solutions to problems compared to other groups that did not follow the scripts.

Collaborative scripts have been employed for many objectives, encompassing the facilitation of collaborative conduct and serving as a mechanism for peer surveillance. The study conducted by Bouyias and Demetriadis aimed to examine and contrast the impacts of compulsory peer monitoring prompts and fading scaffolding scripts in a collaborative argumentation task inside a computer science educational setting. The students participated in an online argumentation learning environment called iArgue [56]. Random pairs were formed between 34 students who belonged to two different classes. In the peer-monitoring group, when student A submitted a task, peer B would receive the submission along with a peer-monitoring prompt asking them to check whether A's argumentation followed the model. In case it does not, student B was expected to indicate where improvements are needed. The monitoring task should be completed before student B can continue their work. Besides the peer-monitoring prompts, consistent scaffolding scripts were also provided (e.g., guidance on argument construction).

Students in the fading group did not receive any prompts for peer monitoring, instead, they were provided with a fading scaffolding script. This means there was less scaffolding after the students' second post. According to the results, the scripts that encourage peer monitoring significantly improved learning outcomes. In another study that investigated the function of collaborative scripts in argumentative learning, these scripts not only served as peer monitoring prompts but also guided students to systematically analyze their partner's argumentation and encouraged their argumentative construction. These activities include guiding students to paraphrase, criticize, ask questions, provide counterarguments, and propose new arguments. Another benefit of using collaborative scripts in argumentative learning is that they encourage students to develop their argumentative constructions [57].

A total of sixty students participated in the study, each of whom worked with a partner on a task that required knowledge from both of their respective fields of study. Specifically, the partnership consisted of two students, one with a background in water management and the other with a background in international development studies. The texts were successful in facilitating the development of argumentation knowledge [58].

The findings indicate that collaborative scripts have the potential to successfully facilitate and improve collaborative processes through the promotion of collaborative behavior, peer monitoring, and knowledge production. This is particularly evident in small-group collaborations. This is true for collaborative scripts used in online environments. Consequently, there should be greater emphasis on incorporating collaborative scripts into educational practice, and more studies on methods that enable collaboration among larger groups.

4.5 Produce, store, and visualize collective knowledge

Besides expanding opportunities for collaborative work and providing direction for collaborative processes, technological tools and platforms have also been developed to facilitate the generation and sharing of collective knowledge by the groups. An example of this is Wikipedia, an online encyclopedia that has grown over the past 20 years and currently has over 56 million entries available in more than 300 different language editions (https://en.wikipedia.org/wiki/List of Wikipedias). It has rapidly developed into one of the most popular websites. A concept behind Wikipedia states that anyone can contribute to the world's most comprehensive encyclopedia. This reflects the latest popular advances in humanity's pool of knowledge.

Several examples of Wikipedia and similar technologies are used in educational settings at various levels. Pifarre and Kleine Staarman investigated the collaborative processes that occur when elementary school students work together on a science book in the context of a wiki. The study focused on student-generated writing from a dialogic perspective, considering the number of paragraphs, words, sentences, and reasoning relationships [59]. This analysis was conducted on student-generated texts, and the conclusion suggests that dialogic methods should be used to examine the process of collaborative interaction. This approach can help create more effective pedagogies related to wiki use in educational settings. Furthermore, a student-centered inquiry learning paradigm has been created and implemented for use in college biology classes as part of the WIKIed Biology project, supported by the National Science Foundation [60].

The students participated by collaboration through the use of Web2.0 technologies, managing and tagging educational resources obtained from the Internet. They collaborated in small groups to create and publish website pages based on their scientific inquiry projects. The results showed that the students significantly improved their understanding of various biological concepts, as well as their ability to think critically and their awareness of the relevance of scientific communication.

Similar to Wikipedia's crowdsourcing model, "citizen science" can be seen as the practice of delegating various scientific endeavors to the general public. The term "citizen science" is used to describe the practice of nonprofessional scientists (such as data collection) by contributing to the generation of professional and scientific knowledge [61]. Digital Earth is an initiative that involves individuals in the collection and conversion of information related to the Earth into digital form [62]. OpenStreetMap (www.openstreetmap.org), an example of Digital Earth, is a collaborative effort to produce a free and editable map of the world populated with geographic data provided by individual users. YouthMapper, a student branch of

OpenStreetMap, which can be accessed online at www.youthmappers.org, is a global network of students who are actively engaged in collaborative mapping projects using OpenStreetMap. Gama et al. [42] examined the collaborative mapping experiences of students from three different institutions in Europe, North America, and Africa. The study found that participating in Mapathons events ("mapping marathons") not only improved students' technical ability and subject knowledge but also enhanced their engagement as socially responsible citizens [42].

Knowledge forums, led by Marlene Scardamalia and Carl Bereiter over three decades of study and development, are a pioneering platform for communal knowledge construction in upper-middle-class spaces [63]. These forums serve as tools to assist and support groups in developing their knowledge. Students compile their knowledge records in relation to specific subjects. One of the strengths of the forum is the provision of multiple scaffolds that students can use to develop their submissions. For instance, students can start a statement by selecting a pre-defined question, such as "My Theory" or "My Understanding Problem," and can also add comments on top to offer a high-level overview. Empirical studies on the topic have shown that students have found the benefits of Knowledge forums in developing their fundamental, domain-specific, and epistemic literacies [64].

The iKOS platform is a web-based knowledge organization tool that enables cross-disciplinary collaborative learning in a classroom environment [65]. The platform allows students to generate, exchange, and organize information. It also has the benefit of incorporating multimodal features [66, 67]. In Wiki mode, students can create knowledge entries similar to those on Wikipedia, while they can tag and annotate images or photos in PicTag mode. In mind mapping mode, they can create concept maps to visualize the relationships between big ideas. In Flipbook mode, they can publish their multimodal pages and interactive flipbook to the general public. This activity supports multiple modes of knowledge representation. According to the results, after completing a lesson on nuclear power, prospective science teachers were able to create a rather voluminous knowledge network [68].

Personal response systems (such as clickers), and polling applications are other technology categories that attract attention. An instructional technique known as Peer Instruction is established to engage students in typically lecture-based classroom conversations. These resources are often effective with Peer Instruction instructional practices. When using clickers or polling software, it is necessary to project and view the distribution of student responses on the screen. This allows students to be motivated to talk to their neighbors about their decisions and thought processes, as the projection of the class' collective state of knowledge engages them in conversation. This strategy has been proven effective in improving student learning and retention through STEM approach [69].

The platforms mentioned above provide a space for students to store, create, and exchange collective knowledge with each other. In addition to improving the community and building knowledge by contributing to the collective knowledge, the students can also benefit from developing their individual knowledge and other personal skills in the process.

5. Evaluation of technology-based collaborative skills through STEM approach

Despite its widespread use, there is a relatively limited amount of empirical study conducted in evaluating technology-based collaborative skills. The assessment of collaborative learning gained can be viewed from two different perspectives, namely collaborative outcomes and processes. Collaborative outcomes refer to what the group produces during the learning process. This collective output can be analyzed to infer the collaborative processes or individual understanding of teamwork. Meanwhile, the processes examine the complex dynamics (such as social interactions among group members), which serve as a link between the individual and the group.

In order to develop products, the discourse documented has been utilized in various technology platforms during the collaborative learning process. For example, synchronous or asynchronous discussion forums can record the discourse that occurs between members of a learning community. This provides a rich data source for evaluating and analyzing the collaboration and knowledge construction among members.

Jimoyiannis and Angelaina investigated Community of Inquiry (CoI) and Social Network Analysis (SNA), using a total of 131 blog posts that addressed the issue of acid rain [52]. Through the use of CoI analysis, community characteristics were examined by classifying publications into three categories, namely social presence (i.e., not involving domain knowledge but for emotional communication and group cohesion), cognitive presence (i.e., involving domain knowledge), and teacher presence. Teacher presence was defined as the presence of a teacher in a publication (i.e., including instruction or scaffolding). The SNA analysis provided a quantitative representation of the engagement level of individuals as well as their social relationships. The amount of direct contact each member has with others may indicate the power distribution among students.

The use of external technologies, such as video or audio recordings, to analyze collaborative processes is frequent. This approach can be used to investigate the collaborative processes. Talentino investigated the use of a mixed reality environment to enhance the study of earth science by high school students. The discourse between students and teachers (e.g., questions and answers) as well as among students, were analyzed and coded using video and audio recordings (e.g., comments, questions, or responses between individual students). The statistics revealed a visible increase in the number of statements driven by student participation. This activity is a usual practice for computer-based learning platforms to track students' activities as they use the platform (e.g., time stamps, clicking on buttons or pages, editing text) [70].

Several previous studies have focused on analyzing log data to gain a deeper understanding of collaborative processes within groups. Altebarmakian and Alterman [68] investigated how elementary school students solve arithmetic problems using an online platform with an intelligent virtual tutor under three different learning conditions, namely collaborative, individual, and combined. They also evaluated the effectiveness of each student's problem-solving approach within the context of their learning conditions by analyzing log data of students' interactions with the virtual tutor. This analysis included students' attempts, errors, and hint requests [71].

Some studies focus on both collaborative procedures and their products. For example, [68] investigated the extent to which 29 university students engaged in collaborative work during a computer science and psychology class by analyzing the data recorded through the system and the substance of the posts made by the students. The students worked on their assignments in groups of three to five using an educational blog as their platform of collaboration. They were required to independently draft and upload their answers to the initial problem, comment on posts made by others, respond to questions posed by others, modify their solutions in light of the group conversation, and eventually submit final responses [68].

The participation of each student in the group project is assessed with respect to their cognitive, social, and behavioral habits during the activity. The measurement of their reading, editing, and commenting behaviors, as tracked by the system, provides insights into their behavioral engagement. The log data also revealed that students' interaction levels depended on whether they referred to previous statements made by their peers. Subsequently, the level of cognitive engagement was evaluated based on the topics covered in the students' contributions.

An important observation regarding the assessment component of the study on collaborative learning is that most studies use assessments to test the effects on individual students' academic or emotional outcomes. However, these tests do not necessarily reflect the development of students' understanding of collaboration or their ability to work together effectively. Therefore, it is crucial they receive timely and useful feedback on their individual and group efforts when participating in learning activities that require collaboration. Continuous assessment of collaborative learning facilitates meaningful communication, not only among collaborative group members but also between students and teachers. Therefore, students have a more positive experience when participating in collaborative learning.

6. Conclusion

The limitations, strengths, and advantages of collaborative learning through STEM approach are outlined in this chapter. The chapter additionally presents instances of technologies that enhance collaboration skills through many means, including facilitating communication among students across different temporal and spatial contexts, fostering written and multimodal exchanges among students, providing guidance for collaborative procedures, and supporting the development of collective knowledge. Further exploration is needed for formative assessment of students' collaborative knowledge and practices. In addition, the assessment of students' collaborative skills should be explored, with a focus on evaluating the effect of collaborative learning on students' academic and affective outcomes (individual) as well as conducting formative assessment of collaborative knowledge and practice.

Author details

Chairil Faif Pasani* and Rizky Amelia
Universitas Lambung Mangkurat, Indonesia

*Address all correspondence to: chfaifp@ulm.ac.id

References

[1] Vygotsky L. Interaction between learning and development. Readings on the Development of Children. 1978;**23**(3):34-41

[2] Dillenbourg P. Collaborative learning: Cognitive and computational approaches. Advances in learning and instruction series. New York, NY: Elsevier Science, Inc.; 1999

[3] Hesse F, Care E, Buder J, Sassenberg K, Griffin P. A framework for teachable collaborative problem solving skills, assess. In: Teach. 21st Century Ski. Methods Approach. Dordrecht: Springer; 2015. pp. 37-56

[4] Chen J, Wang M, Kirschner PA, Tsai C-C. The role of collaboration, computer use, learning environments, and supporting strategies in CSCL: A meta-analysis. Review of Educational Research. 2018;**88**(6):799-843

[5] Kelley TR, Knowles JG, Holland JD, Han J. Increasing high school teachers self-efficacy for integrated STEM instruction through a collaborative community of practice. International Journal of STEM Education. 2020;**7**:1-13

[6] Lai ER. Collaboration: A Literature Review. Vol. 11. Pearson Publ; 2011. p. 2016

[7] Eijkelhof H, Dillon J, Redfors A. Curriculum policy implications of the PISA scientific literacy framework. In: Electronic Proceedings of the ESERA 2013 Conference, Strand 10, Science Curriculum and Educational Policy. ESERA; 2014. pp. 26-33

[8] Michaelsen LK, Sweet M, Parmelee DX. Team-Based Learning: Small Group Learning's Next Big Step. New Directions for Teaching and Learning. Number 116. Vol. 103. United States: John Wiley & Sons; 2008

[9] Wenger E, Trayner B, De Laat M. Promoting and assessing value creation in communities and networks: A conceptual framework. Rapport. Netherlands: The Dutch Ministry of Education, Culture and Science; 2011

[10] Chan T-W et al. One-to-one technology-enhanced learning: An opportunity for global research collaboration. Research and Practice in Technology Enhanced Learning. 2006;**1**(01):3-29

[11] Scardamalia M, Bereiter C. Knowledge Building. Canada: The Cambridge; 2006

[12] Nugraha F, Siahaan P, Chandra DT. The effect of structured individual responsibility on students' achievement increase in cooperative learning science class. Journal of Physics: Conference Series. 2019;**1157**(3):32066

[13] Micari M, Pazos P. Beyond grades: Improving college students' social-cognitive outcomes in STEM through a collaborative learning environment. Learning Environments Research. 2021;**24**:123-136

[14] Froehle CM, Roth AV. New measurement scales for evaluating perceptions of the technology-mediated customer service experience. Journal of Operations Management. 2004;**22**(1):1-21

[15] Ziegler R, Hamker N. The idea of a collaborative competition–a case study. Futures. 2011;**43**(4):441-449

[16] Chen S-Y, Liu S-Y. Reinforcement of scientific literacy through effective argumentation on an energy-related environmental issue. EURASIA Journal of Mathematics, Science and Technology Education. 2018;**14**(12):em1625

[17] Sanders ME. Integrative STEM Education as 'Best Practice'. Queensland, Australia: Griffith Institute for Educational Research; 2012

[18] Kelley TR, Knowles JG. A conceptual framework for integrated STEM education. International Journal of STEM Education. 2016;**3**(1):11

[19] N. R. Council. Monitoring Progress toward Successful K-12 STEM Education: A Nation Advancing? Canada: National Academies Press; 2013

[20] D'Angelo C, Rutstein D, Harris C, Bernard R, Borokhovski E, Haertel G. Simulations for STEM learning: Systematic review and meta-analysis, Menlo Park. SRI International. 2014;**5**(23):1-5

[21] Thibaut L et al. Integrated STEM education: A systematic review of instructional practices in secondary education. European Journal of STEM Education. 2018;**3**(1):2

[22] Sithole A, Chiyaka ET, McCarthy P, Mupinga DM, Bucklein BK, Kibirige J. Student attraction, persistence and retention in STEM programs: Successes and continuing challenges. Higher Education Studies. 2017;**7**(1):46-59

[23] Kyndt E, Raes E, Lismont B, Timmers F, Cascallar E, Dochy F. A meta-analysis of the effects of face-to-face cooperative learning. Do recent studies falsify or verify earlier findings? Educational Research Review. 2013;**10**:133-149

[24] Jeong H, Hmelo-Silver CE, Jo K. Ten years of computer-supported collaborative learning: A meta-analysis of CSCL in STEM education during 2005-2014. Educational Research Review. 2019;**28**:100284

[25] Laal M, Ghodsi SM. Benefits of collaborative learning. Procedia - Social and Behavioral Sciences. 2012;**31**:486-490

[26] Chen X, Soldner M. STEM Attrition: College Students' Paths into and out of STEM Fields (NCES 2014-001). Washington, DC: National Center for Education Statistics, Institute of Education Sciences, US Department of Education; 2013

[27] Lichtenberger E, George-Jackson C. Predicting high school students' interest in majoring in a STEM field: Insight into high school students' postsecondary plans. Journal of Career and Technical Education. 2013;**28**(1):19-38

[28] Romash ZM. Leaving STEM: An Examination of the STEM to Non-STEM Major Change and How the STEM Curriculum Relates to Academic Achievement in Non-STEM Fields. Canada: Seton Hall University; 2019

[29] Gokhale A. Collaborative learning enhances critical thinking. Journal of Technology Education. 1995;**7**(1)

[30] Tolmie AK et al. Social effects of collaborative learning in primary schools. Learning and Instruction. 2010;**20**(3):177-191

[31] Esmonde I. Mathematics learning in groups: Analyzing equity in two cooperative activity structures. The Journal of the Learning Sciences. 2009;**18**(2):247-284

[32] Panitz T. Collaborative versus Cooperative Learning: A Comparison of

the Two Concepts Which Will Help Us Understand the Underlying Nature of Interactive Learning. 1999

[33] Goodyear P, Jones C, Thompson K. Computer-supported collaborative learning: Instructional approaches, group processes and educational designs. In: Handbook of Research on Educational Communications and Technology. 2014. pp. 439-451

[34] Lee VR, Drake JR, Thayne JL. Appropriating quantified self technologies to support elementary statistical teaching and learning. IEEE Transactions on Learning Technologies. 2016;**9**(4):354-365

[35] Pasani CF, Amelia R, Hassanhassan Z. Covid-19 impact in Indonesia's education sector: Challenges and strategy. Journal of Advanced Research in Dynamical and Control Systems. 2020;**12**(7 Special Issue):1722-1731. DOI: 10.5373/JARDCS/V12SP7/20202281

[36] Babaian T, Schiano B. Energize your Online Course with Group Work. In: HBP High. Educ. Inspiring Minds. 2020

[37] Singhal T. A review of coronavirus disease-2019 (COVID-19). Indian Journal of Pediatrics. 2020;**87**(4):281-286

[38] Barab S, Thomas M, Dodge T, Carteaux R, Tuzun H. Making learning fun: Quest Atlantis, a game without guns. Educational Technology Research and Development. 2005;**53**(1):86-107

[39] Cheng Y, Ye J. Exploring the social competence of students with autism spectrum conditions in a collaborative virtual learning environment–the pilot study. Computers in Education. 2010;**54**(4):1068-1077

[40] Tilak S et al. Multi-user virtual environments (MUVEs) as alternative lifeworlds: Transformative learning in cyberspace. Journal of Transformative Education. 2020;**18**(4):310-337

[41] Ketelhut DJ, Clarke J, Nelson BC. The development of River City, a multi-user virtual environment-based scientific inquiry curriculum: Historical and design evolutions. In: Designs for Learning Environments of the Future. International Perspectives from the Learning Sciences. 2010. pp. 89-110

[42] Gama K et al. Mapathons and hackathons to crowdsource the generation and usage of geographic data. In: Proceedings of the International Conference on Game Jams, Hackathons and Game Creation Events 2019. 2019. pp. 1-5

[43] Ibáñez MB, Rueda JJG, Maroto D, Kloos CD. Collaborative learning in multi-user virtual environments. Journal of Network and Computer Applications. 2013;**36**(6):1566-1576

[44] Hwang W-Y, Hu S-S. Analysis of peer learning behaviors using multiple representations in virtual reality and their impacts on geometry problem solving. Computers in Education. 2013;**62**:308-319

[45] Warburton S, García MP. 3D design and collaboration in massively multi-user virtual environments (MUVEs). In: Cases on Collaboration in Virtual Learning Environments: Processes and Interactions. United States: IGI Global; 2010. pp. 27-41

[46] Mantziou O, Papachristos NM, Mikropoulos TA. Learning activities as enactments of learning affordances in MUVEs: A review-based classification. In: Education and Information Technologies. United States. 2018;**23**:1737-1765

[47] Kuyath SJ. The Social Presence of Instant Messaging: Effects on Student

Satisfaction, Perceived Learning, and Performance in Distance Education. The University of North Carolina at Charlotte; 2008

[48] Rosé CP, Ferschke O. Technology support for discussion based learning: From computer supported collaborative learning to the future of massive open online courses. International Journal of Artificial Intelligence in Education. 2016;**26**:660-678

[49] Wang H-C, Rosé CP, Chang C-Y. Agent-based dynamic support for learning from collaborative brainstorming in scientific inquiry. International Journal of Computer-Supported Collaborative Learning. 2011;**6**:371-395

[50] Tegos S, Demetriadis S, Papadopoulos PM, Weinberger A. Conversational agents for academically productive talk: A comparison of directed and undirected agent interventions. International Journal of Computer-Supported Collaborative Learning. 2016;**11**:417-440

[51] Liu H, Spector JM, Ikle M. Computer technologies for model-based collaborative learning: A research-based approach with initial findings. Computer Applications in Engineering Education. 2018;**26**(5):1383-1392

[52] Angelaina S, Jimoyiannis A. Analysing students' engagement and learning presence in an educational blog community. EMI. Educational Media International. 2012;**49**(3):183-200

[53] Gressick J, Derry SJ. Distributed leadership in online groups. International Journal of Computer-Supported Collaborative Learning. 2010;**5**:211-236

[54] Le H, Janssen J, Wubbels T. Collaborative learning practices: Teacher and student perceived obstacles to effective student collaboration. Cambridge Journal of Education. 2018;**48**(1):103-122

[55] Rummel N, Mullins D, Spada H. Scripted collaborative learning with the cognitive tutor algebra. International Journal of Computer-Supported Collaborative Learning. 2012;7:307-339

[56] Bouyias Y, Demetriadis S. iArgue: A web-based argumentation environment integrating collaboration scripting support and flexible fading techniques. Educational Technologies for Teaching Argumentation Skills. Greece: Bentham Science Publisher; 2012:198-224

[57] Noroozi O, Weinberger A, Biemans HJA, Mulder M, Chizari M. Facilitating argumentative knowledge construction through a transactive discussion script in CSCL. Computers in Education. 2013;**61**:59-76

[58] Noroozi O, Biemans HJA, Weinberger A, Mulder M, Chizari M. Scripting for construction of a transactive memory system in multidisciplinary CSCL environments. Learning and Instruction. 2013;**25**:1-12

[59] Pifarré M, Kleine Staarman J. Wiki-supported collaborative learning in primary education: How a dialogic space is created for thinking together. International Journal of Computer-Supported Collaborative Learning. 2011;**6**:187-205

[60] Frisch JK, Jackson PC, Murray MC. WikiED: Using web 2.0 tools to teach content and critical thinking. Journal of College Science Teaching. 2013;**43**(1):70-80

[61] Bonney R et al. Citizen science: A developing tool for expanding science

knowledge and scientific literacy. Bioscience. 2009;**59**(11):977-984

[62] Brovelli MA, Ponti M, Schade S, Solís P. Citizen science in support of digital earth. Manual of Digital Earth. 2020;**18**:593-622

[63] Bereiter C, Scardamalia M. Knowledge building and knowledge creation: One concept, two hills to climb. In: Knowledge Creation in Education. Singapore: Springer Singapore. 23 May 2014. pp. 35-52

[64] Hong H-Y, Chen B, Chai CS. Exploring the development of college students' epistemic views during their knowledge building activities. Computers in Education. 2016;**98**:1-13

[65] Namdar B, Shen J. Intersection of argumentation and the use of multiple representations in the context of socioscientific issues. International Journal of Science Education. 2016;**38**(7):1100-1132

[66] Jiang S, Shen J, Smith BE, Kibler KW. Science identity development: How multimodal composition mediates student role-taking as scientist in a media-rich learning environment. Educational Technology Research and Development. 2020;**68**:3187-3212

[67] Jiang S, Shen J, Smith BE. Designing discipline-specific roles for interdisciplinary learning: Two comparative cases in an afterschool STEM+ L programme. International Journal of Science Education. 2019;**41**(6):803-826

[68] Altebarmakian M, Alterman R. Design heuristics to support cohesion within online collaborative learning groups. In: 2019 IEEE Frontiers in Education Conference (FIE). United States: IEEE; 16 Oct 2019. pp. 1-9

[69] Watkins J, Mazur E. Retaining students in science, technology, engineering, and mathematics (STEM) majors. Journal of College Science Teaching. 2013;**42**(5):36-41

[70] Tolentino L, Birchfield D, Megowan-Romanowicz C, Johnson-Glenberg MC, Kelliher A, Martinez C. Teaching and learning in the mixed-reality science classroom. Journal of Science Education and Technology. 2009;**18**:501-517

[71] Olsen JK, Rummel N, Aleven V. It is not either or: An initial investigation into combining collaborative and individual learning using an ITS. International Journal of Computer-Supported Collaborative Learning. 2019;**14**:353-381

Chapter 10

Improving the Teaching of Mathematical Problem Solving – A Collaborative Research Based on Theoretical Findings from Two PhD Dissertations

Maud Chanudet, Jean-Luc Dorier and Stéphane Favier

Abstract

This text briefly presents two PhDs achieved as part of a larger research project led by the team in mathematics didactics in Geneva on mathematical problem-solving at different school levels. The first PhD, by Maud Chanudet, examines teachers' assessment practices in the context of a mathematics problem-solving course offered at a lower secondary level in Geneva. The second PhD, by Stéphane Favier, aims to document the work of students at different school levels (primary and lower secondary) in problem-solving situations, through detailed analyses of students' work and interactions with each other and with the teacher. These two studies revealed ways to help teachers support students' problem-solving activities. This led us to set up a collaborative project with teachers to study two complementary processes—devolution and institutionalization—with a view to developing a training resource. We'll briefly introduce this new stage of our research.

Keywords: problem solving, assessment practices, heuristics, teacher regulation, collaborative research

1. Introduction

The aim of this text is to present part of a research project led by the team in mathematics didactics in Geneva. The overall project entitled: "Problem solving as an object or means of education at the heart of learning in the mathematics classroom", ResoPro for short, was piloted by Sylvie Coppé and Jean-Luc Dorier and funded by the Swiss National Science Foundation (SNSF – Grant n°100019_173105/1 – Period from 31.08.2017 to 31.01.2022).

The context and main. Results of the whole project are about to be published in a collective book in French [1]. Here we quote a passage from the introduction of this book to help the reader better understand the background of the whole project:

"By presenting innovative activities, these various projects on the inquiry-based teaching approach aim to provide teachers with the tools they need, but few of them question what pupils actually learn, even if this question is always an underlying one. Indeed, the value of problem-solving is often a presupposition that is rarely, if ever, questioned. What's more, questioning the effects of practices and problem types on students' actual learning remains theoretically and methodologically complex. Since the assessments used in classrooms to evaluate learning depend heavily on the type of education provided, what elements should be used to evaluate learning? On what time scale? What are the effects linked to the teacher, in particular his or her skills in engaging and keeping students on task, and in sustaining motivation? That's why we've set ourselves the goal of investigating and evaluating the effects of problem-solving in the mathematics classroom on student learning, using various theoretical frameworks from mathematics didactics and assessment. We draw on the work of our team in which problem solving is used either as a means of teaching thematic knowledge, or as an object of education (to learn how to solve problems). Our aim is to assess how learning classical mathematical themes can be achieved primarily through problem solving, and to better determine what can be learned when the focus is on problem solving independently of mathematical content, then how students can identify this constructed knowledge and know-how (how institutionalization can be managed) and what aids can be provided". ([1], Introduction).

The main aim of the project was to show how the teaching of problem-solving is conceived and constructed, how it is implemented by teachers, and what learning is achieved by students, through various works. The project as a whole has been structured around three PhDs, co-supervised by Sylvie Coppé and Jean-Luc Dorier, and two other works, enabling us to question problem-solving from a variety of angles (school levels, teachers' and students' points of view). In this text, we present only two of the PhD and the new research projects that resulted from them.

Maud Chanudet's PhD [2], defended on October 22, 2019, entitled "Étude des pratiques évaluatives des enseignants dans le cadre d'un enseignement centré sur la résolution de problèmes en mathématiques" (study of teachers' assessment practices in a course centered on problem solving in mathematics), focuses on a course offered at lower secondary level in Geneva centered on problem solving in mathematics. Observation of the practices of teachers involved in teaching and assessing students' problem-solving skills aims to identify their assessment action logics, taking into account both the certificative and formative functions of assessment. The analyses provide a clearer picture of teachers' expectations of problem-solving but also reveal certain contradictions. The theoretical and methodological frameworks, which combine tools from mathematics didactics and the field of assessment, provide a general tool for further study.

Stéphane Favier's PhD [3] was defended on February 8, 2022, under the title "Étude des processus de résolution de problèmes par essais et ajustements en classe de mathématiques à Genève" (Study of trial-and-error problem solving processes in mathematics classes in Geneva). The aim of this work is to document the work of students at different school levels (primary and lower secondary) in problem-solving situations, through detailed analyses of the pupils' work and their interactions with each other and with the teacher. More specifically, this work enables us to characterize students' practices when solving mathematical problems, which may involve trials and adjustments under the usual classroom conditions, leading to a complexity that few other studies on the same subject have tackled, and which therefore constitutes one of its original features.

These two PhDs have led to questions and calls for further research into how to devolve problem-solving and what can be institutionalized at the end of such sessions. However, this work has also revealed ways of supporting students' problem-solving activities. This has led us to develop a new research project, with one strand at the primary level, and the other at lower secondary level, both aimed at setting up collaborative work with teachers to study two complementary processes—devolution and institutionalization—with a view to developing a resource for training, based on the results of our previous research. We will briefly outline the reasons that led us to this new project, and what we are currently doing to implement it.

2. Study of assessment practices in problem-solving

In the canton of Geneva, in the French-speaking part of Switzerland, grade 8 students (13–14 years) in the science stream have a 45-minute period per week dedicated to mathematical problem-solving. As part of this course, entitled "Démarches mathématiques et scientifiques" (mathematical and scientifical approaches (MSA), teachers are required to assess their students' problem-solving skills on a certifying basis. However, the instructions governing this course are vague, leaving teachers a great deal of freedom as to how they wish to organize and manage their teaching and the associated assessment. It is in this context that Chanudet has studied the ordinary practices of teachers, focusing in particular on their assessment practices. Her work focuses on an analysis of the mathematics teachers make their students encounter, and the way they organize and manage such sessions dedicated to problem-solving, with a focus on the assessment dimension of their practices.

2.1 Theoretical approaches to problem-solving and learning assessment

2.1.1 Problem-solving as an object of teaching and learning

A number of approaches, such as "open problems" [4] and "research narratives" [5], have emerged in recent decades with the shared aim of introducing students to scientific practice and transferring mathematical research into the classroom. Georget [6], for his part, proposes to use the term Peer Research and Proof Activities (PRPA) to describe this type of activity "whose main objective is to train students in the mathematical research practice and peer exchanges in the manner of professional mathematicians[1]" (p. 77). He highlights five dimensions that enable the specific features of these activities to be taken into account and characterized: research potential; resistance potential; dynamic resistance potential; debate potential and didactic potential. In our dissertation, we used these potentials to characterize problems designed to make students search, thus enriching the classic a priori analysis tools of mathematics didactics. Indeed, a priori determination of the potential of the problems proposed to students enables us to better understand the effect of the teacher's choices, interventions, and interactions with students during classroom sessions.

From our interest in the teaching and assessment of problem-solving, we came to question the possible learning associated with this object of knowledge. We got inspiration from Houdement [7], who looks at "problems for searching" in elementary schools in France, with a strong hypothesis that research and proof activities

[1] Our translation from French.

between peers can, among other things, lead students to develop learning linked to ways of reasoning and proving in mathematics. The need to characterize this learning in greater detail led us to take an interest in the work carried out by Jeannotte [8] on mathematical reasoning. She highlights the complementarity of two viewpoints: the structural and the processual aspects. The former allows us to take into account the structure of mathematical reasoning and "describe the constituent elements of a step or sequence of steps and the relationships they have with each other[2]" ([8], p. 124). The second allows us to integrate the temporal dimension of reasoning and the goal pursued by the person carrying it out, in connection with the functions of mathematical reasoning. Taking these two points of view into account makes it possible to characterize different types of reasoning (according to the structural aspect of mathematical reasoning, such as hypothetico-deductive reasoning, reasoning by exhaustive study of cases, *etc.*), approaches (according to the processual aspect, such as the experimental approach or the process of adjusting successive trials) and proofs (by ostension, by counter-example, etc.) involved in solving mathematical problems. The variety of ways of searching, reasoning, and proving in mathematics led us to question the way teachers go about organizing and implementing such teaching.

2.1.2 Assessing student learning

The study of the assessment dimension of teachers' practices is based here on a characterization of assessment activity according to four invariants [9]:

- Defining the object of assessment, with particular reference to school curricula;
- Gathering information related to what the teacher is seeking to assess, formally or informally, instrumentally or not, in writing or orally;
- Interpretation of this information, usually using assessment criteria;
- Decision-making and communication, are in line with the function assigned to the assessment.

Research focuses on two functions of assessment: certifying, i.e., whether or not the student has acquired the targeted learning at the end of the associated teaching, and formative, i.e., assessment that serves student learning and aims to advance ongoing learning.

The broader conception of formative assessment emphasizes the possibility of informally gathering information about students' activity, for example during discussions with them [10]. Informal formative assessment practices thus aim to provide information, generated in the course of everyday activities, on student learning [11]. The term interactive episodes, taken from Kiwan Zacka [12], characterizes moments of verbal interaction between students and teacher in the classroom, during which the teacher captures information about students' activity, conceptions, and difficulties. They are an object of analysis in Chanudet's PhD.

Moreover, regulation is a key concept in formative assessment. The term "formative assessment practices" used in this research refers to all teaching practices designed to support and encourage the regulation of student learning. The work of

[2] Our translation from French.

Allal [13] highlights various contextual elements that can be associated with such regulation: the structure of the learning situation, involving in particular the tasks proposed, their articulation and work methods; the teacher's interventions and interactions with students; and interactions between students. The first two levels are examined in particular in the PhD.

2.2 The theoretical framework of the dual didactic-ergonomic approach

The general theoretical framework of Chanudet's PhD is based on the study of practices using the dual didactic-ergonomic approach developed by Robert and Rogalski [14, 15]. This framework invites us to take into account both the generic, by looking at the constraints weighing on the teacher as he or she exercises his or her profession, and the individual, by considering the margins of freedom remaining to him or her and the way in which he or she invests them. Five components are highlighted by the authors to study teachers' practices. The first two, cognitive and mediative, are linked to observations of classroom sessions and possible student activities. The cognitive component relates to the mathematical content presented to students (tasks, their articulation, *etc.*) and provides information on what the authors call the cognitive itinerary proposed to students. The mediative component focuses on the progress of classroom sessions (forms of work, assistance provided by the teacher, exchanges, and interactions with students). The combination of these two components enables us to describe the mathematics that teachers introduce their students to and to identify the logic of their actions. To integrate the fact that teachers' practices are the expression of their work, three other components need to be taken into account: social, institutional, and personal. The social component takes into account the social dimension of the teacher's work (students, parents, colleagues). The institutional component concerns common constraints (programs, timetables, textbooks). The personal component incorporates everything that is specific to the teacher as an individual (his or her representations of the profession, mathematics, knowledge, experience, *etc.*).

2.3 Research question and methodological guidelines

The various theoretical references mentioned above make it possible to formulate the macro-research question: what evaluative logics of action emerge from teachers' practices in the context of problem-solving education in mathematics?

The analyses carried out during the PhD work to answer this question took place on the scale of a school year, with the study of education projects, the study of problems proposed for assessment, their articulation with other problems, and the assessment criteria grids used by teachers; but also on the scale of a sequence and, even more finely, of a session, targeting the actual implementation of problems in the classroom, in connection with informal formative assessment and the study of interactive episodes.

Below are some of the results and contributions of the PhD, opening up new perspectives for research and work with teachers.

2.4 Some results and perspectives

In this research, Chanudet examined how teachers can help regulate student activity [13], in particular through their informal verbal interactions with students

[11, 16] during problem-solving sessions. A typology of the way in which these interactions take place and the subject matter they may address, from the dual viewpoint of information gathered and feedback given by the teacher, has thus been developed. What emerges is that, while all teachers organize discussions that enable them to gather clues about students' activities and difficulties, the extent to which they do so varies greatly from one teacher to another. Moreover, these interactive episodes do not all seem to positively support student activity and, a fortiori, learning. In particular, teachers' interactions with students lead the latter to be more or less active in taking account of the teacher's feedback on their work, results, strategies used, errors, etc. This is in line with the work of Black and Wiliam [17], who question the link between feedback and student involvement in classroom discussions. Moreover, the fact that the teacher seeks to involve students in the feedback he or she gives them does not always seem to be enough to encourage student regulation. In fact, it also seems important that the teacher draws on the students' actual activity and ensures that they are involved in the way this feedback is taken into account, particularly in view of their knowledge. This question of the proximity between students' available knowledge and the relevance of the feedback given by the teacher is also questioned by Robert and Vandebrouck [18]. It thus shows that regulating student activity during a session is not self-evident. This leads us to believe that it would be worthwhile to do specific work with teachers in training (pre- or in-service) to enable them to develop such professional gestures.

Furthermore, this research has identified difficulties when, on the one hand, teachers have a wide margin of freedom when it comes to organizing their problem-solving teaching and, on the other, official instructions in terms of learning objectives remain vague. These difficulties relate to the determination of learning objectives associated with problem-solving, the determination of criteria for selecting problems and organizing their articulation, and finally to assessment. These findings echo the research carried out by Choquet-Pineau [19] on the practices of primary school teachers in France proposing open-ended problems to their students, which led her to identify two teacher profiles, associated with different learning objectives: the first aiming primarily to get students to search and find the solution; the second profile aiming to get students to learn mathematics through investigation. This raises questions about the nature of the learning that can be achieved through the practice of problem-solving and revives the question posed by Hersant [20] about the existence of associated institutional knowledge.

At another level, a second part of this PhD research, based on a large-scale study of teachers' declared practices, reveals that, when selecting problems, teachers often use non-mathematical criteria, sometimes even in opposition to official prescriptions. Complementing this, a detailed study of the actual practices of three teachers in the context of an MSA course focused exclusively on problem-solving shows that, while one of the teachers was careful to choose the problems, she proposed to students by drawing on the strategies involved in solving them, the other two did not seek to articulate the problems selected with possible institutionalizations. The choice of problems and the articulation of these problems over the long term, with a view to learning how to solve problems, are therefore crucial issues for the profession.

Finally, this research has highlighted the difficulties encountered by teachers in institutionalizing knowledge associated with problem-solving. In fact, in the three classes we observed, we found no evidence of institutionalization, either in written or oral knowledge texts. However, this finding seems to go beyond the scope of this research and shows the importance of working on the link between what is institutionalized and what is assessed when problem-solving is taught for its own sake.

In our view, identifying the different types of reasoning, practice, and evidence involved in problem-solving is an interesting way of organizing the choice of problems to be proposed to students, their articulation, but also taking into account the question of learning over a long period of time, or the articulation between institutionalization and assessment.

3. Study of problem-solving processes through trials and adjustments

This PhD work [3] set out to study and characterize the mathematical problem-solving practices used by primary and lower secondary school students. In the scientific literature, various characterizations of problem-solving processes can be found. The seminal work is that of Pólya [21], who proposes a linear model based on four successive stages:

- Understanding the problem;
- Devising a plan;
- Implement the plan;
- Examine the resulting solution.

The linearity of this model was challenged by Schoenfeld (see **Figure 1**) [22]. The latter added an exploration phase to account for the part of the search that moves away from problem appropriation, but which does not yet constitute a plan. In his view, this exploration phase in particular accounts for the unstructured part of the search. This model thus presents a cyclical characteristic in the sequence: Appropriation – Planning – Exploration or Planning – Exploration – Planning.

Recently, Rott [23] has put these models to the test by analyzing the work of 10–12-year-old volunteer problem-solvers in his research laboratory. His results show

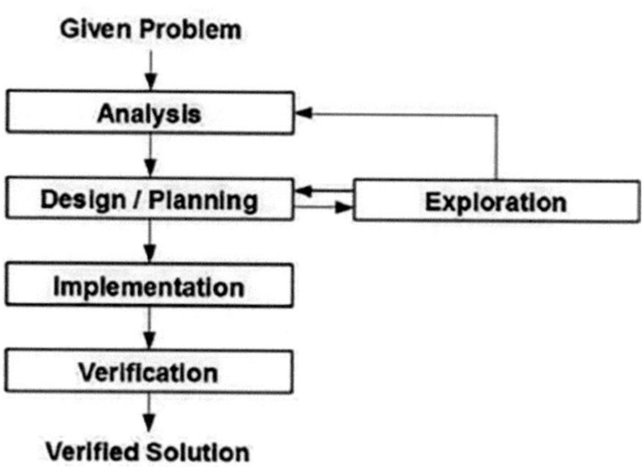

Figure 1.
Schoenfeld's model ([22], p. 110).

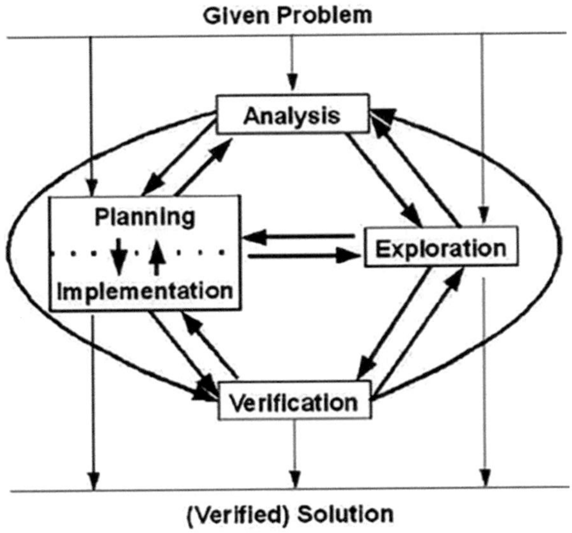

Figure 2.
Rott's descriptive model of problem-solving processes ([23], p. 106).

that Pólya's and Schoenfeld's models are not sufficient to account for the complexity of the phenomena, so he proposes to enrich these models by highlighting the greater complexity of the links between the different phases, as shown in **Figure 2**.

The scientific literature also highlights a second important aspect of the processes involved in problem-solving, namely the role of heuristics. This concept is explored in the field of psychology [24–28], artificial intelligence [29–31], and mathematics education [22, 32–36]. We have adopted Rott's definition, based on a wide-ranging review of the literature in these fields:

"Heuristics is a collective term for devices, methods, or (cognitive) tools, often based on experience. They are used under the assumption of being helpful when solving a problem (but do not guarantee a solution). There are general (e.g., 'working backwards') as well as domain-specific (e.g., 'reduce fractions first') heuristics. Heuristics being helpful regards all stages of working on a problem, the analysis of its initial state, its transformation as well as its evaluation. Heuristics foster problem solving by reducing effort (e.g., by narrowing the search space), by generating new ideas (e.g., by changing the problem's way of representation or by widening the search space), or by structuring (e.g., by ordering the search space or by providing strategies for working on or evaluating a problem). Though their nature is cognitive, the application and evaluation of heuristics is operated by metacognition." ([36], p. 190).

Thus, given the theoretical background we have just presented, we can divide our first research question: "How can we characterize the solving processes implemented by students in the usual conditions of the classroom?", into the following two sub-questions: "To what extent do the various models enable us to describe the work of students solving mathematical problems in the ordinary context of the classroom?" and "What role do heuristics play in the dynamics of the problem-solving process?"

In what follows, we discuss methodological aspects before briefly presenting some of the results of our work.

DOI: http://dx.doi.org/10.5772/intechopen.113258

3.1 Methodology

Our research focuses on three different grades 2, 6, and 8 of compulsory school in the canton of Geneva. For each grade, we proposed a problem that could be solved by trial and adjustment. For example, for two graders, we proposed the Card Game problem: "Each card in my deck represents either a triangle or a square. I pick 15 cards at random. I count all the sides of the figures drawn on the cards I've picked and find 49. How many triangles and squares do you think I picked?

In our research, we preferred that the students work in groups, after a few minutes of individual research, so as to be able to analyze their oral exchanges. However, a major difficulty lies in collecting data as close as possible to the student's work. For this reason, we equipped one student per group with an individual head-mounted action camera. This method of data collection provides first-person images [37]. For his part, the teacher is equipped with a tie microphone, and his interventions are recorded using an external camera controlled by the researcher.

The audiovisual data collected for each group of students were coded using the analysis framework developed by Schoenfeld [22]. This consists of dividing the students' work into macroscopic blocks that he calls episodes: "An episode is a period of time during which an individual or a problem-solving group is engaged in one large task or a closely related body of tasks in the service of the same goal." (p. 292). Schoenfeld thus distinguishes several categories of episodes: *reading, appropriation, exploration, planning/implementation,* and *verification*. In addition, Rott [23, 38] has introduced two further episodes, which we have also taken up: *writing* and *digression*.

Furthermore, the fact that, in our study, students solved problems in class, and not in laboratory conditions, led us to introduce an additional episode that allows us to characterize the moments when teacher and student(s) interact together about problem-solving. We called such episodes: *regulation*.

With regard to analysis in terms of heuristics, we have drawn up a coding manual that lists the various heuristics from the scientific literature that comply with the definition we have adopted. An extract is given in **Table 1**.

This initial list was put to the test by certain experimental data, which led us to supplement it with the following three heuristics (**Table 2**).

In our coding process, we have marked each occurrence of each heuristic as a type of occurrence code [39]. It is therefore possible for the same heuristic to be coded several times throughout the video or, on the contrary, never to appear at all. We also had to deal with a difficulty inherent in the duration of a heuristic, which has an impact on temporal coding. Indeed, some heuristics (e.g. copying or highlighting certain data) have a fairly wide span, while others (e.g. introducing names or

Draw, diagram, figure, graph
Organize data and trials in a particular form (rows, columns, table)
Reformulate the problem
Introduce names or notations
Perform a trial
Generate new data systematically

Table 1.
An extract from the heuristics coding manual.

Recopy or highlight certain data
Introduce artifacts or materials
Search for information (textbook, Internet, etc.)

Table 2.
New heuristics resulting from analysis of our experimental data.

notations) are very punctual. Moreover, for the same heuristic (e.g. Make a trial), we can be confronted with very different durations, depending on how the students implement them. In all cases, we have chosen to code as an occurrence the moment when the students evoke the heuristic, or the start of the action when they implement it without evoking it beforehand, so that duration is not taken into account.

The video data were coded independently by a research assistant and Favier. We analyzed the work of 33 groups in terms of episodes, and 17 groups (of these 33) for heuristics. For each aspect, we compared our coding results, and when these did not coincide, we reached a consensus by recording together. In terms of inter-coder agreement, we calculated a percent agreement (Jacobs et al., 2003) equal to 0.76 for the nature of episodes and 0.82 for time codes. For the coding of heuristics, the inter-coder agreement is 0.73 for their nature and 0.85 for time codes. As Tinsley and Weiss [40], referring to Guttman et al. [41], point out, "here was a 'tacit' consensus that 65% represented the minimum acceptable agreement". Thus, the inter-coder agreement percentages calculated for these different encodings seem quite acceptable.

3.2 Results

In connection with our first research question, we compared the coding results with the linear (Pólya's model) or cyclic (Schoenfeld's model) or neither linear nor cyclic (Rott's model) characteristics. It turns out that the work of only 7 of the 33 groups can be described with these characteristics (2 linear, 3 cyclic, and 2 neither linear nor cyclic).

For the other groups, our analyses reveal numerous episodes of regulation. During these episodes, problem-solving continues to progress from the students' point of view, so it is necessary to take them into account when describing and characterizing their work. However, because of the teacher's intervention, these moments cannot be considered to be of the same nature as the other phases of the model. We, therefore, propose to enrich Rott's model [23] by adding an additional dimension called "Regulation". The 3D representation (**Figure 3**) is an interesting way of symbolizing that this additional dimension lies on another plane. Our observations show that this Regulation can be connected to all the other phases of the model. Double arrows represent these various possible connections.

The distribution of the various solving processes (**Table 3**) shows that the vast majority of solving processes are non-linear, for both primary and secondary school students.

With regard to heuristic analysis, we have operationalized concepts linked to the idea of problem space [42], namely Julo's representation construction processes [43] and the path through semantic spaces [44]. We cannot go into detail here, but we have been able to identify three student profiles. The first, which we have called the *explorer*, shows a certain persistence in the search before considering and changing

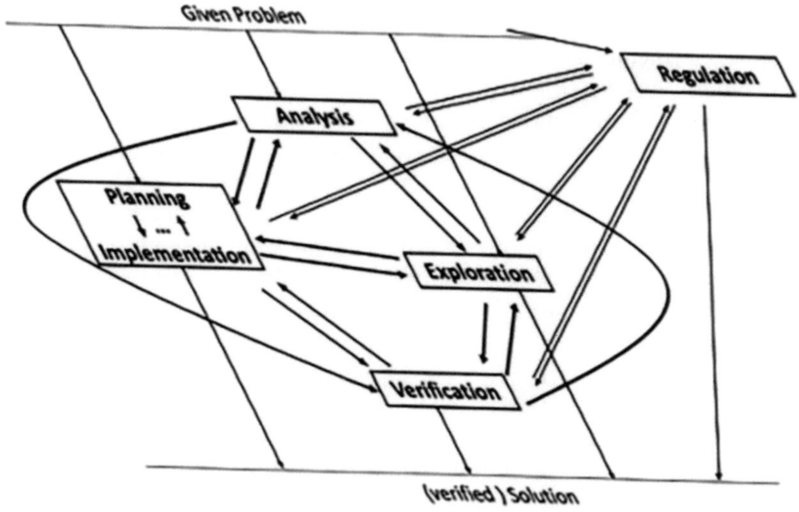

Figure 3.
Proposal for a model to describe students' solving processes in the presence of teacher intervention.

Solving process	Linear	Cyclic	Other	total
Without regulation	2	3	2	7
With regulation(s)	3	6	17	26
Total	5	9	19	31

Table 3.
Distribution of different solving processes.

tracks. Students in this profile have a good track record of problem-solving. In contrast, students in the *butterflyer* profile tend to carry out a more superficial search, with numerous avenues considered without really investigating them. Nevertheless, these students have a good capacity for imagining different ideas. However, these students are not really successful. Finally, a few groups of students combine characteristics of each of the two previous profiles, i.e., they consider different avenues without investigating them before pursuing one. We have labeled students in this profile as *prospectors*, they present a balanced rate of successes and failures.

3.3 Perspectives

These research results show the inadequacy and unsuitability of the models used in education, the main and best-known of which is Pólya's linear model. Moreover, the significant role of heuristics in problem-solving is confirmed, since in our study they enabled us to discriminate between different student profiles. We therefore suggest that heuristics could be an interesting lever, available to teachers, to help students solve problems. Our new research project is a way to transpose the results of our two PhDs into a tool for teachers wishing to teach problem-solving. This is what we will now present.

4. A new project for collaborative research

As we have seen, Chanudet in her PhD [2] characterized various aspects of teachers' problem-solving practices, particularly assessment ones, in the context of a specific problem-solving course at a lower secondary level. On another side, Favier in his PhD [3] provided a detailed analysis of students' problem-solving activities at primary and lower secondary levels. Here, however, the problems proposed were chosen by the researcher, and the sessions observed were ad hoc, making it impossible to take account of problem-solving education over a long period of time. A complementary research work led both by Chanudet and Favier [45]) took up the question of identifying and characterizing possible learning features specific to problem-solving. The starting point for this work was the observation, shared by other authors [7, 46, 47], that in a teaching context designed to get students to practice problem-solving, as well as in the parts of the curricula dedicated to the subject, the targeted learning outcomes are not clearly expressed and do not refer to precise mathematical knowledge. Drawing on the work by Houdement [7], which highlights the fact that problem-solving in elementary school can lead students to develop, in particular, learning related to ways of reasoning and validating in mathematics, the reflection was extended to the secondary school level and led to the identification and characterization of possible learning in problem-solving as pertaining to mathematical practices and reasoning and modes of proof. Chanudet and Favier also drew on the work of Jeannotte [8] to establish a characterization of mathematical reasoning considered from the dual standpoint of its structure and the processes mobilized. It emerges that working on problem-solving can lead students to develop learning linked to trial-and-adjustment or experimental-type approach, hypothetico-deductive reasoning, logical implication, exhaustiveness of cases, disjunction of cases, and proofs by ostension, counter-example or associated with the correct implementation of deductive reasoning. This view of problems seems to us to be a relevant tool for teachers, enabling them to think about and organize problem-solving education.

This has led us to set up a collaborative research project with lower secondary school teachers to study the processes of devolution and institutionalization, and on this basis to develop a resource to equip teachers to help students in problem-solving, covering both the learning objectives targeted and the means for developing this learning. We also plan, through a new PhD, to complement our work with a study of the actual, ordinary practices of primary school teachers, in order to apprehend the professional gestures of these problem-solving teachers and better understand how they organize and manage such classroom sessions. This would be the counterpart to Chanudet's PhD for primary schools. The secondary school component will build on the results of our previous work and enable us to envisage their direct operationalization through collaborative work, while the primary school component will begin by targeting teachers' actual problem-solving practices, before analyzing their potential evolution through collaborative work. In what follows, we present only the secondary school component.

On the basis of the literature review and our previous research project, we put forward a number of hypotheses:

- The devolution and institutionalization processes are always complex for teachers to manage, and are even more so in the context of problem-solving because of the lack of precise identification of the knowledge involved.

- Because of their complexity, problem-solving skills can only be learned over the long term. Similarly, the devolution and institutionalization processes associated with students' research skills also need to be thought through and studied from a long-term perspective.

- For all the above reasons, the implementation of these research findings by teachers is a delicate matter. We hypothesize that only collaborative research can lead to the adoption of such "tools", and their improvement and also enable the processes at play in the classroom to be documented.

To describe and organize the project, we have adopted Desgagné's characterization of the stages involved in collaborative research (see [48]). The first step, before the actual start of the collaborative work, will be to present the project as initially conceived, and the questions that drive it, to the teachers. This "co-situation" stage, central to the collaborative process, may lead us to complete or refine our questioning and our objects of study, in order to take into account the teachers' concerns.

The actual start of the collaborative work will focus on the preparation of problem-solving sessions before they take place in the classroom. The first step will be to present and illustrate to teachers a number of theoretical references drawn from the literature, as well as from our first project. In particular, we will return to the non-linearity of the problem-solving processes highlighted and illustrated in detail in Favier [3], and to the types and articulation of the different episodes (in the sense of [22]) involved. We will then return to the different modes of reasoning, approach, and proof that can be involved in problem-solving. We will then identify these with the teachers in a variety of problems, in order to reflect collectively on the choice of problems to propose to the students, their articulation throughout the school year, and the possible traces of institutionalization. We refer in particular to Julo's work [43, 49] on memory and problem schemas to envisage long-term problem-solving education, and thus think about the articulation of the problems studied.

The following year, collaborative work will focus on the actual implementation of classroom problem-solving sessions. More specifically, the focus will be on the influence of what the teacher does on student activity, and on ways of encouraging devolution and institutionalization processes, with the aim of supporting student activity and learning. This is the "cooperation" stage. Teachers will experiment in their classrooms with the problems chosen above. On the basis of video recordings of the problem-solving sessions, the students' activity, their productions, interactions between students and teachers, and interactions between students during group work phases will be analyzed collectively and discussed during work sessions every two months, which will take place over half-days. It seems particularly interesting to us to be able to monitor students' actual activity in great detail, something that teachers do not usually have access to. This is why the data collection methodology will be based, among other things, on the use of onboard cameras attached to the students' heads. It also seems important to us to bring different types of perspectives to these classroom sessions, focusing specifically on the different issues raised in our previous work. If time allows, we'll try to share our analyses and findings during working meetings with teachers. The idea is to question the evolution of the devolution process, the initiatives taken and left to the students, their share of autonomy in the search for a solution to the problem, and the way in which the associated didactic contract is negotiated. At the same time, the institutionalization process will be studied via the dialectic between two scales of time: local institutionalization, on the scale of one session, and the reinvestment over several sessions of what has been institutionalized previously, and the emergence of knowledge that can only be thought of over the long-term. To take this dynamic into account, we felt it wise to target problems requiring the same type of reasoning or practice. For several reasons, we have chosen to focus on problems involving an experimental approach. Firstly, these problems involve a wealth of mathematical activity on the part of the students (making trials, establishing conjectures, testing, proving). Secondly, we have observed that managing this type of problem is complex, especially given the diversity of student procedures. And lastly, these problems are rarely used in ordinary secondary school mathematics classes in Geneva.

In parallel, the training resource will be developed on the basis of the collaborative work carried out. Its aim will be to equip teachers to design, implement, and manage problem-solving sessions in their classrooms. Teachers who are interested will be able to continue collaborating with the researchers in the development phase of this resource. The content, which will be based on the collaborative work carried out and the results produced, as well as the form of the resource (video capsule or written document accompanied by videos of class sessions and their analyses in particular) will be discussed. The resource will then be distributed to other teachers for testing so that we can gather and analyze their feedback.

The aim of this project is to gain a better understanding of what happens between teaching practices and student activity when problem-solving is widely introduced into the classroom, through the study of the dual process of devolution/institutionalization. In addition, the work will lead to the production of a tool for initial and in-service teacher training. This practice should provide teachers with information on the knowledge involved in problem-solving, and on how to manage such sessions to support student learning.

5. Conclusion

Problem-solving plays an important role in mathematics and in school curricula. However, studies show that its teaching is complex. The aim of all our work is to gain a better understanding of the interplay between teachers' practices and students' learning around problem-solving in mathematics. More specifically, we are looking at how teachers can help students better appropriate a problem, without overdoing it for them.

Chanudet's dissertation has shed light on the practices of secondary school teachers when teaching mathematical problem-solving. We will take up this work again with a study of some primary school teachers. Favier's dissertation has given us a better understanding of how students work when solving problems through trials and adjustments. We intend to extend his work to other types of problems.

Our collaborative project with secondary school mathematics teachers (with plans to collaborate with primary school teachers at a later date) aims to gain a better understanding of what students do in the classroom when they solve a problem, and what can be done to support and encourage their work.

We are using the design and implementation of sessions and the analysis of their progress to devise an in-service training system designed to equip teachers.

Our work is based on a collaborative research approach, which we believe is conducive to a better appropriation of the results obtained. Various theoretical frameworks from the didactics of mathematics are mobilized to construct the research devices and analyze the data collected.

Author details

Maud Chanudet, Jean-Luc Dorier* and Stéphane Favier
University of Geneva, Switzerland

*Address all correspondence to: jean-luc.dorier@unige.ch

References

[1] Coppé S, Dorier JL. La résolution de problème au cœur de l'activité mathématique. Quels enjeux pour l'apprentissage ? Grenoble: GPA Éditions; to be published

[2] Chanudet M. Etude des pratiques évaluatives des enseignants dans le cadre d'un enseignement centré sur la résolution de problèmes en mathématiques. Switzerland: Université de Genève; 2019

[3] Favier S. Étude des processus de résolution de problèmes par essais et ajustements en classe de mathématiques à Genève. Switzerland: Université de Genève; 2022

[4] Arsac G, Mante M, editors. Les pratiques du problème ouvert. Lyon: Scéren édition; 2007

[5] Bonafé F, Sauter M, Chevallier A, Combes MC, Deville A, Dray L, et al. Les narrations de recherche du primaire au Lycée. Brochure APMEP n°151; 2002

[6] Georget JP. Activités de recherche et de preuve entre pairs à l'école élémentaire: Perspectives ouvertes par les communautés de pratique d'enseignants [PhD thesis]. Paris, France: Université Paris Diderot; 2009

[7] Houdement C. Une place pour les problèmes pour chercher. Annales de didactique des sciences Cognitive. 2009;**14**:31-59

[8] Jeannotte D. Raisonnement mathématique: proposition d'un modèle conceptuel pour l'apprentissage et l'enseignement au primaire et au secondaire [Thesis]. Montreal: Université du Québec; 2015

[9] Allal L. Evaluation des apprentissages. In: Van Zanten A, editor. Dictionnaire de l'éducation. Paris: Presses Universitaires de France; 2008. pp. 311-314

[10] Allal L, Mottier LL. L'évaluation formative de l'apprentissage : Revue de publications en langue française. In: L'évaluation formative: Pour un meilleur apprentissage dans les classes secondaires. Paris: OCDE; 2005. pp. 265-290

[11] Ruiz-Primo MA, Furtak EM. Exploring teachers' informal formative assessment practices and students' understanding in the context of scientific inquiry. Journal of Research in Science Teaching. 2007;**44**(1):57-84

[12] Kiwan ZM. Des pratiques d'enseignement de l'algèbre élémentaire aux apprentissages des élèves. In: Cas des expressions algébriques en EB7 et EB8 au Liban. Liban: Université Saint Joseph; 2018

[13] Allal L. Régulations des apprentissages : Orientations conceptuelles pour la recherche et la pratique en éducation. In: Allal L, Mottier Lopez L, editors. Régulation des apprentissages en situation scolaire et en formation. Bruxelles: De Boeck; 2007. pp. 7-23

[14] Robert A. Que cherchons-nous à comprendre dans les pratiques des enseignants ? Quelles analyses menons-nous ? In: Peltier ML, editor. Dur d'enseigner en ZEP Analyse des pratiques de professeurs d'école enseignant les mathématiques en réseaux d'éducation prioritaire. Grenoble: La Pensée Sauvage; 2004. pp. 15-32

[15] Robert A, Rogalski J. Le système complexe et cohérent des pratiques des enseignants de mathématiques: Une double approche. Canadian Journal of

Science, Mathematics, and Technology Education. 2002;**2**(4):505-528

[16] Ruiz-Primo MA, Furtak EM. Informal Formative Assessment of Students' Understanding of Scientific Inquiry. Report No.: 639. Los Angeles: Center for the Study of Evaluation (CSE); 2004

[17] Black P, Wiliam D. Developing the theory of formative assessment. Educational Assessment, Evaluation and Accountability. 2009;**21**(5):5-31

[18] Robert A, Vandebrouck F. Proximités-en-acte mises en jeu en classe par les enseignants du secondaire et ZPD des élèves: Analyses de séances sur les tâches complexes. Rech en Didact Math. 2014;**34**(2-3):239-283

[19] Choquet-Pineau C. Une caractérisation des pratiques de professeurs des écoles lors de séances de mathématiques dédiées à l'étude de problèmes ouverts au cycle 3 [PhD thesis]. Nantes, France: Université de Nantes; 2014

[20] Hersant M. Empirisme et rationalité au cycle 3, vers la preuve en mathématiques [PhD thesis]. Nantes, France: Université de Nantes; 2010

[21] Pólya G. Comment poser et résoudre un problème (Traduit de: How to solve it). Sceaux: J. Gabay; 1989

[22] Schoenfeld AH. Mathematical Problem Solving. New York: Academic Press Inc; 1985

[23] Rott B. Models of the problem-solving process – A discussion referring to the processes of fifth graders. In: Bergqvist T, editor. Learning Problem Solving and Learning through Problem Solving, Proceedings from the 13th ProMath Conference. Umeå, Sweden; 2012. pp. 95-109

[24] Houdé O. Comment raisonne notre cerveau. Que sais-je? Paris: Presses Universitaires de France; 2019

[25] Kahneman D. Système 1/ système 2: les deux vitesses de la pensée. Paris: Flammarion; 2012

[26] Richard JF. La résolution de problèmes. In: Richelle M, Requin J, Robert M, editors. Traité de psychologie expérimentale (vol.2). Paris: Presses Universitaires de France; 1994. pp. 523-570

[27] Richard JF. Les activités mentales. 4e éd ed. Paris: Armand Colin; 2004

[28] Verschaffel L. Realistic mathematical modeling and problem solving in the upper elementary school: Analysis and improvement. In: Hamers JHM, Van Luit JEH, Csapó B, editors. Teaching and Learning Thinking Skills. Lisse: Swets & Zeitlinger. 1999. pp. 215-239

[29] Feigenbaum EA, Feldman J. Computers and Thought. New York: McGraw-Hill; 1963

[30] Romanycia MH, Pelletier FJ. What is a heuristic? Computational Intelligence. 1985;**1**(1):47-58

[31] Tonge FM. Summary of a heuristic line balancing procedure. Management Science. 1960;**7**(1):21-42

[32] Brousseau G. Des dispositifs Piagétiens... aux situations didactiques. Éducation Didact. 2012;**6**(2):101-128

[33] Koichu B, Berman A, Moore M. Patterns of middle school students' heuristic behaviors in solving seemingly familiar problems. In: Novotná J, Moraová H, Krátká M, Stehlíková N, Krátká M, Stehlíková N, editors. Proceedings of the 30th Conference of the International

Group for the Psychology of Mathematics Education. Prague: PME; 2006. pp. 457-464

[34] Pólya G. Comment poser et résoudre un problème (Traduit de : How to solve it). Paris: Dunod; 1957

[35] Posamentier AS, Krulik S. Problem Solving in Mathematics, Grades 3-6: Powerful Strategies to Deepen Understanding. USA: Corwin Press; 2009

[36] Rott B. Rethinking heuristics – Characterizations and examples. In: Ambrus A, Vasarhelyi E, éditeurs. Problem Solving in Mathematics Education – Proceedings of the 15th ProMath Conference. Haxelnyomda, Hungary: Eötvös Loránd University; 2014. pp. 176-192

[37] Andrieu B, Burel N. La communication directe du corps vivant. Une émersiologie en première personne. Hermès Review. 2014;**1**:46-52

[38] Rott B. Problem solving processes of fifth graders: An analysis. In: Ubuz B, editor. Proceedings of the 35th Conference of the International Group for the Psychology of Mathematics Education. Ankara: Türke; 2011. pp. 65-72

[39] Jacobs J, Garnier H, Gallimore R, Hollingsworth H, Bogard Givvin K, Rust K, et al. Third International Mathematics and Science Study 1999 Video Study Technical Report. Volume 1: Mathematics. Washington: National Center for Education Statistics: Institute of Education Statistics, U. S. Department of Education; 2003

[40] Tinsley HE, Weiss DJ. Interrater reliability and agreement of subjective judgments. Journal of Counseling Psychology. 1975;**22**(4):358-376

[41] Guttman HA, Spector RM, Sigal JJ, Rakoff V, Epstein NB. Reliability of coding affective communication in family therapy sessions: Problems of measurement and interpretation. Journal of Consulting and Clinical Psychology. 1971;**37**:397-402

[42] Newell A, Simon HA. Human Problem Solving. Englewood Cliffs, NJ: Prentice-Hall; 1972

[43] Julo J. Représentation des problèmes et réussite en mathématiques: Un apport de la psychologie cognitive à l'enseignement. Rennes: Presses universitaires de Rennes; 1995. p. 255

[44] Poitrenaud S. La représentation des procédures chez l'opérateur. Description et mise en oeuvre des savoir-faire. Paris: Université de Paris; 1998. p. 8

[45] Chanudet M, Favier S. Les démarches et modes de raisonnement en jeu dans les problèmes de « Recherche & stratégies » en 10H. Revue de Mathématiques pour l'école. 2021;**235**:88-98

[46] Hersant M. « Problèmes pour chercher ». Des conduites de classe spécifiques. Grand N. 2008;**81**:57-75

[47] Mercier A. Une question curriculaire de l'enseignement élémentaire des mathématiques : La « résolution de problèmes ». In: Programme national de pilotage Actes du séminaire national L'enseignement des mathématiques à l'école primaire. Paris: Eduscol; 2008. pp. 93-116

[48] Desgagné S. La position du chercheur en recherche collaborative: Illustration d'une démarche de médiation entre culture universitaire et culture scolaire. Recherches qualitatives. 1998;**18**:77-105

[49] Julo J. Des apprentissages spécifiques pour la résolution de problèmes ? Grand N. 2002;**69**:31-52

Chapter 11

Collaboration in Science, Technology, Engineering, Mathematics (Stem) Education and Technical, Vocational Education and Training (TVET) for Graduates' Competence in Nigeria

Edet Okon, Maria B. Onabe, Mary A. Iyam, Ann Dijeh, Francesca Nneoyi-Egbe, Stephen Bepeh Undie, Ita E. Okon, Agnes L. Okute, Mfon E. Usip and Agnes A. Ewuru

Abstract

The demands for skills, especially modern work-related, entrepreneurial and self-reliant skills are necessary in this present twenty-first century. It is clear that a single discipline may not impart comprehensive and required skills, and provide the competencies needed by graduates to remain relevant. There can be no better collaboration than STEM Education and Technical Vocational Education and Training, considering the level at which their contents, mode of delivery and job descriptions relate. The author's opinion is that Nigeria's chronic rate of unemployment among youths can be addressed through right-skilling and up-skilling. Collaboration of these two broad disciplines should urgently be experimented for skill development to reposition youths to fit into the demands of the present era.

Keywords: collaboration, education, vocational education, skill, skill development, graduates

1. Introduction

Science, Technology, Engineering and Mathematics (STEM) Education started gaining prominence in Nigeria's education system with the introduction of the 6-3-3-4 system of education in the 1980s. Therefore, in the formulation of her educational policies, much attention was drawn to this important aspect of education, such as curriculum restructuring from basic to tertiary education, provision of science laboratory equipment to schools, increased funding for STEM-related programmes, including overseas scholarship for STEM-based students, teachers and researchers, among others. The importance attached to STEM Education has not changed as Nigeria, in the last few decades, has witnessed educational reforms in favour of STEM, in view of repositioning the education sector of the economy to deliver its mandate, considering her choice of education as a veritable tool for national development. The recent trends in Nigeria's educational reforms have witnessed the government and stakeholders showing much concern for STEM Education. Relevant stakeholders in education, especially those at the level of formulating, regulating and implementing educational policies, increasingly design programmes that draw the interest of school children to STEM-related academic disciplines and careers. This was equally followed by the establishment of Science schools at the secondary school level, while a good number of Science-related courses were mounted across Nigeria's tertiary education institutions. The idea was to train the younger generations with critical thinking and inventive abilities to grow Nigeria's economy as a country endowed with an abundance of natural resources.

Another area in the education sector that was given top priority in the 1980s, after STEM Education, is Technical Vocational Education and Training (TVET). Interestingly, Nigeria, in response to global trends at the outset of the new millennium, increased the tempo in terms of priority given to TVET. Therefore, one of the action plans to achieve the Millennium Development Goals (MDGs) was to expand TVET delivery systems in order to provide the needed skills to the younger generation in particular and empower them to create wealth. This was to ensure that they were productive and remained job and wealth creators. The emphasis on TVET, a broad-based discipline drawn from technical and vocational educational clusters, covers a broad range of subjects from basic to tertiary education levels.

Significantly, too, between the 1980s and late 1990s, emphasis on STEM was expanded to embrace TVET. However, in the early 2000s, Nigeria's attention to TVET saw the establishment of more technical colleges and the introduction of TVET courses at the tertiary education levels. Thus, regular and recent revisions of the National Policy on Education have been to accommodate these newly adapted sub-sectors of education, having recognised their potency in the production of graduates needed to drive Nigeria's natural and human resources potentials to shower economic development and sustainability.

This chapter further addresses concerns in STEM Education and TVET as cross-cutting courses in Nigeria's education system, Nigeria's educational reforms with STEM and TVET, practical approaches and regulatory agencies required for reforms in STEM and TVET, structural comparison of STEM and TVET, funding as a vehicle to effective STEM Education and TVET, pedagogical approaches, practical inclination and collaborative strategies for the future of STEM Education and TVET are harnessed in this chapter.

2. Historical analysis of Nigeria's education reform and the concern for STEM and technical vocational education and training

Nigeria's education reforms can be deciphered in two phases: the pre-independent and post-independent phases. Unarguably, Western merchants, missionaries and colonial suzerains brought formal education and established schools in Nigeria, with the first primary school established in 1842, long after an attempt by Portugues merchants to do so with restrictions to traditional rulers, who were already exerting so much influence in their respective Kingdom. In other words, European adventurers, whether for business, religious or colonial intentions, brought Nigeria to the limelight of formal/western education. It is often remarked that the type of education handed down to Nigerians was to serve their purposes and not to develop Nigerians. Hence, skilled subjects such as STEM Education and TVET were not introduced into the curriculum. Ossai [1] noted that Nigerians were taught the four Rs – Arithmetic, Reading, Religion and Writing. The curriculum introduced in the school system between 1882 and 1925 marked a significant milestone in Nigeria's educational history. The next phase of curriculum development spanned from 1925 to 1952, marking the establishment of commissions such as the Phelps–Stokes Commission. Interestingly, curriculum development during this period marked a minimal shift from the four Rs to areas of need in Nigeria.

Unfortunately, Nigerians themselves had little understanding of how to adapt to a paradigm shift in this regard. Thus, Afimag.com noted that Nigerians, at the time, favoured academic curriculum above agricultural and technical curriculum because previous students who completed secondary education gained admission into the tertiary institutions with these subjects and saw it as a pathway to securing white-collar jobs. Despite the adherence of Nigeria to the existing curriculum, Phelp-Stoke strongly recommended and made a move intended to match education in Nigeria to the actual needs of Nigerians. This was why the colonial government made efforts towards the provision of agricultural and technical education based on the recommendations of the Phelp-Stoke Commission. The recommendations equally included the need for teacher preparation and registration. Another stage, beginning from 1952 to 1960, marked the birth of examination bodies such as the West African Examination Council (WAEC), which was established in 1952. Notably, emphasis was laid on the development of curriculum, evaluation of academic programmes through services of education inspectors, training of teachers, and funding of schools by colonial masters, missionaries and other proprietors. Meanwhile, during the periods under review, no concerted effort was made to develop STEM Education and TVET.

The second phase of curriculum development witnessed the introduction of STEM and TVET into the curriculum. This followed the recommendations by Ashby Commission that certain vocational and commercial subjects be introduced into the curricula of schools. Subjects introduced include agricultural science, commerce, business mathematics. This is the same period of re-echoed emphasis on technology; thus, Aina [2] noted that technology and engineering education in Nigeria are evolving to date. Aina compared the trends in Nigeria with those of developed countries, where the trajectory of STEM in Nigeria revealed significant achievements in the past and the present critical challenges that require attention. The author further noted that STEM in Nigeria shows substantial achievements in some sectors of the economy such as health, agriculture and telecommunication. Greater parts of the achievement of STEM Education would have been in training students to choose career parts relevant to their aspirations. Rosenzweig and Chen [3] classified career reference into four groups: appealing, unappealing, polarising, and overlooked. They frowned at the poor development of STEM education, which the author reported as hindering national development. Growing out of colonial influence and administration in 1960 gave Nigerians the opportunity to handle their affairs in the education sector based on the peculiarities of Nigeria and global realities.

Basic Education (BE) in Nigeria has STEM offered as Basic Science and Technology with a curriculum segmented into three levels under the 9-year Basic Education Curriculum thus: Basic Science and Technology for Primary 1-3, 4-6, and JSS 1-3 [4, 5].

Subjects are classified in clusters under each stage. For instance, the subjects clustered under Basic Education, according to FRN [6], are Basic Science, Basic Technology, Information Technology and Physical and Health Education. On the other hand, the curriculum structure of STEM at the post-basic education level has Science and Mathematics as a cluster and technology as another cluster. The former (Science and Mathematics) covers Biology, Chemistry, Physics, Further Mathematics, Health Education, Agriculture, Physical Education and Computer studies. The latter (Technology Education) covers Technical Drawing, General Metal Work, Basic Electricity, Electronics, Auto Mechanics, Building Construction, Woodwork, Home Management and Food and Nutrition.

3. Technical and vocational education and training (TVET)

According to FMoE [7], pre-vocation studies are a new Universal Basic Education (UBE) subject curriculum created from the integration of Agriculture and Home Economics in a bid to reduce subject listing at the UBE level. The objectives of pre-vocational studies at the UBE level are to enable the pupils/students to:

1. Develop interest in Agriculture and Home Economics (pre-vocational studies);

2. Acquire basic knowledge in pre-vocational studies;

3. Become aware of the occupational areas in the pre-vocational studies;

4. Acquire basic skills required for entry into the occupational areas of the pre-vocational studies;

5. Appreciate the dignity of labour and entrepreneurship.

This component of education is classified under post-basic education and career development, offered after basic education. Pre-vocational education is offered at the basic level, beginning at the middle basic level (Primary 4-6). Thus, the curriculum is structured as follows: pre-vocational studies for Primary 4-6, JSS 1-3, and Technical and Vocational Education and Training (TVET) for Senior Secondary School.

There is uniqueness in the content of the pre-vocational studies curriculum. The introductory paragraphs, objectives, justification and brief description of the curriculum are uniquely described for both pre-vocational studies 4–6 and pre-vocational studies for JSS 1-3 [4, 5]. A definition advocated for TVET by the Federal Republic of Nigeria [6] refers to those aspects of the educational process involving, in addition to general education, the study of technologies and related sciences and the acquisition of practical skills, attitudes, understanding and knowledge relating to occupations in various sectors of economic and social life.

4. The nature and role of STEM and TVET education in Nigeria

It is pertinent to emphasise here that both STEM and TVET have more practical components attached to little theoretical enunciations at the ratio of 70 and 30 per cent, respectively. Practical approaches are adopted for enhanced skill development and competence among graduates in these two broad areas (STEM and TVET). In the course of reforms, STEM Education and TVET have been given preference with practical approaches, including but not limited to establishment and separation of relevant regulatory and supervisory authorities responsible for STEM Education and TVET programmes; approval of students' industrial work experience schemes for students from these two clusters; establishment of vocational enterprise institutions for continuous training through collaborations; among others.

The importance attached to STEM Education and TVET warranted the establishment of relevant agencies for regulating these two significant areas of education. In the UBE and Post-UBE categories (Primary and Secondary Education), STEM Education and TVET are directly regulated by Science and Technical Education Board. At the Federal level, the National Business and Technical Education Board (NABTEB) oversees STEM Education TVET at UBE, Post-UBE and National Board for Technical Education at the Higher Education (HE). This independent identity allows for effective monitoring of academic programmes in these two education clusters, development techniques for effective evaluation and ensures quality graduates' output.

Approval and administration of students' industrial work experience scheme (SIWES) for students from STEM Education and TVET clusters.

Students undergoing studies in STEM Education and TVET courses are made to undergo compulsory training in industries to acquire practical work experience and relate theoretical enunciations obtained in the classroom with hands-on experience in work and business environments. Classroom enunciation may be as effective as it is but does not replace 'work ready' state of graduates. Thus, Obi [8] posited that SIWES prepares 'work ready' graduates.

During the most recent review of the National Policy on Education, a remarkable innovation was the establishment of vocational enterprise institutions (VEIs) for continuous training through collaborations. The need to solve perennial problems in Nigeria led to the review of the National Policy on Education (NPE), where innovations were introduced that led to the establishment of Vocational Enterprise Institutions (VEIs). The mandate of VEIs is to redirect people into creativity and innovativeness in line with global trends. According to the Federal Republic of Nigeria, VEIs will be offered to graduates of Senior Secondary Schools who could not transit to tertiary institutions. The period of training can last for two years. Summary of the goals of VEIs according to the Federal Republic of Nigeria [6] focuses on:

a. Training for specific skills and knowledge with a certificate issued to trainees on successful completion;

b. promote a training environment where trainees can think creatively to enhance knowledge transformation through technological processes, thereby creating wealth for economic development;

c. provide intensive training for the acquisition of functional skills using modern technologies for the achievement of national goals;

d. avail increase access to technological education at the tertiary level;

e. promotes the acquisition of skills in craftsmanship for global competition;

f. equip the trainee with competences to explore life opportunities; create opportunities for self-reliance, and make trainees wealth creators and providers of employment.

Emphasis on STEM Education and TVET in Nigeria is further premised on the need to engage graduates in productive venture as captured in the goals of VEIs highlighted above. There are other practical approaches to enhance the functionality of these two education clusters, including expansion through the approval and establishment of more academic programmes, technical colleges, polytechnics, and university of technologies, among others. Acquisition of skills will enable graduates to be productive and contribute to national development. In our era, the acquisition of generic skills become significant. Obi [8] observed that generic skills build self-esteem.

5. Structural comparison of STEM education and TVET in Nigeria education system

It is pertinent to note that STEM Education, TVET, and pre-vocational studies are interrelated as broad areas. The interrelationship provides an advantage for collaboration. Both broad areas have many things in common: Some courses in STEM Education are offered in TVET. The areas of interrelatedness are discussed generally in line with the basic education curriculum. The interrelationship is presented in tables. It is also pertinent to consider the interrelatedness based on the most recent structure of STEM Education and TVET (either pre-TVET) under the 9-year compulsory universal basic education, 3-year post-basic education, and post-senior secondary education.

Comparison of STEM Education and TVET (either pre-TVET) under the 9-year compulsory universal basic education

Table 1 identifies the interrelatedness between the two broad and very important areas of the entire education. Therefore, both STEM Education and TVET at the basic education level are interrelated. The integral objective of these academic programmes at the basic level is to equip learners with basic skills for progression to post-basic education level with interest in TVET, which covers other related areas such as STEM offered in technical colleges, Vocational Enterprise Institutions (VEIs) offered through collaborations, and National Vocational Qualification Framework (NVQF). Again, a concise comparison of STEM Education and TVET under the 3-year post-basic education is necessary at this stage. The structure of TVET at the post-basic education is largely merged. Thus, FRN [6] maintained that TVET should cover the following: Technical Colleges and Vocational Enterprise institutions and National Vocational Qualification (NVQF). In other words, TVET is broadened at the Post-Basic Education Level to cover vocational subjects, technical subjects, science subjects, engineering at the foundational level, and, of course, mathematics. Some related areas may not necessarily be done in formal setting but regulated and formalised for certification under VEIs and NVQF. The provisions of Nigeria's National Policy on Education on these educational aspects, according to FRN [6], are described in a tabular form and presented in **Table 2**.

The analysis in **Table 2** shows the interrelatedness of the components embedded in TVET, which includes technical/technology education and other related programmes such as VEIs and NVQF. This means that TVET is the umbrella academic programme and can be delivered using formal and informal approaches. This implies that TVET and associated areas highlighted in **Table 2** have the cardinal objectives for training for skills acquisition.

Parameter	STEM	TVET
Subjects' composition	Subject and course composition at all levels covers sciences, engineering, and mathematics, among others	Subject and course composition in technical education include engineering; agricultural science and home economics education include the sciences; computer education includes technology; and business education includes mathematics, especially commercial mathematics
Practical orientation	Delivery is largely practical, which is why large classes should be grouped at the ratio of 1:20 in line with the provisions of the National Policy on Education	Delivery is largely practical, which is why large classes should be grouped at the ratio of 1:20 in line with the provisions of the National Policy on Education
Laboratory/Workshop	Practical for all STEM courses are delivered either in the laboratory or workshop. For instance, science practical is conducted in the laboratory, engineering in workshops, and mathematics and technology both in laboratory and workshop	Subjects and courses related to agricultural science education are in the laboratory, field or farm; technical education in workshops; business education in computer/ICT laboratory
Pedagogical strategies	Delivery strategies are mainly activity-based and other scientific teaching methods	Delivery strategies are mainly activity-based and other scientific teaching methods
Model Curriculum Structure of STEM Education and TVET	Formal STEM (broad-based) curriculum under the 9-year basic education cuts across Primary 1–3, Primary 4–6, and JSS 1–3	Formal TVET (broad-based) curriculum begins at Primary 4–6 and continues with JSS 1–3
Objectives (JSS 1–3)	a. Provides activities that motivate students for preference in science and technology b. Train students to acquire basic skills and knowledge in science and technology c. Train students to apply skills acquired to meet contemporary and societal needs d. Create awareness of diverse career opportunities in science and technology e. Prepare students to advance their education in science and technology f. Inculcate positive attitudes that drive away social vices. g. Train students on safety and security measures	a. Develop interest in agriculture and home economics (pre-vocational studies) b. Acquire basic knowledge in pre-vocational studies c. Become aware of occupational areas in the pre-vocational studies d. Acquire basic skills required for entry into the occupational areas of the pre-vocational studies e. Appreciate the dignity of labour and entrepreneurship

Source: Developed by Okon [9], as adjusted from FGN, 2023.

Table 1.
Comparison of STEM education and TVET (either pre-TVET) under the 9-year compulsory universal basic education.

A comparison of STEM Education and TVET under post-senior secondary education reveals that tertiary education in Nigeria is complex, with different types, levels and ownership. These include universities, polytechnics, mono-technic and colleges.

Basis for comparison	TVET	Technical Colleges	Vocational Enterprise Institutions (VEIs)	National Vocational Qualification (NVQF)
Goals/ Features	Workforce training and specialised areas related to science, technology and business; Acquisition of vocational and technical skills for the productive sectors of the economy; and Training to provide personal and self-reliant skills	The curricula promote training in different trades with learning outcomes structured in a systematic manner that allows for mastery of skills; The learning outcomes are grouped into general education; theory and related courses; workshop; industrial training/ production work; entrepreneurial training; teacher–student ratio shall be kept at 1: 20	Provides training for the promotion of creativity. Opportunity to accommodate diverse groups within a short-duration of 1 to 2 years	Provides the basis for the identification of diverse skill areas with trainees certification, thereby adding formal status to informal training with government recognition and acceptance of such certificates.

Source: Modified from FRN [6].

Table 2.
TVET under the 3-year post-basic education and related vocational programmes.

Again, within each type of institution, there may be variation based on programmes and regulatory bodies. For the purpose of this research, colleges of education and universities are modelled for comparison between STEM Education and TVET. College of Education in Nigeria has the following categories: Traditional College of Education, College of Education (technical) and Distance Learning. The uniqueness of STEM Education and TVET at the college of education and university levels is based on the fact that STEM Education is offered in faculty domains, namely: STEM Education in the Faculty of Education/Vocational Technical/Technology Education; STEM Education in the Faculty of Sciences; and STEM Education in the Faculty of Engineering.

In most cases, courses offered under the auspices of TVET are offered in cognate faculties. For instance, courses offered in Technical Education are offered in Engineering; those offered in Agricultural Science Education are offered in the Faculty of Agriculture and Forestry, among others.

Again, in most tertiary institutions in Nigeria, Technical Education (a major component of TVET), according to Federal Republic of Nigeria [10], has various areas of specialisation such as Automobile Technology; Building Technology; Electrical/Electronic Technology; Metalwork Technology; and Woodwork Technology.

Similarly, Business Education, another major component of TVET, has areas of specialisation such as Accounting, Office Management, Marketing, and Management found in other cognate faculties such as management sciences and business administration, among others. Home Economics Education as an aspect of TVET that has options in Textile and Clothing, Food and Nutrition, and Home Management are also in other cognate departments.

6. Educational funding on STEM education and TVET

Globally, education funding is critical to effective training of learners. Education funding has been a perennial issue in Nigeria and pre-dates independent Nigeria. For instance, Okon [9] noted that the Ashby Commission crystallised 'Investment in Education' with projection in the development of Higher Education. Okon further explained that the Commission was not just making proposals but making a programme for educational development that would be massive and expensive. Beyond this, the Federal Government of Nigeria has constantly been on the side of improvement in funding education with policy trust that equally attracts alternative funding. Notably, increased funding of education has been voiced for the development of STEM Education and TVET in Nigeria.

Therefore, funding was captured and accorded a section in the 6th edition of the National Policy on Education (NPE) that was developed and subjected to regular revision to match pace with changing global and indigenous realities occasioned by digital transition and application in virtually all facets of human endeavours. Funding, according to FRN [6], can effectively be addressed through the development and articulation of policies and provided in the 6th edition of the National Policy on Education, involving collective and wide consultation by the Nigeria Educational Research and Development Council (NERDC). This is to ensure that the persistent gaps in education policy provisions and implementation in Nigeria are bridged. The current edition of the Policy highlights emphasise critical areas such as consolidating the 10-year Basic Education in line with Universal Basic Education; emphasis on quality assurance; development of the National Education and Management Information System (NEMIS); improvement strategic planning; teacher quality; and functional collaboration with relevant stakeholders.

In terms of funding, the policy acknowledged that education is capital intensive and requires adequate financial provisions and the need for these stakeholders, especially all tiers of government and private sector to jointly fund education in Nigeria. The Policy acknowledged interventionist programmes from Federal Government agencies such as Universal Basic Education Commission (UBEC); Tertiary Education Trust Fund (TETFund); international donor agencies such as Word Bank, USAID, UNIDO, UNICEF, UNDP, as well as NGOs, among others. Aina [2] asserted that funding is globally critical to sustainable STEM Education. The author further observed that inadequate funding has been a problem of STEM Education in Nigeria, noting that the Nigerian Government has not done enough to make funds available for STEM Education. Aina further reported that most schools are without modern learning resources as a result of inadequate funding, a situation that the author described as creating an appalling learning condition.

Therefore, the quest for improved funding for STEM Education is being voiced by stakeholders within and outside the purview of STEM. According to Adeniran [11], academic experts have called on the government at all levels to allocate special funding for STEM Education in Nigeria as a means to encourage students and boost the working condition of services for academics in the country. In the course of funding STEM Education in Nigeria, the government plays a major role. Therefore, the government at all levels needs to provide a good stimulus for pragmatic science and technology education and equally map out strategies to promote interventionist programmes that attract appropriate special funding for STEM Education and TVET programmes in Nigeria. Lawal [12] acknowledged the importance of STEM Education worldwide and its ability to stimulate development, especially in developing countries, but observed that Nigeria has little or no investment in STEM Education. Equally, Asaolu [13], in an inaugural lecture, advocated special funding for STEM Education in Nigeria. According to Asaolu, such development is borne out of the need to fasten the rapid industrialisation that Nigeria needs in her quest for national development. Okoli et al. [14] expressed dissatisfaction with the poor funding of TVET, noting that the nonchalant attitude dates back to the colonial days.

Nigeria can leverage collaborations with other economies across the globe. Evidence from the East Asia and Pacific (EAP) can be useful, where training is collaboratively handled by both government and private organisations. In most cases, private organisations identify skills and offer training to members of the immediate communities, thus contributing optimally to the development of the workforce. According to Palmer [15], countries in the region face strong pressures to expand their technical and vocational education and training (TVET) systems and enhance their quality, while also facing spending pressures on basic and higher education, adding that as the demand increases, the need for sustainable financing for TVET becomes more urgent. Furthermore, Palmer argued that this is not only about ensuring that sufficient and predictable revenue streams exist to fund training programmes but perhaps just as importantly about how financing mechanisms themselves can be strongly linked to achieving policy objectives of making TVET systems more accessible, equitable, efficient, demand-driven, responsive and relevant. From the preceding, therefore, the following areas of funding strategies highlighted by Palmer are summarised as follows:

a. Diversification through enterprises, individuals and other innovative sources of funding and through private-public partnerships (PPP)

b. Efficiency and equity in financing policies and mechanisms

c. Effective utilisation of resources through a sound administrative system

d. Funding from private resources in the form of students' fees

1. Pedagogical approaches and practical inclination

Simply put, Nigeria's experience in education funding requires a paradigm shift. Funding education has remained a frontline issue in Nigeria over the decades. This view is widely reported by scholars and people from other walks of life, including government functionaries who are in a position to reverse the trend. One could sum up such reports by concluding that education is grossly underfunded. This is amidst several sources of funding education, such as reports by researchers. For instance, Nnamdi and Nwite [16] identified sources of funds such as School fees and levies on school children in some states and other monies raised within the school; fund yield ventures such as lunching of school magazines and almanacks, sports' complex, inter-house sports' appeal fund; revenue from agricultural yield and home economics exhibitions; donations from philanthropic individuals; corporate funding from local and international organisations such as World Bank with interest in STEM and TVET funding through school building renovation and rehabilitation and supply of school equipment (science and technology education post-basic project (STEP-B) World Bank Assisted), endowment funds, local community aid through parents and teachers' association (PTA) and some wealthy members of the community; among others.

Education suffers poor funding in Nigeria. Consequently, the slow rate of development it portends to the education system and economic development beckons for a more aggressive and enterprising approach to funding education. It is equally pertinent to advocate for divorce and independent funding of different components in education based on segmentations provided in the National Policy on Education, where Technology and TVET education can be funded independently.

Again, heavy funding of STEM Education and TVET under long-term investment in the next two decades is urgently required. Therefore, the Federal Ministry of Education, in conjunction with State ministries of education, can draw a funding plan for commitments by all tiers of government. Thereafter, the government at all levels should introduce initiatives for private sector engagement and commitment in contributing to the funding of education in Nigeria.

Furthermore, STEM and TVET institutions should map out models for funding sustainability through venture development during the 20 years long-term investment through heavy funding of education by the Government of Nigeria. Venture enterprises by STEM and TVET institutions should take into consideration, and utilise the global business environment through e-business opportunities in a knowledge economy.

Importantly, too, prudence in the management of funds and accountability are necessary qualities school managers must possess. Personnel in charge of appropriating educational funds should be made to account for funds at their disposal, and heavy sanctions should be meted out on defaulters.

7. Collaborative strategies for the future of STEM education and TVET

The importance attached to STEM Education in Nigeria is attracting collaborations. Lawal [12] was emphatic on the benefits of collaboration of Science, Technology, Engineering and Mathematics (STEM) education, and TVET to the development of Nigeria. Collaborative strategies can take the following forms:

Exchange programmes: These include students' exchange programmes for 1 year internship between STEM and TVET students/institutions globally; instructors' exchange programmes through sabbatical leave, adjuncts and visiting teachers and lecturers; direct training of students in advanced economies through Federal and State Governments' sponsors.

7.1 Research collaboration in topical issues in STEM and TVET

The authors' collaboration to enhance published works as good sources of content that lecturers in STEM Education and TVET can use and enrich content delivery; institutions' collaboration in research among authors can be across institutions within a country. Lathabai et al. [17] are of the opinion that institutions should shift from what they call trust-based funding to performance-based funding. Of course, this implies that funds made available would be put to proper use. Again, Tim and Leuprecht [18] recognised the globalising world and the need for cross-border collaborations. Also, Yao [19] noted that international collaborative research involves cross-country teams that share research interests, conduct research and promote research results to advance knowledge and promote positive shifts in practice. These are lessons that institutions offering STEM Education and TVET need to enhance their collaborative mechanisms to enhance economic and sustainable development.

1. Nigeria can seek World Bank sponsorship in the development of STEM and TVET
2. International exhibition through STEM and TVETFair for fundraising
3. Endowment funds from wealthy individuals in the society
4. Scholarships and capital projects by industries
5. Increasing education endowment funds paid by companies through education tax

8. Conclusions

The nature of STEM and TVET reveal striking similarities, with both having the mandate of producing skilled-bound graduates that can proffer solutions to Nigeria's ailing economy through problem-solving and critical thinking and the acquisition of functional skills for self-reliance, productivity and poverty eradication. This chapter reveals that the initial effort by the Western world to establish Western/formal education witnessed very little or no emphasis on STEM and TVET education but provided a platform for education development. Thus, different commissions instituted by pre-independent education bodies made recommendations that provided the roadmap for what later evolved as STEM and TVET. The growing needs of Nigeria, as they took over the affairs of their country, at independence, yearned for an educational system that could address emerging realities. Thus, Nigeria saw the need to revise the previous curriculum that seemingly addressed colonial and other foreign needs to reflect the needs of the Nigerian Society. This yielded results as Nigeria laid and is still laying much emphasis on STEM and TVET. This has resulted in increased education funding with a reasonable proportion voted to STEM and TVET, development of a curriculum that addresses the concerns of STEM and TVET that have proven to be of collaborative relevance in addressing the needs and aspirations of Nigeria in terms of national development.

This chapter therefore concludes that the mandate of education to produce graduates with functional skills can best be achieved through effective and continuous STEM and TVET collaboration. Unarguably, therefore, collaborations among disciplines can be very useful in the production of graduates with competencies who can efficiently fit into the demands of the twenty-first century. Considering the similarities of STEM Education and TVET in structure, approach and delivery, educational institutions cannot afford to ignore such collaborations. In order to uphold their mandate, therefore, practical activities can be better achieved through collaboration strategies. This, therefore, requires training to enhance skill development.

Author details

Edet Okon[1*], Maria B. Onabe[1], Mary A. Iyam[1], Ann Dijeh[1], Francesca Nneoyi-Egbe[1], Stephen Bepeh Undie[1], Ita E. Okon[2], Agnes L. Okute[1], Mfon E. Usip[3] and Agnes A. Ewuru[1]

1 University of Calabar, Calabar, Nigeria

2 Cross River State College of Education, Akamkpa, Nigeria

3 University of Port-Harcourt, Port-Harcourt, Nigeria

*Address all correspondence to: okon_e@ymail.com

References

[1] Ossai C. History of curriculum in Nigeria. Africa Investigative Magazine. 2022 [Accessed: June 13, 2023]. Available from: https://afimag.com/history-of-curriculum-development-in-nigeria/

[2] Aina JK. Stem education in Nigeria: Development and challenges. In: Aina JK, Abdulrahman A, editors. Current Research in Language, Literature and Education 3. Lafiagi: BP International; 2022. DOI: 10.9734/bpi/crlle/v3/2258C

[3] Rosenzweig EQ, Chen XY. Which STEM careers are most appealing? Examining high school students' preferences and motivational beliefs for different STEM career choices. International Journal of STEM Education. 2023;**10**:40, 2-25. DOI: 10.1186/s40594-023-00427-6

[4] Federal Republic of Nigeria – FRN. 9-Year Basic Education Curriculum, Pre-Vocational Studies 4-6. Lagos: NERDC Publisher; 2012a

[5] Federal Republic of Nigeria – FRN. 9-Year Basic Education Curriculum, Pre-Vocational Studies JSS 1-3. Lagos: NERDC Publisher; 2012b

[6] Federal Republic of Nigeria – FRN. In: National Policy on Education. 6th ed. Lagos: NERDC Publisher; 2014

[7] FMoE. Pre-Vocational Studies. Lagos: NERC Publishers; 2012

[8] Obi OC. Generic skills acquisition and the realisation of the goals of business education in Nigerian universities: Implications for curriculum innovation. Nigerian Journal of Business Education. 2021;**8**(2):222-230

[9] Okon EE. Historical development of business education. In: Duruamaku-Dim GCE, Chukwurah CC, Akeke MNG, Okute AL, editors. Contemporary Issues and Concepts in Business Education. Calabar: Sunny-Tech Link & Logistic Concept; 2020

[10] Federal Republic of Nigeria – FRN. Nigeria Certificate in Education Minimum Standards for Vocational and Technical Education. Lagos: NERDC Publisher; 2012c

[11] Adeniran N. Academic experts seek special funding for STEM education. The Tribune. 2021 [Accessed: November 18, 2021]

[12] Lawal I. STEM education as catalyst for national development. The Guardian. 2017 [Accessed: October 24, 2017]

[13] Asaolu O. UNILAG Don Advocates Special Funds for STEM. Being 18th Inaugural Lecture in the 2021/2022 Academic Session of the University of Lagos. Lagos: University of Lagos Press; 2023. Available from: https://unilag.edu.ng/?p=21780

[14] Okoli NJ, Wejinya SA, Agam PO, Asufi MR. Funding of technical and vocational education and training in Nigeria: Issues and challenges. Merit Research Journals. 2016;**4**(6):75-78. Available from: http://www.meritresearchjournals.org/er/index.htm

[15] Palmer R. Financing TVET in the East Asia and Pacific Region: Summary Report, Current Status, Challenges and Opportunities. Korea World Bank Partnership Facility. Washington DC: World Bank; 2017

[16] Nnamdi SE, Nwite O. Education funding in Nigeria: Issues and prospects. Journal of Teacher Perspective. 2014;**8**(4):1-8

[17] Lathabai HH, Nandy A, Singh VK. Institutional collaboration recommendation: An expertise-basded framework using NLP and network analysis. Expert System with Applications. 2022;**209**:118317. DOI: 10.1016/j.eswa.2022.118317

[18] Tim L, Leuprecht C. Securing cross-border collaboration: Transgovernmental enforcement networks, organized crime and illicit international political economy. Policy and Society. 2021;**40**(4):565-586. DOI: 10.1080/14494035.2021.1975216

[19] Yao B. International research collaboration: Challenges and opportunities. Journal of Diagnostic Medical Sonography. 2021;**37**(2):107-108. DOI: 10.1177/8756479320976130

Chapter 12

Cross-Cultural Experiences of Canadian Science Educators Visiting the Sister Schools in Chongqing, China: A Cross-Cultural Experience

Emilia M. Iacobelli and Geri Salinitri

Abstract

The benefit of cross-cultural learning is two-fold: first is recognizing the practices and foundations that build education in another country. The second is reflecting on the educational practices within one's home country. Cross-cultural learning also allows one to put the practice of education into perspective by having another side to which they can compare their experiences. This research is a narrative inquiry case study of the Canadian perspective and experience of two in-service science teachers who visited sister schools in China. This case study explores Canadian teachers' perceptions of teaching science, what inquiry-based teaching looks like, what equipment aids in the process, and what experiences and teaching methodologies can be shared between Canadian schools and the Sister Schools in Chongqing, China. This research was funded by the Social Sciences and Humanities Research Council (SSHRC) Partnership Grant.

Keywords: reciprocal learning, science education, Canadian teachers, Chinese teachers, inquiry-based teaching

1. Introduction

The benefits of cross-cultural learning are two-fold: first is recognizing the practices and foundations that build education in another country. The second is reflecting on and analyzing the educational practices within one's home country. Cross-cultural learning also allows one to put the practice of education into perspective by comparing methods. "When researchers from two countries collaborate on research involving teaching and learning in two countries [such], research not only provides a researcher with an understanding of science education in another country but also sharpens insights into science education in his or her own country" [1]. In addition, using cross-cultural experiences as comparisons for pedagogical research can help in-service teachers evaluate their teaching styles and philosophies. This reflective practice works to strengthen the practice and effectiveness of educators in the classroom. Viewing education as a social construct that is not bound by geographical limitations provides educators who may have never had the chance to learn about classroom practices in other countries an opportunity to do so.

This research is a narrative, inquiry, case study of the experience and Canadian perspective of two in-service science teachers who visited sister schools in China. The case study explores Canadian teachers' perceptions of teaching science, what inquiry-based teaching looks like, what equipment aids in the process, and what experiences and teaching methodologies can be shared between Canadian schools and the Sister Schools in Chongqing, China. When looking at China and Canada, there are foundational, curricular, and cultural differences that affect how teachers convey information and how students are expected to learn. This study highlights connections between Canadian and Chinese teachers and teaching practices to identify similarities and differences within the two countries' educational systems and the best practices that can be adopted from Chinese teaching into a traditional, mainstream Canadian classroom and vice versa.

Reciprocal learning is a process that can improve the practices in two areas or countries by introducing one participant's best practices to the other and vice versa [2]. Reciprocal learning is beneficial to both parties, as each side is able to take strengths from outside programs, in order to make connections to their own. For example, due to the Reciprocal Learning Partnership between the University of Windsor, in Canada, and Southwest University, in China, teachers are provided with opportunities to share experiences and learn from one another. This partnership aims to share knowledge, appreciate cultural and educational differences, and learn from one another. By opening the lines of communication and inviting teachers with varying experiences and perspectives into one another's classrooms, educators in both countries can draw upon each other to improve their practice. Specifically, reciprocal learning in science classrooms is important as it provides educators with opportunities to draw upon the experiences of teachers from varying backgrounds, in educational systems with potentially different values [3].

In the context of this research, reciprocal learning can and has occurred between Canadian and Chinese classrooms concerning instruction, planning, and student learning, leading to the improvement of student engagement, lesson delivery, and learning. As a result of a Social Sciences and Humanities Research Council (SSHRC) Partnership Grant, two Canadian science teachers visited Chinese science classrooms to explore and investigate standard Chinese educational practices and compare and analyze findings based on similarities and differences that can be used to strengthen science education in Canada and China. Using cross-cultural experiences as comparisons for pedagogical research can help in-service teachers to evaluate their teaching philosophies.

2. Definition of terms and relevant literature

2.1 Definition of terms

Reciprocal learning is learning that occurs between two parties who learn from one another. In this process, each party is able and encouraged to learn from the other by adopting new ideas that may improve or strengthen their current methods while also sharing core aspects of their practices that may be adopted by the other party to improve their programs.

Canadian teachers are defined as those trained in Canada and predominantly have experience in the Canadian school system. Canadian classrooms are classrooms in Canada governed by the educational laws and mandates of Canada's provincial and federal governments. Similarly, *Chinese teachers* have been trained in China and mainly have experience in Chinese schools and classrooms. Chinese classrooms are those found in China and are governed by the Chinese government's laws and mandates.

*Inquiry-based teac*hing involves teaching lessons that allow students to explore multiple scenarios and make predictions and connections based on their previous learning. Teachers can provide inquiry-based lessons to guide and scaffold student learning, rather than directly telling students the information. While inquiry-based lessons can be carried out across various subject matters, the teachers in this study focused explicitly on its uses and applications in science classrooms.

2.2 Literature review

Canadian and Chinese schools have different teaching models, but something can be learned from each of these models. Cross-cultural learning allows one to put the practice of education into perspective by comparison. As Aldridge and Fraser [1] reported, "when researchers from two countries collaborate on research involving teaching and learning in two countries, such research not only provides a researcher with an understanding of science education in another country but also sharpens insights into science education in his or her own country (p. 102)." This literature review will explore several differences and preferences, including teaching styles, the dynamics of Chinese and Canadian classrooms, and the use of inquiry-based teaching practices in science classrooms.

2.2.1 Science education in Canada and China

Science education in Canada places a large emphasis on problem-solving skills and using technology effectively to understand and decode problems [4, 5]. Specifically, the Ontario curriculum recognizes that while not all students will become scientists, the principles of scientific literacy and understanding media are core skills that are transferrable to numerous professions [4, 5]. In recent years, Canadian science education has also made attempts to decolonialize its science curriculum, by including opportunities to learn and acknowledge the teachings and knowings of Indigenous Peoples in ways that become meaningful to a greater population of students [6].

Science education in China aims to develop competent citizens that have a high level of scientific literacy [7]. As of 2011, the Ministry of Education in Ontario [7] outlined four key methods that would be employed to reach these goals:

1. Extending Science Learning Time;

2. Integrating Engineering and Technological Content into Science;

3. Phase Based Learning Progressions; and

4. Using Big Concepts [7].

There are some key similarities in the underlying values and goals for science education in both Canada and China. The remainder of this literature review will seek to explore the nuances, similarities, and differences in the approaches and methodologies employed to reach these educational goals.

2.2.2 Teaching styles and classroom design

Research found that many Chinese science educators view exam performance as one of the most important factors in determining good teaching [8]. In addition, they tend to equate good classroom teaching with fostering good conduct and learning attitudes, compared to Western teachers who aim to promote an interest in learning.

It was reported by Chen [9] that science education in China tends to be teacher-centred, focused on theory and exams, and supplemented by homework problems. This information closely aligns with many of the teacher participants' interview responses and contrasts with many of Canada's current educational views. Many Canadian classroom teachers are moving toward teaching models that are more learner-centred and focus on learning more holistically, offering several options for learning and understanding topics. Inquiry-based teaching is one such method of curriculum delivery. Chen [9] also reports on the importance Chinese educators place on developing engaged students who become lifelong learners. The research also notes a shift in some Chinese middle school teaching practices to student-centred classes and lessons that align with values present in Canadian classrooms.

2.2.3 Inquiry-based teaching

Inquiry-based teaching is a practice that has become more common in classrooms over the past 20 years [10]. However, Orpwood and Souque [11], on behalf of the Science Council of Canada, were producing work involving deliberate inquiry wherein deliberations of policymakers and inquiry of researchers were being consulted to create a science curriculum that reflected the learning needs of the students, as early as 1985. In the inquiry-based teaching model, the teacher is no longer the source of the knowledge but rather a facilitator and guide of the student learner's discovery [10], which aligns with constructive learning theory that encourages students to think critically and creatively.

As of 2011, the Ministry of Education in Ontario identified four key teaching elements using an inquiry-based approach:

1. Student engagement and focus;

2. Student communication, explanations, and reflections to share learning;

3. Student exploration and investigation; and

4. Student analysis and extension of thinking, further supporting this student learning model [12].

2.2.4 Professional development and collaboration in teaching

Professional development is a vital tool in the growth and learning journey of teachers and their teaching practices. As teachers become more experienced, they tend to participate in more professional development seminars but place a lower-level value on collaboration [13], despite the known benefits of collaboration among educators. Collaboration is more highly regarded by teachers and educational staff members who voluntarily choose to do so, as their belief systems value working together and sharing viewpoints [14]. Rigler [15] states there are many benefits to international professional development and collaboration. In being awarded opportunities to become fully immersed in Chinese classrooms, there were chances for meaningful collaboration, further understanding of the teaching practices in other countries, discussions, lesson development, and creating global learning opportunities for students. Additional research also finds collaboration, specifically in a Reciprocal Learning Program, leads to positive effects for students and teachers, including deeper respect for culture, understanding global citizenship, and a greater appreciation for varying teaching practices [16].

Reciprocal learning across international boundaries has been shown to benefit teachers in both countries. In addition to professional development seminars and opportunities, professional learning communities where educators can engage with and reflect on the differences in educational systems allow for the strengthening of teacher performance and personal practical knowledge [17].

3. Purpose and methods

3.1 Purpose

This research aims to highlight the connections between Canadian and Chinese teaching practices to identify similarities and differences within the educational systems of Canada and China. Such research will identify best practices in Chinese classrooms and teaching adopted within traditional, mainstream, Canadian science classrooms.

3.2 Methods

The following subsections outline the participants and procedures engaged in order to obtain the data that was analyzed for key and emerging themes.

3.2.1 Participants

In this narrative inquiry case study, the participants were two in-service science teachers from a local Windsor school board. Purposeful sampling was used to select these two individuals from all qualified individuals interested in participating in the Reciprocal Learning Project. These teachers visited science classrooms within the Sister Schools in Chongqing, China and observed science teachers and practices within science classrooms. For purposes of confidentiality the following pseudonyms were given to the teachers: Teacher A and Teacher B. Teacher A is a male science teacher who has been a Canadian classroom teacher for 15 years. Teacher B is a female science teacher who been a Canadian classroom teacher for 3 years.

3.2.2 Procedure

Two in-service teachers were allowed to observe science teachers and classrooms in Chongqing, China. These teachers were provided with opportunities to attend Chinese classes, meet students, and speak with Chinese educators to learn more about their classroom and education style in China. Once the teachers returned to Canada, they were individually interviewed and asked questions related to their definitions of student engagement, if and how inquiry-based teaching practices in China compared to those in Canada, and what types of learning equipment were available in classrooms.

4. Results and discussion

Qualitative data was obtained from participating teachers who were individually interviewed upon returning from their visits to secondary science classrooms in Chongqing, China. The interviews were transcribed and sent to participants to member check, amend, and approve their final responses before analysis. The interview data were coded and analyzed for general, common themes. Four main themes emerged: inquiry-based lessons and teaching opportunities in Canada and China, types of available equipment for lab experiments, benefits to Canadian science classrooms based on Chinese practices, and benefits to Chinese science classrooms based on Canadian practices. Each of the four themes are discussed in greater detail within this section.

4.1 Inquiry-based lessons and teaching opportunities

Both of the interviewed teachers perceived greater opportunities for inquiry-based lessons and teaching within Canadian classrooms. However, previous research within the Reciprocal Learning Project suggests no significant difference in inquiry-based teaching applications in the two countries [18]. In this current study, teachers attributed these perceptions to the smaller class sizes within Canadian classrooms and the increased availability of classroom spaces and lab equipment to students and teachers. One significant note was the effectiveness of the inquiry-based teaching practices that were being implemented in Canadian classrooms. Teacher A noted that even though many teachers are using or attempting to use inquiry-based teaching practices within their classrooms, the overall success of these endeavors has not been measured. It is unclear with current measures and assessments if inquiry-based practices are useful to students and how inquiry-based lessons are standardized among different classrooms.

Additionally, there are teachers in Canadian schools who do not utilize inquiry-based teaching methods. Teacher B noted, she witnessed mainly Chemistry classes which consisted of

> *"The teachers just providing lectures to the students...and the teachers would ask students questions on the board or have the students finish equations for them. The teacher would experiment [at the front of the class], and then the students will all be sitting at their desks watching. I was told that the students will have a lab maybe a couple of times a year but very infrequently, and they would have to go to a different classroom to perform the lab."*

The frequency of laboratory experiences for students brings up interesting questions about the lab equipment and space available to students in Chinese classrooms.

4.2 Types of available lab equipment

When asked about laboratory equipment availability, both teachers felt they had access to the necessary equipment to run safe and effective lab experiments in their

Canadian classrooms. They also noted that it was relatively easy to borrow, share, or purchase the necessary equipment if the equipment was not available within their respective schools. In the Canadian teachers' experiences, there was also adequate space within classrooms, in the form of wet labs and dry labs, for students to conduct laboratory experiments. The Canadian perception is that increased lab space gives students and teachers more opportunity to engage in inquiry-based learning practices.

In addition to physical space, Teacher B noted Canadian classrooms had complete sets of glassware and access to enough reactant chemicals that classroom demonstrations were easier to present. The classroom setup itself was also conducive to demonstrations and hands-on laboratory experiences due to the number of students in the class and the physical space, including a lab bench for the teacher and various lab stations around the classroom for the students. Canadian teachers reported that they witnessed some classroom demonstrations while in China and heard that Chinese students also had opportunities to conduct laboratory experiments during their courses, just more infrequently than Canadian students, with the significant limitations being time, space, and availability of large volumes of material.

4.3 Benefits to Canadian science classrooms based on Chinese practices

One of the Canadian teachers' views of best practices is consistency in planning and lesson delivery within Chinese schools and classrooms. Teacher A, who visited the Sister School, noted,

> *"Every lesson is coordinated like a massive, massive synchronized swimming meet. They [the Chinese teachers] all teach at the same time, so they all work on the lessons together."*

All of the Chinese teachers who were teaching the same course worked together to create course content and create lectures and activities common for all students, regardless of the teacher. This level of consistency could help Canadian students within multiple sections of a course, as the curriculum is the same, and students should be learning the same content across sections. Teacher A also noted,

> *"Their [the Chinese teachers] teaching is like an art form in the sense that there are very, very prepared and they are going from a script, and they work together and they collaborate, and it's amazing."*

Given the opportunity, co-planning can also help teachers, especially those who are either new to the practice, new to teaching the course, or more experienced educators looking for new ways to disseminate information and impact student learning. Many teachers within Canadian school systems partake in co-planning opportunities, but it is not mandated in the same ways as Chinese schools.

Finally, Teacher A noted that since the lesson plan developed for each Chinese science class is so detailed, pinpointed, and well-planned, there is more time for teachers to monitor and assess the students while in the classroom, rather than frantically trying to fit in the course material, leading to more opportunities to collect pieces of data that can be used to triangulate assessment practices. A significant difference that could impact lesson delivery is that Chinese students are in school for an extended time each day compared to Canadian students. From a time standpoint alone, there is more time to assess and monitor students, specifically while in the classroom.

4.4 Benefits to Chinese science classrooms based on Canadian practices

One of the main best practices reported by Canadian teachers that can be shared in Chinese classrooms is implementing a greater focus on student collaboration and providing opportunities for students to work together, share ideas, and learn from one another. It was noted that while Chinese students spend a great deal of time problem-solving and developing the skills and knowledge outlined in the curriculum and necessary for future academic success, they do much of this work independently. Chinese students could benefit from more opportunities to collaborate with others and learn about different thought processes and strategies. Teacher A felt

> *"The biggest thing missing in Chinese education classrooms, science classrooms, but all classrooms probably, is the level of feedback they receive during the learning process. I don't think the feedback is enough...and the response that I got from the Chinese teachers is... there's just too many people. That this is what we have to do."*

Teacher A stated it seemed like Chinese science classrooms were filled with a sense of urgency to learn as much as possible, as fast as possible, as independently as possible. One of the standout features of the Chinese classroom that differed from Canadian classrooms was the idea of self-study, which encompassed three or four hours of studying at the school each night. In addition, the Canadian teacher's understanding was that students were not allowed to talk, whereas Canadian classrooms tend to encourage dialog and collaboration among classmates.

5. Final considerations

5.1 Conclusion

In conclusion, Canadian and Chinese science classrooms can gain valuable insight into best practices from one another. Both Canadian and Chinese science classrooms can take steps to focus on inquiry-based lessons and teaching and an increased level of consistency in the delivery of such practices. Since lab equipment is available to students and teachers in both countries, it would benefit students and their learning to provide more lab experiment opportunities. Collaboration between Canadian and Chinese teachers concerning types of experiments would also be beneficial. One main takeaway Canadian science classroom teachers can learn about planning and the shared teaching processes used by teachers in Chinese science classrooms. Conversely, Chinese science classroom teachers can learn from students' collaboration activities and opportunities within Canadian science classrooms.

5.2 Implications

The teachers' interviews and experiences within this study prove that Canadian and Chinese educators can share their experiences and perspectives to give the students a more well-rounded learning experience in secondary science classrooms. Canadian teachers shared that a greater focus can and should be placed on the consistent collaboration of teachers who teach the same courses to provide consistent lessons and course content no matter what teacher is presenting the material, as educators do in Chinese schools and classrooms. They also shared that Chinese science teachers may want to encourage greater collaboration among students to increase their ability to work together and solve problems as a group. This practice may lead to improved understanding by considering other individuals' viewpoints and processes in their class. This research speaks to the significance of educators from different countries, cultures, backgrounds, and learning experiences sharing to create better learning opportunities for students in their respective classrooms, schools, and countries.

5.3 Limitations and future work

One of the major limitations of this study was the sample size of the participants. Only two Canadian teachers visited the sister schools in China and provided an analysis of their experiences. They were not in China for an entire semester, but rather a shorter time, so they could only view a snapshot of science education in China rather than the overall program experienced by the students. Furthermore, Canadian teachers have an understanding and comfort within their school system that they likely do not have within the Chinese school system. As such, in both Canada and China, the classroom experiences were viewed through a Canadian teacher who went through a Canadian school system and continues to teach within that system.

Reciprocal viewpoints were not directly obtained in this study. Still, through other portions of the Reciprocal Learning Project, the views and experiences of Chinese science educators have been received and shared. Given the opportunity, Chinese educators have shared they enjoyed having opportunities to engage students in real-world learning outside of the classroom [19]. Continued comparative research is a promising avenue for future research.

Further work in this area could implement changes within Canadian and Chinese classrooms based on reciprocal learning visits and suggestions from teachers in different contexts. Results will be presented at conferences and in peer-reviewed journals to increase the number of inquiry-based teaching and learning opportunities for students and teachers within Canadian and Chinese contexts.

Acknowledgements

The authors wish to acknowledge the funding support provided to the Reciprocal Learning Project by the Social Science and Humanities Research Council (SSHRC) Partnership Grant.

Conflict of interest

The authors declare no conflict of interest.

Notes and thanks

The authors wish to thank the administrators and educators in our Sister School in Chongqing, China for allowing our Canadian in-service teachers to visit and learn within their classrooms.

Author details

Emilia M. Iacobelli* and Geri Salinitri
University of Windsor, Windsor, Ontario, Canada

*Address all correspondence to: iacobel1@uwindsor.ca

References

[1] Aldridge J, Fraser B. A cross-cultural study of classroom learning environments in Australia and Taiwan. Learning Environments Research. 2000;**3**(1):101-134. DOI: 10.1023/A:1026599727439

[2] Xu S, Connelly M. Reciprocal learning between Canada and China in teacher education and school education: Partnership studies of practice in cultural context. Frontiers of Education in China. 2017;**12**(2):135-150. DOI: 10.1007/s11516-017-0013-6

[3] Xu S. Reciprocal learning in teacher education between Canada and China. Teachers and Teaching. 2019;**25**(6):703-729. DOI: 10.1080/13540602.2019.1659766

[4] Ontario Ministry of Education. The Ontario Curriculum Grades 1-8: Science and Technology (Revised). Ottawa, Ontario: Canada. Queen's Printer for Ontario; 2007

[5] Ontario Ministry of Education. The Ontario Curriculum Grades 9 and 10: Science (Revised). Ottawa, Ontario: Canada: Queen's Printer for Ontario; 2008

[6] Aikenhead GS, Elliott D. An emerging decolonizing science education in Canada. Canadian Journal of Science, Mathematics and Technology Education. 2010;**10**(4): 321-328. DOI: 10.1080/14926156.2010.524967

[7] Pei W. Curriculum reform of science in elementary schools in China. Beijing International Review of Education. 2019;**1**(2):573-578. DOI: 10.1163/25902539-00102007

[8] Gao L, Watkins D. Conceptions of teaching held by school science teachers in P. R. China: Identification and cross-cultural comparisons. International Journal of Science Education. 2002;**24**(1):61-79. DOI: 10.1080/09500690110066926

[9] Chen J. Teachers' conceptions of approaches to teaching: A Chinese perspective. The Asia-Pacific Education Researcher. 2015;**24**(1):341-351

[10] Ismail N, Alias S. Inquiry-based learning: A new approach to classroom learning. English Language Journal. 2006;**2**(1):13-24

[11] Orpwood G, Souque JP. Toward the renewal of Canadian science education. Ii. Findings and recommendations. Science Education. 1985;**69**(5):625-636. DOI: 10.1002/sce.3730690505

[12] Ministry of Education. Capacity building series: Getting started with student inquiry. 2011. Available from: http://www.edu.gov.on.ca/eng/literacynumeracy/inspire/research/CBS_StudentInquiry.pdf

[13] Ritcher D, Kunter M, Klusmann U, Ludtke O, Baumert J. Professional development across the teaching career: Teachers' uptake of formal and informal learning opportunities. Teaching and Teacher Education. 2011;**27**(1):116-126. DOI: 10.1016/j.tate.2010.07.008

[14] Barfield A. Collaboration. ELT Journal. 2016;**70**(2):222-224. DOI: 10.1093/elt/ccv074

[15] Rigler A. Reciprocal learning: One teacher's narrative. Frontiers of Education in China. 2017;**12**(1):219-232

[16] Howitt C. Building a bridge between Western and eastern worlds: Reciprocal learning programmes that create reflective practice, hope and prosperity in education. Teachers and Teaching. 2019;**25**(6):743-751. DOI: 10.1080/13540602.2019.1680358

[17] Huang X. How teachers learn and change in reciprocal learning space. Frontiers of Education in China. 2017;**12**(2):151-179. DOI: 10.1007/s11516-017-0014-5

[18] Salinitri G, Palazzolo S, Nahaiciuc R, Iacobelli E, Li Y, Zhou G. Analysis of Canadian inquiry-based science teaching practices and its implications for reciprocal learning. Universal Journal of Educational Research. 2018;**6**(10):2280-2293. DOI: 10.13189/ujer.2018.061027

[19] Zhou G, Wai-Ying Ho S, Li Y, Luo M, Freedman H, Luo J. Reciprocal learning between Canadian and Chinese schools through the 24 nature notes project. The Journal of Teaching and Learning. 2019;**13**(1):7-24. DOI: 10.22329/jtl.v13i1.5988

Chapter 13

Enhancing Educators' Cultural and Digital Literacies through Makerspace Development Activities

Jennifer Miller

Abstract

Makerspace communities enhance STEM literacy while valuing assimilation of Hispanic and Indigenous culture, language, traditions, and ways of knowing that connect the purpose of STEM disciplines to real world uses by honoring past funds of knowledge. This study investigates the La Frontera mobile makerspace program's impact on educators' self-efficacy toward integrating technology and culturally responsive teaching using pop-up activities. Early outcomes indicate increased interest of emerging technologies and understanding of culturally responsive teaching. The program includes strong community partnerships connecting K12 schools, community college, university, museums, and industry to an isolated area spanning a vast region along the Texas-Mexico border.

Keywords: STEM11, makerspace, digital literacy, education technology, Mobile

1. Introduction

Mobile makerspaces assist with building capacity beyond the university to assist with educator professional development toward using technologies and assist with providing connections to engineering outreach and engagement [1]. The dynamic nature of makerspace activity may help sustain faculty curiosity/intellectual growth through the inherent challenges. Mobile makerspaces offer strong internal outreach opportunities, with educators' responding positively to professional development centered on digital and design literacies offered in a mobile makerspace environment perhaps due to the creative and flexible approaches during pop-up activities. Mobile makerspace pop-up activities increase awareness and understanding of digital literacies in an active learning atmosphere [2]. This provides a mobile benefit to remote institutions serving underrepresented minority (URM) communities. Digital literacies in mobile environments often incorporate use of creative mobile technologies: 3D modeling, fabrication, robotics, laser cutting and lendable technologies. In addition to providing increased access, mobile makerspace has shown to positively connect to engineering outreach, increasing diversity within engineering education tasks [1]. Mobile makerspace have shown to excite diverse audiences but often lack robust participation in curricular models. Research needs include contributing mobile makerspace curriculum approaches in which improved participation through pop-up challenge layout and instructional foster participation beyond excitement to strong engagement [3]. The La Frontera mobile STEM pilot project provides for the delivery of makerspace popup activities to allow for personalized expressions of interests and values, promoting higher engagement and interest. Mobile makerspaces will mitigate isolation issues in STEM education. Advancing post-secondary success of under-represented minority communities through the availability to Indigenous and/or Hispanic serving institutions (HSI) in rural communities can significantly positively influence and increase the number of URM attaining postsecondary STEM degrees. [4]. Current post-secondary institutions serving underrepresented minorities often lack strong STEM mentorship systems along with many students feeling culturally isolated in STEM undergraduate programs. HSI's represent a critical component serving the "marginalized majority" by focusing on service to an many first generation Hispanic students, targeting individual student needs [4]. This pilot project examines how mentorship experiences within a makerspace community may contribute toward the development of self-efficacy among URM and address gaps to refine the scientific communities' understanding on how engagement within makerspace environments is not only established within URM remote communities but maintained to establish a culture of inquiry and hands-on approaches beyond informal learning. Establishing and supporting a makerspace culture continues to gain momentum and attention from policymakers, leaders, and scholars, but literature lacks a deep understanding of the maker communities and how they can be leveraged among families to support learning [5]. Broadening the use of evaluation instruments beyond measuring the impact of middle school and high school STEM programs would provide invaluable insights as to how informal learning approaches might serve to measure informal learning approaches on URM remote students as well as larger community efforts connecting K-12, community college, and university programs through a makerspace community effort.

2. STEM engagement through mentoring in a makerspace

Enhancing the relevance of STEM through stronger STEM programming that values assimilation of Hispanic and Indigenous culture, language, traditions, and ways of knowing through a transactional process could benefit underrepresented communities. A mentoring approach through "shared interactions, experiences, and mutual exchanges" with the La Frontera Mobile STEM lab is introducing STEM faculty and educators to culturally appropriate outreach strategies. Strong STEM mentorships over a two-year period have shown to benefit underrepresented students with improving scientific efficacy and scientific identity, with students feeling as if they belong to a community of scientists [6]. Gaps in achievements, representation and interest in STEM are persistent across all levels of educational achievement, with African Americans, Latinos, and Indigenous populations representing only 6% of the total STEM labor workforce, with URM representing 25% of the total population in the United States [7, 8].

Outcomes from the pandemic presents an opportunity to improve technology training postsecondary institutions for many Hispanic and Indigenous students. Enhancing the relevance of STEM through stronger STEM programming that values assimilation of Hispanic and Indigenous culture, language, traditions, and ways of knowing through a transactional process could benefit underrepresented communities. A mentoring approach through "shared interactions, experiences, and mutual exchanges" could assist STEM faculty in identifying culturally appropriate outreach strategies [9]. Transformative STEM (science, technology, engineering and mathematics) learning spaces have grown rapidly in schools, universities, libraries, and museums as "Learning Labs" or "Makerspaces." These spaces are designed to encourage deep engagement with STEM-integrated content, critical thinking, problem solving, and collaboration while sparking curiosity [10].

Makerspaces are informal sites for creative expressions in STEAM (sciences, technology, the arts, engineering, and mathematics) in which learner's blend digital and physical technologies in an informal learning environment to share with a wider community [11]. Makerspace emphasizes learning through direct experiences, hands-on projects, inventions, and is based on a constructionist learning theory [12]. The La Frontera project outcomes would assist in addressing a gap in literature investigating how to establish a STEM culture through faculty development beyond grant funding. Beavers et al. [13] project leveraged 3D printing in a library to first engage and create interest in fabrication programs. STEM faculty previously viewed the library as an independent unit or place, disconnected from academic programs. However, providing a fabrication printer formed a grassroots effort to engage the community in pop-up challenge activities, creating interest groups and encouraging administration to support a formal makerspace area. Faculty worked with support staff and student workers to create pop-up makerspace activities, hosted a mini-maker faire, provided financial incentives to students to participate in pop-up activities and the maker faire, supported robotics and makerspace camps, which created a culture of STEM engineering engagement, educator professional development, and resulted in improved STEM undergraduate programing. Outcomes brought about increased engagement, use of student support library services by faculty and students, and improved teaching practices [10]. It has been observed through a strong scientific community mentoring plan, quality STEM engagement has increased along with opportunities for educator development to STEM faculty, K-12 teachers, and preservice teachers. Program outcomes contribute toward understanding how remote mobile makerspace programs may impact isolated URM STEM enrollment, retention, self-efficacy and persistence in STEM. Examining persistence through a lens of social influence will address gaps exploring relationships between the scientific community, makerspace community, industry, faculty, and mentoring and how relationships might be leveraged as an agent approach toward social change. Literature has shown the development of efficacy, identity, and values predicts URM persistence but the impact of makerspace on engagement in STEM have limited support in literature [6].

Mobile makerspaces assist with building capacity beyond the university to assist with educator professional development toward using technologies and assist with providing connections to engineering outreach and engagement [1]. The dynamic nature of makerspace activity may help sustain faculty curiosity/intellectual growth through the inherent challenges. Mobile makerspaces offer strong internal outreach opportunities, with educators' responding positively to professional development centered on digital and design literacies offered in a mobile makerspace environment perhaps due to the creative and flexible approaches during pop-up activities. Mobile makerspace pop-up activities increase awareness and understanding of digital literacies in an active learning atmosphere [2]. Digital literacies in mobile environments often incorporate use of creative mobile technologies: 3D modeling, fabrication, robotics, laser cutting and lendable technologies. In addition to providing increased access, mobile makerspaces have been shown to positively connect to engineering outreach, increasing diversity within engineering education tasks [1] while exciting diverse audiences, but often lack robust participation in curricular models. Research needs include contributing mobile makerspace curriculum approaches in which improved participation through pop-up challenge layout and instructional foster participation beyond excitement to strong engagement [3].

3. STEM literacy divide

Accessibility issues can no longer be defined in terms of physical access. The pandemic has shown that education organizations must approach the use of distance education and digital learning from an equity lens. Age does not define digital fluencies or abilities to apply digital skills to communicate effectively. The term 'digital divide' traditionally was used to describe inequities in access to devices and broadband, but this definition fails to capture gaps in educational experiences, curricula, social, cultural, and economic realities of many [14]. Texas ranked highest in the nation, with 34% of students lacking broadband connection and 24% of students lacking access to a device [15]. Post COVID-19 presents the immediate need to rethink digital leadership training approaches to begin to address the digital divide presented by lack of quality training, equity and accessibility issues in isolated areas. This project addresses the digital divide challenges facing remote underrepresented and underserved Indigenous and Hispanic communities with limited access to technology and faculty development opportunities. This would be accomplished through a community mobile makerspace program aimed at addressing the digital divide and digital literacy needs of STEM educators. The term 'digital divide' traditionally has been used to describe inequities in access to devices and broadband, but this definition fails to capture gaps in educational experiences, curricula, social, cultural, and economic realities of many [14]. This issue has long been debated in literature but school closures in isolated regions during the pandemic magnified preexisting socio-economic and education disparities on a massive scale, revealing large inequities in access to resources and learning quality [16]. Schools must begin to truly identify gaps in accessibility to include physical access, technology mediums used, equity, resources, digital learning, and digital literacy experiences.

While remote areas of Texas face a digital divide, national employment projections in STEM reported by the Bureau of Labor forecast overall growth in STEM fields to increase by 11% with Austin, Texas leading national STEM indexes over the next 5 years [17]. Nearly 28% of all occupations in the Austin metro area require a bachelor's degree, with salaries of occupations requiring a bachelor's degree double the salary of a position requiring a high school diploma [18]. Seven out of 10 job occupations identified in the Bureau of Labor Statistics report were computer related, with mathematical science occupations expected to grow by 28.2%, with top occupation growth among statisticians, research analysists, cartographers, forensic scientists, biomedical engineers, mathematicians, computer systems analysists, software developers, actuaries, and information security analysts. In addition, in that same report, over 73% of STEM occupations required a bachelors' degree, while only 21% of overall occupations required a bachelor's degree [17].

Main factors contributing to the digital inequities may not be identified as physical access only but rather a divide in digital literacy skills to the extent in which populations are divided and excluded [19]. The pandemic has presented educators with a renewed insight on the importance of participation in digital learning development, as now they are aware of their own digital skills, abilities, and limitations [20]. This presents an opportunity for organizations to foster stronger development approaches and to identify gaps in digital competencies, teacher capabilities, and abilities to use technology to communicate well, or digital fluencies. Schools can look to the International Society of Technology Education (ISTE) standards, Common Framework for Teaching Digital Competence or TPACK as a starting point. Investment in technology training cannot be an afterthought during the strategic plan process and is a required component to equip students with future-ready skills. Measuring the quality of teacher training and resources can provide insights on how financial allocation of professional development influence URM student engagement, STEM literacy, use, and academic performance. Makerspaces provide an informal space for active learning outside of the classroom to facilitate self-directed learning opportunities for faculty development in which faculty collaborate with others to learn about digital resources and technology approaches to improve integration efforts in the classroom [2]. Makerspace environments provide a safe place for educators to participate in trial-and-error experimentation of new technology tools, for learning and investigating within a makerspace community [21]. The La Frontera Mobile STEM program expands upon Miller's study that investigated the effect of professional development on educator perception of integrating technologies with STEM. The NASA funded Makers' Guild study incorporated 2D and 3D technologies during STEM makerspace pop-up activities in which participants reported a statically significant increase in self-reported competence in technology integration, the use of the World Wide Web, and Emerging Technologies for Student Learning as measured by the TPSA C-21 [11].

4. Methodology

The La Frontera mobile STEM program is addressing the following question: How does informal learning through mobile makerspace pop-up activities impact self-efficacy of URM undergraduate students and STEM faculty from isolated communities? The La Frontera Mobile Makerspace pilot project provides for the delivery of makerspace workshops and speakers from STEM industries and STEM organizations using either a mobile lab or a makerspace classroom in which pop-up activities allow for personalized expressions of interests and values of faculty, promoting higher engagement and interest. The La Frontera Mobile STEM program works in conjunction with the Noyce en la Frontera National Science foundation Track 1 program, which provides scholarships to math and biology majors seeking a teaching certification at the middle school or high school level. Both programs represent a collaborative partnership with a local community college, surrounding high needs K12 schools located on the Texas Mexico border, the Science Mill Museum, and additional community outreach programs. The Noyce en la Frontera program pairs URM preservice STEM teacher with a highly qualified STEM mentor teacher or faculty member. The program provides faculty development toward becoming a mentor, culturally responsive teaching approaches, and digital literacy training incorporating the Universal Design for Learning framework and employs tactile technologies to assist preservice teachers in their self-efficacy of teaching with technology using informal learning popup activities through the mobile program. The program introduces preservice teachers, other educators, and mentors to the possibilities of incorporating stronger digital learning approaches to improve engagement and expression of knowledge. After participating in a series of workshops, Noyce scholars along with mentors utilize the mobile STEM program with collaborative schools or public libraries located in isolated areas along the southwest between El Paso and Eagle Pass, a 400-mile or 640-kilometer area. The La Frontera program financially supports faculty development by providing stipends to faculty and K12 certified secondary educators enrolled in a mentoring program to lead co-developed makerspace mobile events, lectures, and workshops. Mentoring is facilitated through a partnership between STEM and Education faculty at both the university and community college; the Science Mill Museum; and the Texas Workforce Commission. Program activities have been designed to address future-ready skills and to address achievement gaps by improving relationships between STEM K-12 programs, community college STEM student populations, STEM majors attending the university, faculty teaching STEM or STEM related topics and industry in rural Southwest Texas.

The La Frontera pilot program hypothesizes that facilitating improved faculty development that is focused on diversity, equity, and inclusion (DEI) intervention approaches using an informal learning creative space will serve to improve retention, self-efficacy, STEM identity, and values of undergraduate STEM preservice teachers and STEM faculty as measured by the Teacher Beliefs and Attitudes Toward STEM (T-STEM) Survey pre to post using Qualtrics [22]. The TPSA C-21 is used to measure the impact of technology integration training on teacher self-efficacy, investigating six factors: Email, World Wide Web, Integrated Applications, Teaching with Technology, Emerging Technologies, for Student Learning, and Emerging Technologies for Teacher Development, and reports a Chronbach's Alpha between .75 and 1.00, with .88 reported for the entire scale [23–25]. The TPSA-C-21 is a 34 item survey used for Miller's NASA funded Makers' Guild program, which reported that makerspace development significantly impacted teachers' self-efficacy ($p < .05$) in emerging technologies and teacher professional development programs [11].

In addition, activities focused on investing in faculty development that integrates DEI to improve culturally responsive intervention would increase self-efficacy of STEM faculty to integrate technology in STEM content courses as measured by Technology Proficiency Self-Assessment for twenty-first Century Learning (TPSA C-21) [22]. Faculty and STEM preservice teachers participating in the makerspace workshop development receive training in Culturally Responsive Intervention, Emerging Technologies to Support Teaching and Learning, and Project Based-Learning through Makerspace Pop-Up Activities. The mentor program includes faculty development focused on the following topics through a face-to-face summer training program and additional online training made available through Canvas: Advancing Equity & Inclusion, Cultural Literacy in Teaching and Learning, The Importance of Family Support, Resources for Makerspace Mentoring Programs, and Incorporating Digital Techniques in Teaching STEM. Extended faculty development is offered within the mobile program that incorporates advanced technology to include virtual reality, augmented reality, robotics, artificial intelligence, social media, creation technologies, and fabrication technologies.

Data is collected using Qualtrics to include the following evaluation components: makerspace participant surveys, TPSA C-21 preservice and mentor pre to posttest survey responses, T-STEM preservice and mentor pre to posttest survey, Culturally Responsive Teaching survey pre to posttest survey response, attendance, STEM enrollment in the teacher preparation program, retention rates, and graduation rates [21, 22, 24–26]. Pretest data collection of preservice and mentor participants is administered during an orientation session prior to receiving training and participating in makerspace activities over a 2 year period of time. During the last semester of the 2-year program, posttest data collection of preservice and mentor participants occurs during an exit interview.

5. Results

Early outcomes of the La Frontera Mobile STEM program provide insight as to how makerspace pop-up activities might improve faculty development approaches and to improve faculty self-efficacy and abilities to integrate educational technology into formal learning experiences serving remote and isolated URM populations. Activities designed in the Summer and Fall of 2022 by museum STEM teachers and the La Frontera research team resulted in 8 popup stations designed to introduce participants to future ready skills needed in the agricultural industry All activities encompass a career awareness component, which is available in both Spanish and English. Tactile components included in the makerspace program incorporated a cardboard challenge, fabrication printing, Arduino watering station, wind wattage challenge, drone cattle search, microbit weather station, cacti catastrophe, and a citizen scientist challenge. All activities do require electricity but do not require broadband access, as many areas located along the Texas Mexico border lack connectivity. After participating in a 6-month training, Noyce scholars began teaching informally through the mobile program. For many scholars, this was their first introduction to teaching.

The La Frontera Mobile STEM program has reached over 3348 participants since programing began in January 2023. The mobile van has traveled 2380 miles or 3830 kilometers serving several isolated communities with additional programing planned in the summer of 2023. The mobile van has visited several middle schools, the community college, university, and surrounding public libraries. Noyce scholars and mentors are learning how to incorporate coding, fabrication printing, drones, and multimedia while teaching in a fun and low stress environment. Participants are asked to participate in a program survey to investigate their attitude or engagement of activities. Seven Noyce preservice teachers, six mentors, and additional faculty have participated in training events. All seven Noyce preservice teachers along with STEM faculty have facilitated mobile makerspace activities.

During the last academic year, the program is lending considerable support toward achieving the goal of improving current support systems for STEM teaching enrollment, retention, and faculty development by strengthening relationships between institutions, departments, and the community at large. All participants were invited to participate in a pretest data collection. Although sample sizes continue to be small, the pre-test portion included 13 participants, 7 Noyce scholars and 6 mentors. Not all mentors have participated in hosting mobile pop up activities but all have participated in the faculty development training. All Noyce scholars have participated in hosting a mobile STEM lab event at either the community college, university, public library, or secondary school. T-STEM analysis revealed that twenty-first Century Learning Attitudes found a significant finding ($p < .05$) that preservice teacher reported a higher confidence toward teaching in the twenty-first century skills compared to mentor teachers as measured by the T-STEM survey. TPSA C-21 analysis determined that mentor self-perceived competence toward using the World Wide Web, teaching while integrating technology to be higher than preservice teacher perceptions, with results educationally significant ($p < .05$) in the area in two areas spreadsheet and newsletter. Data analysis indicated that Teacher Efficacy and Attitudes toward STEM, pretest data reported preservice teachers' efficacy and beliefs toward having the necessary skills to teach as significant ($p < .05$) compared to faculty. Data indicated that males self-reported a higher efficacy toward the World Wide Web while females reported a higher efficacy toward teaching with technology. Analysis of culturally responsive teaching revealed a significant finding ($p < .05$) suggesting that science content teachers report a higher culturally responsive attitude toward inclusion and attitudes compared to math content teachers. Further, science content teachers reported a higher use of culturally responsive practices compared to math content teachers. The teacher preparation program has experienced a strong increase in enrollment of math majors seeking a teaching certification. It is expected that the pilot project will exceed the Noyce en la Frontera grant goals of increasing the number of highly qualified STEM teachers by 10% as the program has doubled in size over the last academic year. Excitement of the La Frontera Mobile STEM van may be a contributing factor toward improving interest, enrollment, and retention of STEM majors seeking a teaching certification.

6. Conclusion

The La Frontera Mobile STEM Program is contributing toward understanding how remote mobile makerspace programs may influence isolated faculty serving unrepresented minority populations. Examining persistence through a lens of social influence will address gaps between the scientific community, makerspace community, industry, faculty, and mentoring and how relationships might be leveraged as an agent approach toward social change. Literature has shown the development of efficacy, identity, and values predicts URM persistence but the impact of makerspace on engagement in STEM have limited support in literature [5]. This pilot project examines how mentorship experiences within a makerspace community may contribute toward the development of self-efficacy among URM and address gaps to refine the scientific communities' understanding on how engagement within makerspace environments is not only established within URM remote communities but maintained to establish a culture of inquiry and hands-on approaches beyond informal learning. Broadening the use of evaluation instruments beyond measuring the impact of middle school and high school STEM programs would provide invaluable insights as to how informal learning approaches might serve to measure informal learning approaches on URM remote students as well as larger community efforts connecting K-12, community college, and university programs through a makerspace community effort. Hands-on learning experiences that incorporate literacies within a makerspace may be less intimidating and abstract for underserved students, families, community and faculty. Refocusing mobile makerspace outreach efforts to provide faculty development to support active learning and discovery would begin to address gaps identified in literature [2]. Outcomes of the pilot program consider how culturally responsive makerspace programs might impact teaching intervention approaches among underrepresented and remote populations. The impact of the pilot project has great potential toward improving faculty development so that diverse participation and persistence in STEM careers increases, improved STEM education and faculty development, increased public scientific literacy and engagement in STEM, and could lead to a broader collaboration and partnership among industry, community colleges, Hispanic serving institutions assisting isolated populations.

Acknowledgements

Dr. Jennifer Miller currently serves as the PI for the Noyce en la Frontera grant. The Noyce en la Frontera program, award 2050173, has received positive evaluation results from both the external evaluator and program officer. The PI, Dr. Miller, has received several NASA funding awards and STEM education grants and served as a graduate research assistant for the NSF funded Middle Schoolers Save the World project with the University of North Texas while earning her Ph.D.

Conflict of interest

The author declares no conflict of interest.

Notes/thanks/other declarations

The author wishes to thank the National Science Foundation, Sul Ross State University Texas State System administration, the Science Mill Museum, and countless educators who continue to embrace STEM through makerspace experiences.

Author details

Jennifer Miller
Sul Ross State University, Alpine, United States

*Address all correspondence to: jennifer.miller@sulross.edu

References

[1] Compeau S. Designing a mobile makerspace: A strategy for increasing diversity by offering engineering outreach workshops to underrepresented youth. Proceedings of the 18th Canadian Engineering Education Association (CEEA). 2018 June 3-6, University of British Columbia; 2018. pp. 1-5

[2] Purpur E, Radniecki T, Colegrove PT, Klenke C. Refocusing mobile makerspace outreach efforts internally as professional development. Library Hi Tech. 2016;**34**(1):130-142

[3] Sinha S, Rieger K, Knochel AD, Meisel NA. Design and preliminary evaluation of a deployable mobile makerspace for informal additive manufacturing education. In 2017 International Solid Freeform Fabrication Symposium 2017; Austin, Texas: University of Texas; 2017

[4] Núñez AM. Commentary: Centering the "marginalized majority": How Hispanic-serving institutions advance postsecondary attainment. American Educational Research Journal. 2017;**54**(1_suppl):135S-139S

[5] Litts BK. Making Learning: Makerspaces as Learning Environments [dissertation]. Madison: University of Wisconsin; 2015

[6] Estrada M, Hernandez PR, Schultz PW. A longitudinal study of how quality mentorship and research experience integrate underrepresented minorities into STEM careers. CBE—Life Sciences Education. 2018;**17**(1):9.1-9.3. p. 13

[7] Board NS. Revisiting the STEM Workfource: A Comparison to Science and engineering Indicators 2014. Alexandria, VA: National Science Foundation; 2015

[8] Barton AC, Tan E, Greenberg D. The makerspace movement: Sites of possibilities for equitable opportunities to engage underrepresented youth in STEM. Teachers College Record. 2017;**119**(6):1-44

[9] Bonny SM. Effective STEM outreach for indigenous community contexts-getting it right one Community at a Time! International Journal of Innovation in Science and Mathematics Education. 2018;**26**(2)

[10] Koh K, Abbas J. Competencies for information professionals in learning labs and makerspaces. Journal of Education for Library and Information Science. 2015;**56**(2):114-129

[11] Miller J. Effect of makerspace professional development activities on elementary and middle school educator perceptions of integrating technologies with STEM [dissertation]. Denton, TX: University of North Texas; 2016

[12] Stager GS. Papert's prison fab lab: implications for the maker movement and education design. Proceedings of the 12th International Conference on Interaction Design and Children. New York, NY; June 24, 2013. pp. 487-490

[13] Beavers K, Cady JE, Jiang A, McCoy L. Establishing a maker culture beyond the makerspace. Library Hi Tech. 2019;**37**(2):215-228

[14] Gorski P. Education equity and the digital divide. AACE Review (Formerly AACE Journal). 2005;**13**(1):3-45

[15] Chandra S, Chang A, Day L, Fazlullah A, Liu J, McBride L, et al. Closing the K–12 Digital Divide in the Age of Distance Learning. Boston, MA, USA: Common Sense and Boston Consulting Group; 2020

[16] Vishkaie R. The pandemic, war, and sanctions: Building resilience for the digital divide in education. Interactions. 2020;**27**(4):36-37

[17] Watson SF, Alan Lacey, and Audrey. Science, Technology, Engineering, and Mathematics (STEM) Occupations: Past, Present, and Future: Spotlight on Statistics: U.S. Bureau of Labor Statistics. www.bls.gov. Available from: https://www.bls.gov/spotlight/2017/science-technology-engineering-and-mathematics-stem-occupations-past-present-and-future/home.htm

[18] Ker B. Job growth & unemployment [Internet]. Austin: Austin Chamber of Commerce; 2020. Available from: https://www.austinchamber.com/blog/09-22-2020-job-growth-unemployment

[19] Li S, Li E. The impact of digital divide on education in USA amid COVID-19 pandemic. InHCI International 2021-Posters: 23rd HCI International Conference, HCII 2021, Virtual Event, July 24-29, 2021, Proceedings, Part III 23 2021. Online: Springer International Publishing. (pp. 571-576)

[20] Laufer M, Leiser A, Deacon B, Perrin de Brichambaut P, Fecher B, Kobsda C, et al. Digital higher education: A divider or bridge builder? Leadership perspectives on edtech in a COVID-19 reality. International Journal of Educational Technology in Higher Education. 2021;**18**:1-7

[21] Moorefield-Lang HM. When makerspaces go mobile: Case studies of transportable maker locations. Library Hi Tech. 16 Nov 2015;**33**(4):462-471

[22] Friday Institute for Educational Innovation. Teacher efficacy and beliefs toward STEM survey. Raleigh, NC; 2012. Available from: https://stelar.edc.org/sites/default/files/T-STEM%20Survey.pdf

[23] Christensen R, Knezek G. The technology proficiency self-assessment questionnaire (TPSA): Evolution of a self-efficacy measure for technology integration. In: Proceedings of the KEYCIT2014: Key Competencies in Informatics and ICT Conference, KEYCIT. Potsdam, Germany: University of Potsdam; 2015. pp. 311-318

[24] Cronbach LJ. Coefficient Alpha and the Internal Structure of Tests. Psychometrics. 1951;**16**(3):297-334

[25] Christensen R, Knezek G. Validating the technology proficiency self-assessment questionnaire for 21st century learning (TPSA C-21). Journal of Digital Learning in Teacher Education. 2017;**33**(1):20-31

[26] Rhodes CM. A validation study of the culturally responsive teaching survey. Universal Journal of Educational Research. 2017;**5**(1):45-53

Chapter 14

Creating a Culturally Responsive Mathematics Education: The Case of Gebeta Game in Ethiopia

Solomon A. Tesfamicael and Danyal Farsani

Abstract

This chapter presents an example of creating a culturally responsive mathematics education in Ethiopia using a locally available game called Gebeta. The study is framed using two theoretical frameworks: funds of knowledge and cultural commognition. To this end, an ethnographic study has been employed, which helps to describe, analyze, and investigate a particular group, culture, or community. The findings show that the game is well situated as a developed body of knowledge and skills in the culture. It has remarkable potential to foster mathematical thinking and communication at different levels of the school context. Specifically, in terms of mathematical thinking, the game has affordances to foster early numerical and algorithmic thinking. Educators and stakeholders involved in designing tasks and activities for the curriculum and syllabus should consider incorporating the Gebeta game and other culturally available activities to embed them as part of formal school mathematics in a meaningful way.

Keywords: culture, communication, cognition, funds of knowledge, Gebeta

1. Introduction

The content of the mathematics curriculum and classroom practices in schools used to be dominated by the view that mathematics is a value-free body of knowledge disassociated from its cultural and social context [1–4]. According to Stephen Lerman, the social turn in mathematics education has challenged such practices [5] since, for sociocultural theory learning theorists, learning is a social phenomenon, that is. knowledge is acquired in social settings [4–6]. While emphasizing the importance of a child as an independent learner, Vygotsky defined the "more knowledgeable other (MKO) as anyone who has a better understanding or a higher ability level than the learner, particularly regarding a specific task, concept, or process," [6]. Hence, learning happens with the presence of an MKO in a given social context, creating the opportunity for discourse.

Prahmana reflected that Freudenthal and Ubiratan D'Ambrosio had reacted to the movement in the USA in the teaching and learning of mathematics, called "back to basics," which focused solely on mathematical procedures and rules [7]. As a result, D'Ambrosio initiated the idea of Ethnomathematics, while Freudenthal introduced realistic mathematics education (RME). The former focuses on "mathematics exploration from ideas, methods, and techniques a society uses to respond to its environment," and the latter "emphasizes the students' thinking levels and the mathematization process from surrounding phenomena" ([7], p. 257). RME uses real context from the students' surroundings. At the same time, ethnomathematics emphasizes values while reinventing the concept of mathematics, and both somehow consider the reality and the social and cultural context of the learners. If school mathematics should be rooted in these principles and values mentioned briefly, the curriculum (intended-implemented-attained) plays a vital role in helping teachers and learners realize it [4]. The challenge has been how to implement such ideas in practice. This chapter aims to contribute to such endeavors within Ethiopia and other countries where such emphasis is needed across the globe. Hence, we first define the Ethiopian context and the game under consideration in this study, which can be used in the teaching and learning of mathematics at different school levels in Ethiopia.

2. Setting and research question

In Ethiopia, one of the oldest countries in the world, there are 294 local games, according to the Cultural Sports Office of the country. Only 11 have been considered official cultural sports for competitions in the country. The Gebeta (ገበጣ) game is one of these 11 games that have obtained official status. It is played across towns and rural areas in the country. It has many different versions with various local rules. However, the nationally recognized game follows standard rules, similar to the internationally recognized version of Mancala. Studying such cultural activities for education, particularly mathematics education, is an ongoing concern for many educators in the country [4, 8, 9].

Such games and activities, which are locally available and accessible, provide out-of-school mathematical activities [10]. Kids engaged with such games at an early age are already exposed to different mathematical concepts that will be learned formally within the school curriculum. If school mathematics utilizes such opportunities to learn the subject, student learning could be optimized. The school mathematics curriculum and instruction must systematically embed such games [11]. The challenge is finding example activities and tasks rooted in the culture and situating those activities to fit within a given curriculum. Thorough investigation and rigorous work are needed if such locally available games and activities are to be integrated into the formal education system *via* the curriculum, syllabus, and textbook to inform the teaching and learning of mathematics. Partly, this need motivates us to fill the gap in Ethiopia's education system's curriculum and instruction. Therefore, we wanted to explore and investigate the potential of the Gebeta game to support mathematics teaching and learning in the country. So, we asked the following questions: a) To what extent is Gebeta an available cultural game in society? b) Which mathematical concepts, operations, and procedures are embedded in the Gebeta game? c) How can this be integrated into the education system? These questions shape the present research at hand, and this paper shows examples of the affordances found in the culture to attain a bigger picture.

3. Theoretical perspectives

The study about the availability and relevance of the Gebeta game in the culture and the opportunity embedded in the game for the teaching and learning of mathematical concepts and procedures should be grounded theoretically. For this reason, two different theoretical frameworks that complement each other for the chosen study are presented as follows.

3.1 Funds of knowledge (FoK)

Moll coined the term funds of knowledge (FoK) to refer to "the historically accumulated and culturally developed bodies of knowledge and skills" ([12], p. 133). They studied household knowledge to foster a participatory pedagogy that would enable educators to facilitate the transfer of knowledge between home and school contexts. As a result, they developed a framework with broader and more diverse knowledge, including agriculture, mining, medicine, economics, religion, and others. Given that FoK emanates from a given household or community, the list can also include different plays and games embedded in that household or community.

According to Bishop, playing is one of the six mathematical activities found in almost every culture [1]. It is possible to argue that the games available at home or in their community that interest children can be counted as their FoK [2]. Games provide opportunities for community participation, which is crucial to social learning and cognitive development [13]. Chesworth indicated a FoK approach to strengthening curriculum and pedagogical decisions informed by children's play choices and interests [14]. Of course, what interests children in play is a complex matter. Chesworth argues that "FoK offers an alternative mode for understanding children's interests within the context of their participation in the activities for their homes, classrooms, and communities" ([14], p. 296). Out-of-school mathematics is another terminology that emphasizes the nonformal mathematical-related activities learners engage in daily life [10].

3.2 Cultural commognition

Commognition is another interpretation of sociocultural learning theory focused on education. It comprises two words, communication and cognition, signifying the unity of interaction and thinking [15, 16]. This means that doing mathematics in this theoretical framework includes both thinking and communicating it. In this regard, culture can serve as a mediator for forming and developing competencies, such as problem-solving and critical thinking. Radford argued for the connection of mathematical thinking (cognition) to its historical and cultural situatedness and further stated: "I will suggest that thinking in general, and mathematical thinking (MT) in particular, are forms of reflective, mediated social praxis where the organization of individuals' sensuous cognitive processes is related to the meaning of things as they become objectified in practical and theoretical activity" ([17], p. 440).

Further, it was inspired by scholars, such as Brenner, who added: "Cognition to the formula for culturally relevant instruction in mathematics" ([18], p. 214). Turner synthesizes findings that integrate children's mathematical thinking with cultural FoK (CFoK) in mathematics instruction [19]. Therefore, cultural commognition refers to the mathematical thinking and communication situated in the given cultural

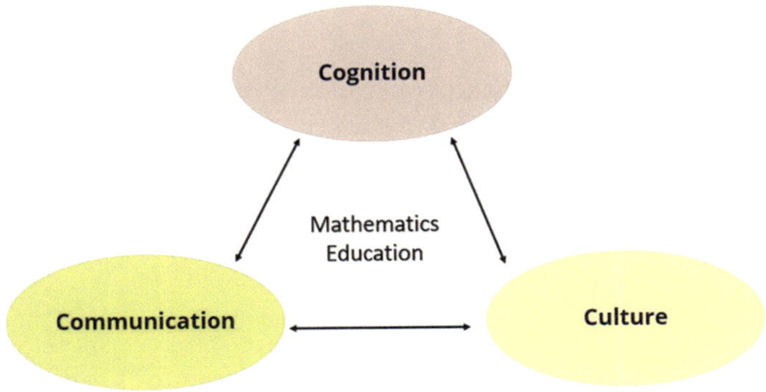

Figure 1.
Cultural commognition in mathematics education is inspired in Ref. [20] work of interconnection that involves society, cognition, and culture.

setting and context. Inspired by those who argued that mathematical production is subsidized by an interconnection involving society, cognition, and culture, this work presents the interconnection between communication, cognition, and culture [20], as shown in **Figure 1**.

3.3 Introducing Gebeta as an artifact

Culture-specific tools are integral in how learners organize and think about mathematical concepts and procedures [11]. "Gebeta" is one of the ancient African board games. It originated from ancient Ethiopians and was discovered by archeologists in Eretria, a place called Matara, which was then part of Ethiopia (see **Figure 1a** & **d**). Generally, it is also referred to as Mancala in other locations. However, the type of game described in this article is unique to Ethiopia. Tesfamicael argued that "it is difficult to find structured literature about the Gebeta game in the country." Nevertheless, some evidence was uncovered by archeologists who searched for relics in the northern part of the country.

They found a Gebeta-playing artifact and estimated that the game was practiced in Ethiopia from the sixth to seventh BC. The game can be played in both rural areas and towns. Players can dig 12 or 18 equal-sized small half-sphere-shaped holes in the ground and collect 48 small stones, beads, or seeds, placing four in each hole. Alternatively, the game can be made on a wooden material. **Figure 2** provides examples of 12-hole (homes) Gebeta artifacts, including both old and modern versions. Additionally, a digital version created using block programing, Scratch, is included.

3.4 Mathematical thinking (MT) *via* Gebeta game

MT is a complex matter to deal with. If one uses ChatGPT to get an answer to the question, "What is MT"? then ChatGPT provides a comprehensive definition as follows [21]:

> *Mathematical thinking is a cognitive process that involves analyzing, reasoning, problem-solving, and making connections using mathematical concepts and principles. It goes beyond simply performing calculations and involves deeper levels of understanding and manipulation of mathematical ideas.*

Figure 2.
a) Gebeta playground made on the ground b) virtual Gebeta made of scratch block programming. c) Commercial Gebeta (mancala) game boards available www.kodkod.com. d) Gebeta game board made in Zege monastery in Lake Tana.

To highlight the process aspect further, the ChatGPT definition provides ten aspects of MT: abstraction, logical reasoning, pattern recognition, problem-solving, creativity, symbolic manipulation, critical analysis, visualization, generalization, and communication. Stacey emphasized the importance of MT as it is a goal of schooling, a way of learning, and teaching mathematics [22]. Furthermore, Stacey explained MT using two pairs of processes: specializing vs. generalizing and conjecturing vs. convincing [22]. Additionally, aside from the process aspect, there is a content-specific aspect of MT, such as numerical, algebraic, geometrical, statistical, stochastic, and computational (algorithmic) thinking. The curriculum mainly encompasses these content aspects of MT. A robust curriculum should also consider the process aspect. In connection to the Gebeta game, early numerical thinking (ENT) and algorithmic thinking (AT) are evidenced and highlighted. Therefore, this study primarily focuses on these two aspects.

According to van de Walle and colleagues, developing early number concepts and number sense involves dealing with several activities and concepts such as counting, one-to-one correspondence, cardinality, comparing and ordering quantities, subitizing, counting on and counting back, developing verbal counting skills, numeral writing, recognition, and thinking about zero [23]. Furthermore, algorithmic thinking (AT) is being embedded into curricula in STEM courses [24–26] as AT has become necessary for every child to experience twenty-first century skills [25]. Of course, the broader concept is computational thinking (CT), and in recent years, Bocconi argues that educational stakeholders have advocated CT and related concepts, such as programing and AT, as abilities for all that is as fundamental as numeracy and literacy [26]. They provided ways AT and programing can be integrated into the Nordic curriculum. Lockwood presented the construct of AT in mathematics education comprehensively and used NCTM's AT definition as follows: "AT is a method of thinking and guiding thought processes that use step-by-step procedures, require inputs and produce outputs, require decisions about the quality and appropriateness of information coming in and information going out, and monitor the thought processes as a means of controlling and directing the thinking process. In essence, algorithmic thinking is simultaneously a method of thinking and a means for thinking about one's thinking" ([24], p. 7). In this sense, the Gebeta game can foster some aspects of AT.

4. Methodology

This is a qualitative study, and it follows an ethnographic research approach as a methodology. According to Creswell & Poth, ethnographic research is "a qualitative design in which the researcher describes and interprets the shared and learned patterns of values, behaviors, beliefs, and language of a culture-sharing group" ([27], p. 90). In this study, the researchers engaged in extracting the shared values, experiences, and engagement with the Gebeta game in the Ethiopian context. Using cultural commognition and the theory of funds of knowledge as a lens, the researchers studied the affordances of the game through fieldwork, primary interviews, observations, questionnaires, documents, and video recordings to find answers to the research questions posed.

4.1 Data collection

4.1.1 Entry and organization access

During our previous engagement in investigating the curriculum implications of ethnomathematics to promote student learning in Ethiopia, one of the researchers approached different ministries and offices to gather data on the subject [4]. Fortunately, we found relevant information about activities considered a cultural sport officially for computation with the Ethiopian Cultural Sports Federation. Capitalizing on this networking during the summer of 2021, one of the present researchers reached out to this office. Fortunately, the people were more open and collaborative, and they showed interest in our work. After a few telephone conversations, we held the first meeting with the head of the cultural sports division and his staff. They explained their work about cultural sports and provided us with a soft copy of the guiding documents about cultural sports in the country. This document is used as a resource for the data, partly.

4.1.2 Observation and video data

On the second day, two players aged 15 and 17 from the Gebeta game team in one of the subregions of Addis Ababa city were invited to demonstrate the game at the national level. An official referee was also present to officiate the game between the two young players. Initially, the official explained the rules and procedures with great enthusiasm. Subsequently, the two young players were allowed to compete. We were granted permission to video record the game, and after its completion, we conducted a follow-up questionnaire with the players and the referee. The researcher documented the session and made field notes.

Another video data is recorded in the northern-western city called Gondar to identify and locate a culture-sharing group for this ethnographic study [27]. The kids were asked to play the Gebeta game as they usually do. Otherwise, Gebeta is a game played spontaneously during the summer holidays. The video is recorded by a volunteer who is originally from the area. The goal is to situate the study within the context of the culture and investigate if Gebeta is one aspect of CFoK for the learners in the country.

4.1.3 Small-scale survey on social media

To further strengthen the investigation of whether Gebeta is one of the accessible and meaningful cultural FoK for learners in Ethiopia, an additional small-scale survey was conducted on social media using Google Forms. Such an online survey will result in a sample of unknown representativeness, but it is more cost-effective and time-saving than in-person surveys and postal questionnaires [28]. The participants were not from a particular region or part of the country. They were recruited through personal networks, a mathematics education network in Ethiopia, *via* Telegram and Facebook Messenger. Most of those who participated and responded were highly educated (minimum master's degree in one field).

4.2 Data analysis

Ethnographic research follows three types of data analysis, according to Creswell & Poth: description, analysis, and interpretation of the culture-sharing group [27]. As a starting point, the researchers described the cultural context of the Gebeta game in Ethiopia. The story is told in a way that helps us understand the game within the given cultural settings in the community, society, and country at large. Furthermore, we interpret the data by examining different aspects of the game's influence on mathematics education. Finally, analysis, the quantitative side of qualitative study, is employed to highlight specific material introduced in the description of this cultural game [27]. In search of patterns of evidence for CFoK, the data will be triangulated by comparing and testing information from different sources. The conceptual frameworks presented above, CFoK and cultural commognition, generally guide the data analysis.

4.3 Ethical consideration

The study was conducted by obtaining informed consent and ensuring confidentiality [28]. The researchers obtained informed consent from the participants by providing information about the research process. Before collecting the data, permission to responsibly collect personal data was obtained from Norwegian Center for Research Data (NSD) in Norway. The participating authorities and persons in Ethiopia were also duly informed, and efforts were made to minimize or avoid using personal data. The images used were carefully selected to ensure they did not harm the participants despite their consent to participate in the research.

5. Findings

5.1 Gebeta is part of the cultural in the family, community, and society

During the visit to the Ethiopian Cultural Sports Federation office, the officials claimed that there are 293 registered cultural games in the country. This excerpt from the document shows the office's mandate to organize and structure scientifically. Furthermore, the document reveals that only 11 of these games have been adopted as national cultural sports games. Among these 11, the Gebeta game is included. This information is depicted in **Figure 3**, which includes a text excerpt from the guiding document about cultural sports in the country.

During the videotaped Gebeta game at the head office, two young players from the sub-city of Arada in Addis Ababa played against each other, officiated by a referee who has been working on cultural sports for 30 years. They were all dressed according to the cultural context of the game. The head of cultural sports opened the demonstration. He said, “Today, as promised, we have invited players from one of the sub-cities in Addis Ababa called Arada. They are here with a referee, who is also a game instructor. He leads a cultural sports project in the Arada area. He will take responsibility for officiating and explaining the rules and cultural interpretations of the game today. The two young people are here because they are among the best, and they are the ones who know all the four types of Gebeta game (see **Figure 4**).”

After that, the referee took over and introduced himself first. Then, he continued with a deeper explanation of the country’s history and development of the cultural sport. In the Room was also another officer who has facilitated this meeting.

Because Ethiopia's cultural sports games are growing over time, the need to organize and synchronize these games scientifically has increased. This is to ensure that they can be widely distributed and enjoyed by society. In 1967, the Sports Commission was established by a decree, and it was decided that the cultural sports games would fall under this commission. Mr. Tefera Mekonnen, a member of the commission's staff, was responsible for traveling across the provinces to gather approximately 293 cultural sports games. Among these, the commission selected games that were considered nationally acceptable and established rules and regulations for the competitions. Currently, these games are played throughout Ethiopia in competitive situations.

Figure 3.
Extract from the cultural sports federation guiding document (Amharic text translated to English by one of the researchers).

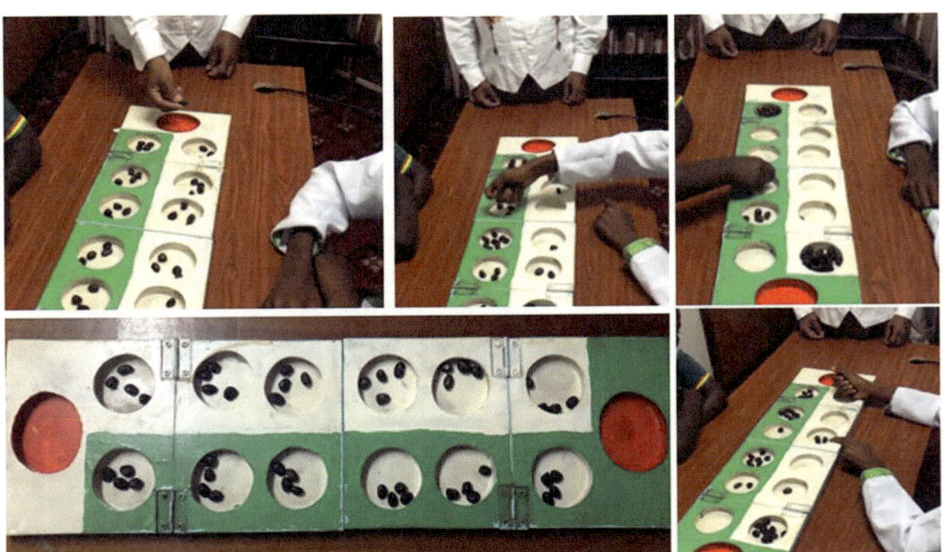

Figure 4.
The referee, the two teenage players from the Arada sub-city of Addis Ababa, and the official game board of 84 cm x 20 cm with 12 holes/homes and two storing holes (homes).

> *"I am ...; I have contributed to a cultural sport for over 30 years. Whenever I mention cultural sports in Ethiopia, I always want to recognize the contribution of the late Mr. Tefera Mekonnen, who tremendously contributed to establishing this work in the country. He is the one who traveled throughout the country around 45 years ago, sitting on a horse and mule, and traveling on foot, collecting information about these games. I am always grateful that he showed me the way and recognized and entrusted me with this work to continue to the next generation; that is why I am working hard to realize that. I am training, teaching, as well as leading projects that involve young people. Today, two young teenagers from Arada are here to demonstrate the game."*

Furthermore, the referee stated, "the game usually starts with greetings and handshakes. However, due to coronavirus concerns, the players will now begin with an elbow bump greeting." Both the players and the referee have already donned cultural clothes. As the game commences, he also explains which versions of the game will be demonstrated, focusing on grammar, spelling, and punctuation errors.

> *"The 12-homes Gebeta game has four different types: 1-kinchibosh (ቅንጭበሽ), 2-kinchibosh, 3-kinchibosh, and 4-effisosh (እፍሶሽ). There are different versions Gebeta game but are only focused on those that are nationalized by establishing rules and regulations of the game scientifically. Of course, we could have worked harder to recognize and register these cultural sports at the Olympic level. Even we need people who work on it so that it is played using technology."*

Hence, first, 1-kinchibosh was demonstrated. Mr. Yaregal said, "First, we should decide who to start. We do it using a draw by flipping a coin. What do you prefer, Anbesa (አንበሳ meaning lion) or Sew (ሰው meaning man)?" Then, he flipped the coin, and the one dressed green got the chance to start the game. After he took his round, the second player chose a home (hole) to start and moved the stones further. We asked him why he chose to start there. He explained, and Mr. Belayneh interjected in between and provided the different strategies to win the game. After the first round, they continued demonstrating the other three game versions. The detail is not the focus of this article, and we return to it later on the mathematics, language, cultural representations, and interpretations involved in the game. In addition to this official data, other evidence collected to establish that Gebeta is among the CFoK was from the survey. The result of the small-scale survey is summarized in **Table 1**.

5.2 Scaffolding 'Gondar's Gebeta' game

Gebeta game has many versions. Even the 12 homes Gebeta game has four different ones, as mentioned above. The kind of Gebeta game that is the focus of this work is different from the official cultural sports, but the one that one of the researchers used to play before he started elementary school. It is played around Gondar, the northwestern part of the country; we call it Gondar's Gebeta game. It might also be known in other parts of the country; further, investigation is needed. The rationale for presenting this version of the game is the following: i) its simplicity to be understood and played, ii) its contribution to mathematical cognition, iii) it is known to us at an early age before school, and iv) its availability within our proximity and how we enjoyed it.

Questions	Response
Do you know the game called Gebeta (ገበጣ)?	95% of the 22 who participated in the survey responded YES.
Have you played it before?	81.8% of the 22 who participated in the survey responded YES. 18.2% said NO.
Did you play it by making holes in the ground and collecting pebbles (small stones), or you played on a material made for the Gebeta game?	By making holes in the ground = 63.6%. On a material made for the Gebeta game, 34.4%
Why did you play the game?	Entertainment (most answers), for leisure, Competition, fun and interesting, to enjoy, interesting to play, it was a tradition in my place I was born, just to spend my time, it is one type of game that I used to play with my siblings because it a good game.
When did you start playing it? At which age?	12, 10, almost at 7 when I was a kid, when I was a kid, 5, 7, 12, 14, at 7, around 10, maybe in early 20, 15, before 35 years, I think as a kid, when I was 5, 14.
Where was the first time you played it?	At birthplace at rural area, in the rural area, in rural areas, northwest & central, Ethiopia (two times), in my birthplace, anywhere, home, in my home town Gondar (town), in my birth village, Lay Armachiho, Tikildingay, Dessie, Ethiopia, cannot remember, neighborhood, Debrezeit (town).

Table 1.
Summary of the small-scale survey questions and responses about the Gebeta game in Ethiopia.

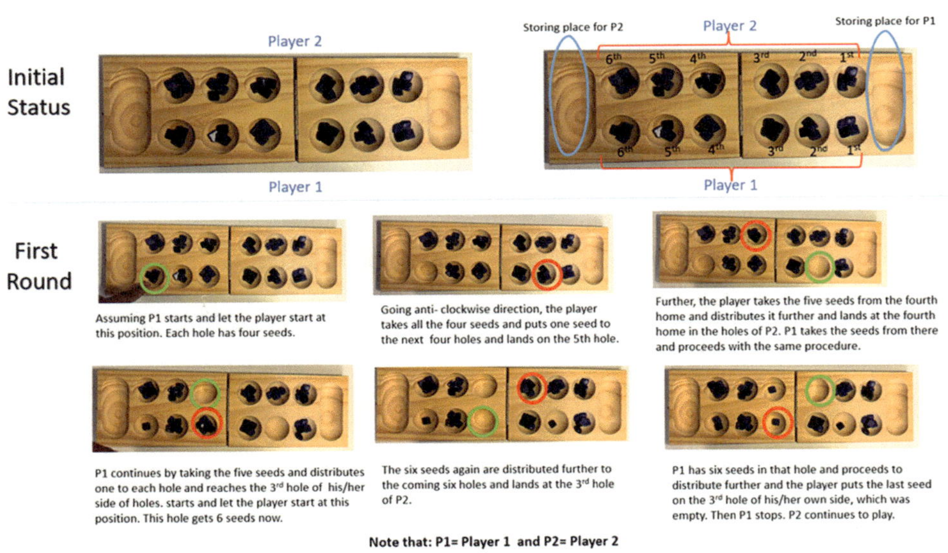

Figure 5.
The initial status of the two-player playing the Gebeta game and the first round play by player 1.

Assume there are only two players (see **Figures 5–7**). At the start, each player has six equal holes called homes (in the local language, ቤት bet). Each hole contains four seeds, which count as one completed home. The players aim to own as many holes/homes as possible. To achieve this, players must empty their holes and then rebuild them to contain four seeds again, thus completing a home. The game is played in multiple rounds until one player loses some or all of their initial homes. The first player

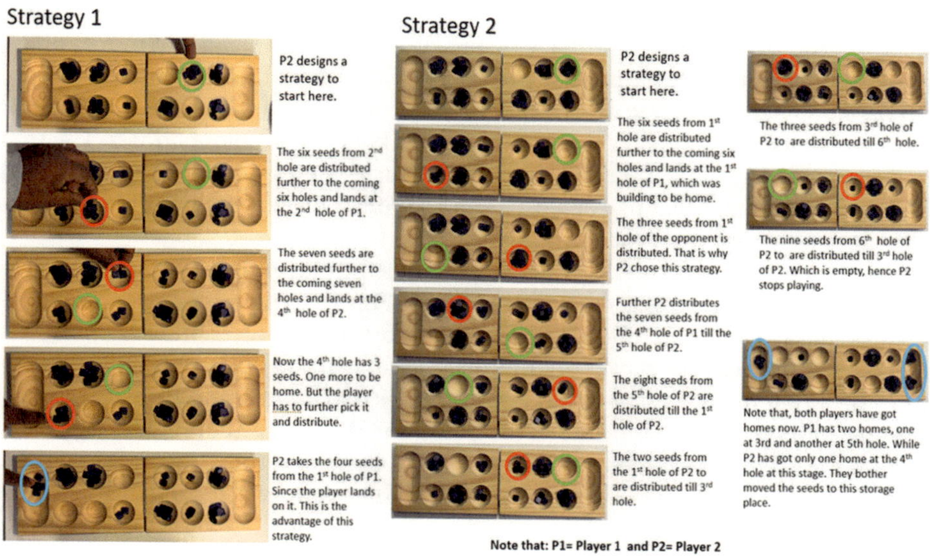

Figure 6.
Two possible strategies by player 2.

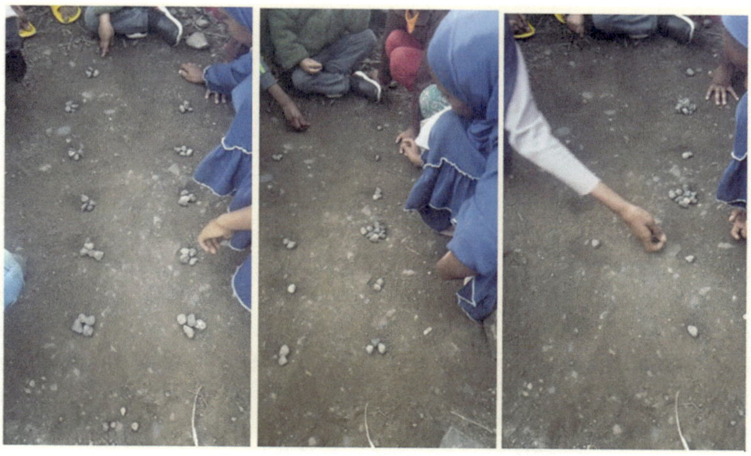

Figure 7.
Kids playing the Gebeta game on the ground in Gondar.

begins at one of their homes, picks up all four seeds, moves around in an anticlockwise direction, and distributes one seed to each subsequent hole until all seeds have been distributed. The player then collects all the seeds from the last hole they placed seeds in. This process continues until the player reaches an empty hole, at which point they stop playing and pass the turn to the other player.

The second player follows the same procedure starting with a hole of their choice and picking up all the seeds in that hole. As the players continue to go around and around, if a hole becomes filled with four seeds again, the player who owns that hole immediately collects the four seeds and claims their first home. However, if a player accidentally adds another seed to a hole that already contains four seeds, they are unable to pick up the four seeds. On the other hand, if a player can put his last pebble on the hole/homes of the other side, which has three, then that player collects that home for him/herself. Actually, that is one of the ways a player can win more homes. Another strategy is to be able to attack the holes, which are on the way to four seeds (complete homes). The game can be completed if one of the players has more or even all homes. Hence, the game involves strategic thinking to gain more homes. The first round for each player is simulated as follows. Let us assume two players, players 1 and 2, are playing. Assume that player 1 starts and **Figure 5** provides the simulation of the player's steps starting from the initial step.

Player 2 starts his/her turn. The game demands the player to design strategies to optimize the win, that is. to acquire more homes. **Figure 6** provides two different strategies that Player 2 can follow, resulting in differing outcomes. In strategy 1, at the end of the first round, the player obtains one home, while the other player gets nothing. On the other hand, in strategy 2, Player 1 gains two homes, and Player 2 only gets one home. Which strategy is better in the long run? Was there another strategy that surpasses the aforementioned two? The answer to these questions depends on various variables. However, the Gebeta game undeniably requires critical and strategic thinking.

Figure 7 showcases versions of the Gebeta game played by kids in Gondar. In the video, the kids surrounding the two players can be heard suggesting different strategies. Sometimes, competitive players may not allow external suggestions, while in other cases, they may tolerate them.

6. Discussion

6.1 Gebeta game as a cultural funds of knowledge

The Gebeta game is considered one of the cultural games, as evidenced in both the national document of the cultural sports federation in Ethiopia and the response from participants in a small-scale survey conducted *via* social media. Additionally, children playing the Gebeta game in Gondar, as described above, further supports its cultural significance. The survey results showed that 95% of respondents were aware of the game, and 81.2% had played it between the ages of 5 and 15. The game is predominantly played in rural areas and towns, usually on the ground, for entertainment, fun, and competition. This indicates that the Gebeta game is a prevalent form of cultural knowledge in most communities in Ethiopia. Children engage in this game at an early age, regardless of whether they live in cities or rural areas. This aligns with Moll and colleagues' definition of FoK (Funds of Knowledge) as a historically accumulated and culturally developed body of knowledge and skills within a household or community [12]. Clark demonstrated how nonschool activities that draw on children's FoK can enhance problem-solving abilities [29]. In this context, we argue that Gebeta is one of those cultural activities that can be considered as part of learners' FoK. The game incorporates various concepts and procedures of school mathematics. If adequately studied and integrated into the school curriculum and instruction in Ethiopia, it has the potential to enhance mathematical learning.

The Vygotskian perspective of learning advocates learning as participation in cultural activities [6, 13]. The Gebeta game is inherently suited for such learning since it is accessible to kids at an early age. It can be played by more than two players (if three players, they can share four homes at the start). One knowledgeable individual can train the other(s). That is to say, the more knowledgeable other (MKO), according to Vygotsky, helps the novice to learn the game [6, 13]. The MKO can be a father, mother, sibling, or friend with whom the village kids frequently interact. The MKO and those learning the game communicate different strategies, which occur in the most natural context of the culture. In the video, the kids from Gondar demonstrate that others suggest the best strategies for the players. This demonstrates participation and communication of winning strategies at every level as the game can progress in various ways at any given level. Finding the best ways to optimize the chances of winning requires abstraction, pattern recognition, critical and strategic thinking, conjecturing, and convincing, which are the process aspects of MT [21, 22].

6.2 Gebeta game as a cultural commognition artifact

Commognition signifies the unity of interaction and thinking [15, 16], and a cultural artifact, such as the Gebeta game, could stimulate it. In particular, the Gondar Gebeta game presented in this work can potentially mathematically engage kids at age 5 and 6 [4, 11]. We are doing more profound research with kids in Ethiopia and Norway to see which aspects of mathematical thinking, language, and communication are visible in both contexts. In this regard, as depicted in **Figure 8**, the potential of the Gebeta game to facilitate both the mathematical thinking and the communication aspect could be investigated very well. The question could be which mathematical thinking is most pronounced. Mathematical thinking (cognition) is a cognitive process by which students come to understand mathematical concepts and procedures. The process includes problem-solving, reasoning and proof, communications, connections, and representation, according to [30]. If mathematics should be thought of as a human activity, games, such as Gebeta, can facilitate such a learning process. In the game, many mathematical activities are embedded. These are presented as follows:

6.2.1 Gebeta and early numerical thinking (ENT)

Counting: For starters, it is basically a counting activity. One has to count how many seeds a home has to play the game till the end of the game. A home can have 1, 2, 3, 4, ..., 11, or more seeds. It can be empty also, representing the concept of zero. As counting is the basis of numerical thinking [11, 23], playing this game at an early age repeatedly could boost skills in early numeracy.

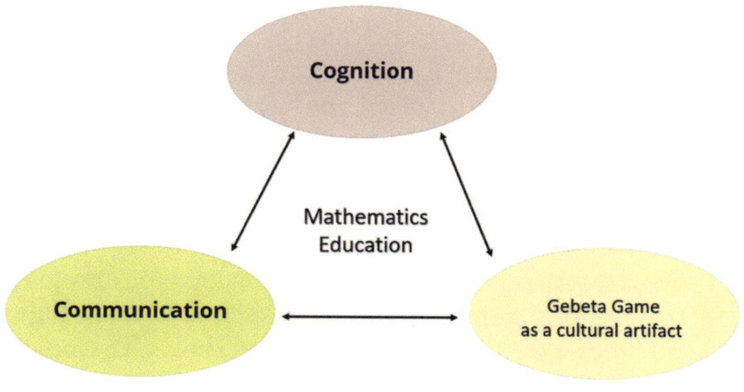

Figure 8.
Gebeta game as an artifact for commognition in mathematics.

One-to-one corresponding: During playing, one first counts how many seeds each home holds and decides to start at one of the homes. Then the players grab the seeds in one home and distribute the seeds to consecutive homes (holes). This is done using the one-to-one corresponding principle: For each home, you put one seed. Of course, this is a subtraction activity too. From the number of seeds in the hand, you reduce one each step forward. At the same time, it is an additional activity to the number of seeds in each hole/home.

Concept of zero: While at the start, one grabs all four seeds (see **Figure 1**). That home now has 4–4 = 0 seeds, and the following four homes gain one more seed, 4 + 1 = 5. Then, one empties the fifth hole/home and adds one seed for the next five holes. Making homes empty is an integral part of the game. In doing that, one communicates the concept of zero several times in a single game round. Hence, the concept of zero is visibly demonstrated *via* an empty hole/home so that preschoolers understand its meaning.

Subitizing, cardinal, and ordinal numbers: These skills are highly involved if one intends to win more homes in the game. Each player continuously engages in subitizing (perceptual or conceptual) activity from the start. For example, a home can have 5, 6, or 7 seeds. Then one has to update as the game proceeds instantly. For instance, in strategy 2 (see **Figure 2**), Player 2 begins at that house with six seeds, intending to annul the potential of being a home for the opponent player. It means that by counting the number of seeds to reach that home, Player 2 can instantly recognize the six seeds and use the one-to-one correspondence principle with the order of the homes to reach that particular home of Player 1. In addition, the concept of the ordinal number, the sixth home, is embedded in the thinking process.

The 12-home Gondar's Gebeta game involves a maximum of 48 seeds, which is 12 times 4. Repeated addition of four represents multiplication. Each player has six homes in this game, so 6 x 4 = 24 seeds. When you win a home, you have 1 x 4 = 4 seeds. If you have two homes, you have 2 x 4 = 8 seeds, and so on. One can also think that if there are two players, they divide the 12 homes by two, so 12 ÷ 2 = 6 homes each. The 48 seeds are divided into 12 homes, so 48 ÷ 12 = 4. To win a home, the players must first empty the hole (zero seed) and then put in four seeds to build it. This game uses a base-five counting system in addition to the base-ten counting system. Different arithmetic operations are embedded from the outset. If played many times with a grown-up person, kids can develop advanced numerical thinking even before they start formal school.

6.2.2 Gebeta and algorithmic thinking

Adopting NCTM's definition of AT [24], we can say that the Gebeta game is inherently algorithmic. It is a game in which one has to decide where to start and where to land for many rounds, following step-by-step procedures, and requires a strategy to win using the game's rules. According to Bishop, playing is considered one of the mathematical activities [2]. This game challenges players to be creative to win homes. One should have good counting strategies for repeatedly determining where to start and finish. The game can even be programmed using block programming tools, such as Scratch (See **Figure 9**) or text programing, such as Python. **Figure 8** shows the adoption of the Mancala game, which one can find through a Google search, to a Gebeta game used in this study.

There are two different activities here. First, it is possible to create a programing activity that simulates the game. Another, it is possible to make applications that can

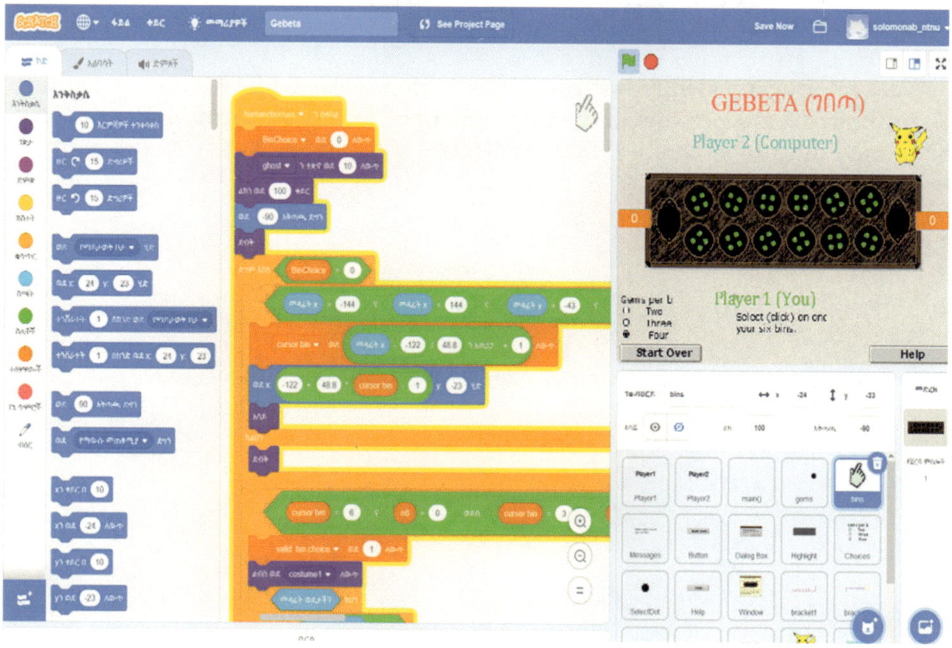

Figure 9.
Example of block programming to Gebeta game to foster algorithmic thinking.

help play the game on computers and mobile applications. That makes the game to be available virtually, and kids can access it on their mobile or PC.

6.3 Conclusion and implications

The Gebeta game is played in towns and rural areas and is integrated nationally as a cultural sports competition, as evidenced above. However, there needs to be clear evidence that it is used as part of the mathematics curriculum in the country. Gebeta games can be played without a proper game board, as shown in **Figure 6**. The availability and accessibility of the Gebeta game for everyone could be considered an advantage in considering it as part of the resources for mathematics education. The variations of the game, from simple to complex (such as the Gondar Gebeta game, 1-kinchibosh, 2-kinchibosh, 3-kinchibosh, and 4-kinchibosh), allow for learning the different mathematical concepts embedded in the game, as discussed above. Hence, the Gebeta game can be used as a CFoK (Cultural Form of Knowledge) and for cultural cognition. What is left is to design activities and tasks that can be integrated into the mathematics curriculum and syllabus in the country. This demands those who are engaged in designing tasks, instructional materials, and textbooks to search, for example, the Gebeta game and others, such as the 293 cultural games mentioned in the National Cultural Sports Federation document and embed the activities and tasks that can foster mathematical cognition and communication across the school mathematics curriculum. There has been a movement among education stakeholders that indigenous knowledge and practices have to be part of the formal school system, as evidenced in the new curriculum of the country [31]. However, implementing it is challenging, and this work provides one particular example within mathematics education.

Acknowledgements

We want to acknowledge our fruitful discussions with the IMED research group and its international network at the Department of Teacher Education, Faculty of Social and Educational Studies (SU) at NTNU.

Author details

Solomon A. Tesfamicael* and Danyal Farsani
Faculty of Social and Educational Sciences, Department of Teacher Education, Norwegian University of Science and Technology (NTNU), Trondheim, Norway

*Address all correspondence to: solomon.a.tesfamicael@ntnu.no

References

[1] D'Ambrosio U. Society, culture, mathematics and its teaching. Educação y Pesquisa. 2005;**31**(1):99-120

[2] Bishop AJ. Mathematical Enculturation: A Cultural Perspective on Mathematics Education. Dordrecht: Kluwer Academic Publishers; 1998

[3] Ernest P. The Philosophy of Mathematics Education. London: Falmer Press; 1991

[4] Tesfamicael SA, Nakken AH, Tumebo T, Gray P. Curriculum implications of using Ethnomathematics to promote student learning in Ethiopia. Universal journal of educational research, 7(3), 754-765. Journal of Mathematics and Culture. 2021;**15**(3):86-123

[5] Lerman S. Learning and knowing mathematics. In: Rowland T, Andrews P, editors. Master Class in Mathematics Education: International Perspectives on Teaching and Learning. London: Continuum Publishers; 2014. pp. 15-26

[6] Vygotsky L. Mind in Society: The Development of Higher Psychological Processes. Cambridge, MA: Harvard University Press; 1978

[7] Prahmana RCI. Ethno-realistic mathematics education: The promising learning approach in the city of culture. SN Social Sciences. 2022;**2**:257

[8] Nair BS, Abera T. Indigenous knowledge in Ethiopian school curriculum. Social Sciences. 2017;**18**(2):89-95

[9] Ayalew Y, Areaya S. Sociocultural mathematics in Ethiopia curriculum: A Glocal view of Dire Dawa context. Journal of Mathematics and Culture. 2021;**15**(3):66-85

[10] Bose A, Subramaniam K. Exploring school children's out of school mathematics. In: Ubuz B, editor. Proceedings of the 35th Conference of the International Group for the Psychology of Mathematics Education. Vol. 2. Ankara, Turkey: PME; 2011. pp. 177-184

[11] Tesfamicael SA, Nakken AH, Tumebo T, Gray P. Curriculum implications of using ethnomathematics to promote student learning in Ethiopia. Journal of Mathematics and Culture. 2021;**15**(3):85-123

[12] Moll L, Amanti C, Neff D, González N. Funds of knowledge for teaching: Using a qualitative approach to connect homes and classrooms. Theory into Practice. 1992;**31**(2):132-141

[13] Vygotsky LS. The Collected Works of L. S. Vygotsky, Vol. 4: The History of the Development of Higher Mental Functions (R. W. Rieber, Vol. Ed; M. J. Hall, Trans.). New York: Plenum Press; 1997. (Original work published 1941)

[14] Chesworth L. A funds of knowledge approach to examining play interests: Listening to children's and parents' perspectives. International Journal of Early Years Education. 2016;**24**(3):294-308

[15] Sfard A. Learning, commognition and mathematics. In: Scott D, Hargreaves E, editors. The Sage Handbook of Learning. London: Sage; 2015. pp. 129-138

[16] Sfard A. Commognition. In: Lerman S, editor. Encyclopedia of Mathematics Education. Cham: Springer; 2020. DOI: 10.1007/978-3-030-15789-0_10003

[17] Radford M. Complexity and truth in educational research. Educational Philosophy and Theory. 2008;**40**(1):144-157

[18] Brenner ME. Adding cognition to the formula for culturally relevant instruction in mathematics. Anthropology & Education Quarterly. 1998;**29**:214-244

[19] Turner EE, Drake C. A review of research on prospective teachers' learning about Children's mathematical thinking and cultural funds of knowledge. Journal of Teacher Education. 2016;**67**(1):32-46

[20] Mendes IA. Society, culture, and cognition: Interconnections In mathematics education. Paradigma. 2022;**43**:870-897

[21] Open AI. ChatGPT (Aug 22 version) [What is Mathematical Thinking?]. 2023. Available from: https://chat.openai.com

[22] Stacey K. What is mathematical thinking and why is it important? In: APEC Symposium. Innovative Teaching Mathematics Through Lesson Study II, 3-4 December 2006. 2007

[23] Van de Walle JA. Elementary and Middle School Mathematics: Teaching Developmentally. Tenth edition, global edition/. (Bay-Williams JM, Karp KS, red.). Upper Saddle River, New Jersey: Pearson; 2020

[24] Lockwood E, DeJarnette A, Thomas M, Mørken K. Exemplifying algorithmic thinking in mathematics education. In: Twelfth Congress of the European Society for Research in Mathematics Education (CERME12). Bozen-Bolzano, Italy; Feb 2022. hal-03748468

[25] Lytle N, Cateté V, Boulden D, Dong Y, Houchins J, Milliken A, et al. Use, modify, create: Comparing computational thinking lesson progressions for STEM classes. In: Proceedings of the 2019 ACM Conference on Innovation and Technology in Computer Science Education. ITiCSE '19. Aberdeen, Scotland, UK; 15-17 July 2019. pp. 395-401. DOI: 10.1145/3304221.3319786

[26] Bocconi S, Chioccariello A, Earp J. The Nordic Approach to Introducing Computational Thinking and Programming in Compulsory Education. Report Prepared for the Nordic@ BETT2018 Steering Group; 2018. DOI: 10.17471/54007

[27] Creswell JW, Poth Cheryl N. Qualitative Inquiry and Research Design: Choosing among Five Approaches by Cheryl N. Poth, John W. Creswell. London: SAGE Publications, Inc; 2018. 9781506330211R180

[28] Clark T, Foster L, Bryman A, Sloan L. Bryman's Social Research Methods. United Kingdom: Oxford University Press; 2021

[29] Razfar A. ¡Vamos a Jugar counters! Learning mathematics through funds of knowledge, play, and the third space. Bilingual Research Journal. 2012;**35**(1):53-75

[30] National Council of Teachers of Mathematics (NCTM). Principles and Standards for School Mathematics. Reston, VA: Author; 2000

[31] Ministry of Education (MOE). Ethiopian Education Development Roadmap (2018-30). Ethiopia: Ministry of Education; 2019

Chapter 15

The Importance of Activity-Based Learning for Mastering Mathematical Tasks during Distance Learning in Austria

Sabine Apfler

Abstract

The conversion of teaching to distance learning was a major challenge for all those involved in education. For elementary school-aged students, in particular, the transition was enormous, especially in mathematics. A research project in Austria investigated which features of activity-based, tools-supported mathematics instruction were helpful for mastering mathematical tasks during distance learning. The results of the research project show that action-oriented, tools-supported teaching, in which Maria Montessori's didactic mathematics tools were used, was helpful for mastering mathematical tasks during the time of distance learning. By working with tools, the students were able to build up inner images of mathematical content and contexts. These inner images favored the process of abstraction, and they could fall back on these ideas during the time of distance learning.

Keywords: elementary school, hands-on-activities, research project, inner images, Montessori education

1. Introduction

As in many other countries, the COVID-19 pandemic led to school closures throughout Austria. To ensure that all students were taught during this time, classes were switched to distance learning without teachers and students learning how to deal with this situation. For three school years, teachers, students and parents had to react flexibly due to the constantly changing regulations. This caused a number of problems for all of them. One of the main problems arose from the organization of school, teaching and education, which was still based on the nineteenth-century factory model [1].

Another problem arose from the technical equipment of Austrian elementary schools, although media education is included in the Austrian curriculum for elementary schools [2]. Therefore, neither the teachers nor the students had the necessary skills to cope with distance learning at the beginning.

Especially in the subject of mathematics, the implementation of distance learning was a challenge for many teachers. In mathematics, the curriculum calls for activity-based, tool-supported instruction through which children can develop an understanding of mathematical principles and relationships [3]. However, during distance learning, in many cases, there was a lack of opportunities to provide didactic materials for action-oriented instruction. This could have negative effects on the development of mathematical skills in elementary school children.

Therefore, a research study addressed whether students who could work with didactic materials in an action-oriented way in mathematics classes before the school closures could benefit from the acquired skills during distance learning.

After an examination of the theoretical background, the results of this research are presented in this article.

2. Literature review

In order to consider the theoretical foundations of the research field, it is necessary to deal with mathematics education in Austrian elementary schools. This is constantly confronted with changes and new requirements. For example, the implementation of educational standards and competence orientation took place in 2012. Starting in the school year 2023/24, there will be new curricula that are based on the current state of research and are designed to focus on competence orientation and action orientation [3]. The current state of research and the events resulting from the necessary school closures during the COVID-19 pandemic will be discussed in detail subsequently.

2.1 Mathematics instruction in elementary school in Austria

International comparative tests [4] have shown for years that the mathematical competencies of Austrian students do not match the level of students in other countries. The question, therefore, arises as to what mathematics instruction students need in order to improve their competencies in mathematics. Research like that of Krajewski [5] emphasizes the importance of mathematical precursor skills already at preschool age and of basic mathematical competencies in the first years of schooling. For example, Krajewski [5] was able to show a connection between children's understanding of number concepts in preschool and performance in mathematics as a subject several years later. In order to be able to build up sustainable basic ideas in all mathematical areas, but especially an idea of numbers and quantities, children need many experiences with didactic materials. For Kuntze [6], it is of great importance that the children actively deal with these materials and can act with them. These materials, in turn, can be mapped in the next step, translated into a pictorial representation, and later into a symbolic representation. The connection of action, illustrations, linguistic expressions and symbolic representations enables the development of sustainable ideas and insight into mathematical laws [7, 8]. It is essential that the children develop inner conceptions of the actions through the active, acting engagement with didactic materials, which are essential for further abstractions. For Lorenz [9], mathematical learning processes are based on these actions and mathematical thought structures are precisely these internalized actions. In order to successfully build up these mental images, it is necessary to select suitable materials. As an example of didactic materials for the development of a sustainable understanding of numbers and place values, decimally structured materials can be mentioned, such as those that Montessori made available to children as early as preschool age [10].

The development and training of the child's senses was of great importance to Montessori because, with them, the child grasps its environment and learns to find its way in it. In the course of his development, the child is confronted with an infinite number of sensory impressions and objects. However, the properties of these objects are limited, which makes it necessary for the child to learn to distinguish and order these properties. For Montessori, there is a connection between this ordering of properties and the development of a mathematical understanding. Montessori refers to this need for structure and order as the "mathematical mind" [10]. She developed didactic materials to best support and encourage children's mathematical development and education. Central to these materials is that the children are given the opportunity to actively engage with mathematical content, which can lead to a deeper understanding of mathematical content.

The materials developed by Montessori cover all mathematical areas of arithmetic and geometry. They are designed in such a way that the level of abstraction increases, allowing children to eventually succeed in detaching from the material [4, 10, 11].

Current research shows that students who have the opportunity to explore mathematics with materials designed by Montessori in the mathematics classroom show positive skill development [4, 12–14]. In a longitudinal study, Apfler [4] investigated whether there is a difference in the level of competence in mathematics when students in public elementary schools are taught by teachers with or without Montessori training. At the end of the third school year, those students taught by teachers with Montessori training showed significantly higher levels of competence in the competence areas of working with numbers and communicating. Whether there is a relationship between the higher level of competence and Montessori education cannot be clearly demonstrated with this study. Nevertheless, it was found that the teachers with Montessori training used a variety of Montessori materials, especially for building a viable understanding of numbers in small number spaces and for an understanding of the decadal system.

As an example of a material for building a sustainable understanding of numbers and operations, the Golden Bead material can be mentioned. In this material, which represents the decadic system, the ones is represented as a bead. The ten is represented as a bar with ten individual beads joined together. The number 100 is represented by ten tens bars joined together. The thousand is represented by a cube consisting of ten plates of hundreds. The beads can also be replaced by small cubes, as shown in the following **Figure 1**.

In the active confrontation, the students place quantities and numbers with this bead material and thereby recognize the decadic structure of the number system under the guidance of the teacher (see **Figure 2**).

However, not only numbers and quantities but also arithmetic operations can be placed and understood with this material. **Figure 3** shows how students use the bead material to place a division. Here, the total quantity of 642 is divided into two figures. The simultaneous notation provides insights into the algorithm of written division.

Due to the freedom of methods that the legal framework in Austria allows, teachers who have completed Montessori training can incorporate elements of Montessori education into their teaching. Studies show that in public elementary schools, the concept of Montessori education is not implemented in its entirety, but individual elements are used. The materials in mathematics are often used by Montessori teachers to enable the children to actively engage with the materials. Comparing the findings of current research in the field of mathematics didactics, it can be seen that the methodological approach of Montessori education is consistently confirmed [4, 15].

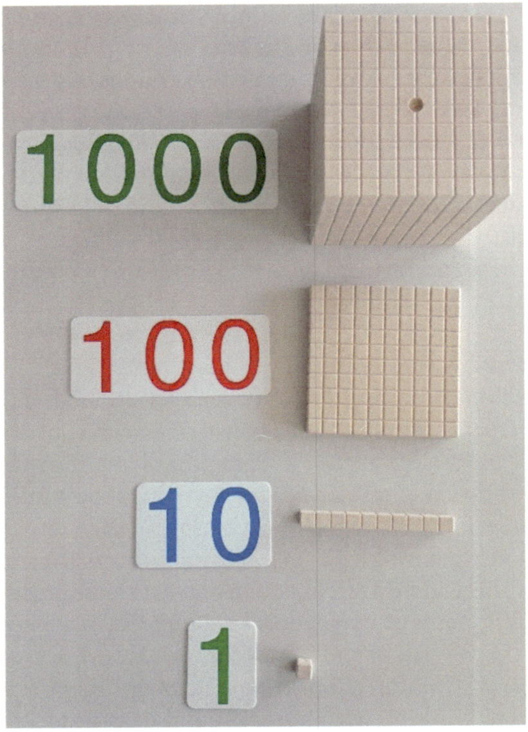

Figure 1.
Material for building the decadic number system.

Figure 2.
Number sets with the decadic material.

Of course, digital media do not yet play a role in Montessori's writings. Nevertheless, it must be questioned whether the use of digital media can also be justified in Montessori classes in the 21st century. Thus, Burow [16] also pleads to supplement the prepared learning environment, which in Montessori classes consists of analog materials, with digital offerings and virtual learning spaces, but not to replace them. The importance of action orientation, active doing, and acting engagement with the environment is thereby maintained, but expanded.

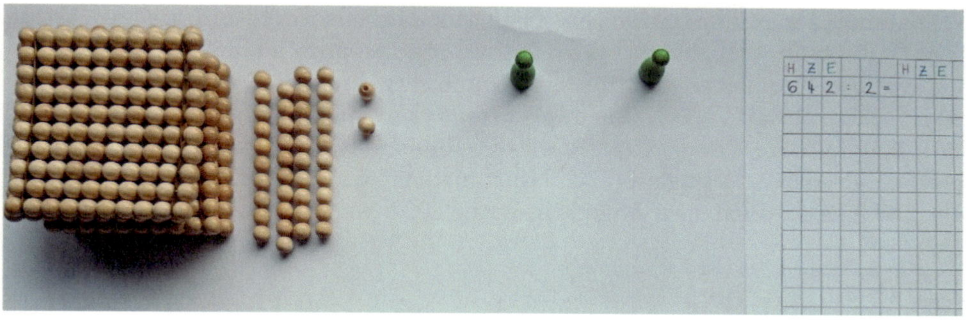

Figure 3.
Written division with the decadic material.

2.2 Distance learning in Austria

As in many other European countries, schools in Austria were closed in March 2020 due to the COVID-19 pandemic. Teachers only had the weekend to change lessons to distance learning. The Ministry of Education gave instructions not to develop any new subject matter during the first weeks of the school closures but only to practice, repeat and consolidate already known content [17]. This situation, unique in Austrian history [18], was an attempt to fulfill the educational mission of schools even during the pandemic [19]. In the course of the next almost 3 years, schools were permanently confronted with new demands and framework conditions and were challenged to always adapt to the circumstances in a new and flexible way.

Schools were repeatedly closed for several weeks in 2020 and 2021, and students were cared for at home by parents. Teachers in the elementary schools primarily prepared workbooks in print format, which parents picked up from school, worked on with the children at home, and brought back to school at the end of the week. Teachers corrected students' work over the weekend and prepared the papers for the next week. Especially in elementary school, few digital offerings were used [17, 20].

Teachers played a special role in the design of distance learning. As has been confirmed in studies on face-to-face learning [21], teachers also had the most effective influence on student learning in distance learning [22]. Teachers were particularly challenged by the ever-changing environment. They had to completely change how they had been teaching, which meant significantly more time spent on school-related work [17, 23]. Despite the high workload, which resulted mainly from the lack of experience with distance learning, elementary school teachers showed a high level of commitment [24]. In addition to the high workload, teachers perceived the feeling of permanent accessibility and the dissociation of professional and private life as stressful. How teachers coped with the transition to distance learning was a function of their digital competencies and previous teaching styles. A large proportion of teachers tried to take their face-to-face teaching styles and map them digitally during distance learning. This was especially difficult for those teachers who typically work in a reform pedagogical, holistic, and student-centered manner [23].

The impact of school closures on children and young people has been enormous, but the long-term effects are not yet even foreseeable [25]. In addition to social and psychological symptoms among the children and adolescents, negative effects on learning development were evident early on. Particularly, educationally disadvantaged children achieved hardly any learning gains during distance learning. In the

Netherlands, for example, which has a high standard of schools compared to Austria, "The findings imply that students made little or no progress while learning from home, and suggest losses even greater in countries with weaker infrastructure or longer school closures" [26]. There are currently few studies for Austria, but they show that there are scissor effects in the development of basic cognitive skills and that many children have large gaps [27]. Unfortunately, the results of the IKM Plus study, which will be conducted nationwide in spring 2022, are not yet available.

3. Research project

The research findings presented in the theoretical part of the article show that there is already a lot of research on the conversion of teaching to distance learning and its effects on children and adolescents. It shows that some children have come through the pandemic well, while others have major deficits [26–28]. In mathematics, it is shown that students benefit from activity-based, material-supported instruction and especially from activity-based engagement with Montessori materials to achieve high levels of competence in mathematics [4, 12, 13, 29, 30]. When teachers have completed Montessori training, they implement elements of the concept in their teaching and use Montessori didactic materials, especially in mathematics [4, 31, 32].

During distance learning, the teaching methods had to be changed and adapted; the lessons took place at home and the students could not work with the materials in a proven way. Research in the field of mathematics is scarce for the German-speaking countries. The question arises whether students who were already able to build up mathematical competencies during face-to-face teaching through active engagement with Montessori materials could benefit from this during distance learning [11].

The research desideratum gives rise to the following research question for the study presented here: What features of action-based, materials-supported mathematics instruction do Montessori teachers judge to be conducive to developing mathematical competencies during distance learning in times of the COVID-19 pandemic? [11].

The purpose of this study is to find out how adaptable Montessori's educational concept is to the needs of children in the 21st century, especially in the area of mathematics, and whether recommendations for school and classroom development can be derived from it.

Within the framework of the study, a second question concerning the self-organization of learning processes was also addressed. However, this cannot be addressed in this article.

3.1 Method

A qualitative research design is best suited to answer the research question in order to ensure the subject-relatedness and subjective assessments of teachers on this issue. An expert survey was chosen as the research method, which was conducted in the form of guided interviews. A guideline enables a partial standardization and comparability of the individual interviews [33, 34]. The quality criteria of qualitative research, according to Mayring [34], were observed in the presented study.

The study was conducted in spring 2022, i.e., after the end of the distance learning phases. [11].

3.1.1 Study participants

To answer the research question, an attempt was made to select a representative sample. According to Hug and Poscheschnik [33], a sample is representative if it reflects the population in terms of essential characteristics. This resulted in several characteristics related to the sample.

- The selected teachers had completed Montessori training.
- The selected teachers implement individual principles of Montessori's concept in their teaching.
- The selected teachers were leading an elementary school class or elementary school from September 2019 to the time of the study.
- The selected teachers taught at different Austrian elementary schools in different regions of the country.
- The selected teachers were willing to participate in the study [11].

Due to the limitations of the research field based on the aforementioned criteria, nine female teachers from four Austrian provinces finally participated in the study.

What is interesting about the sample is that one teacher manages a public school and leads a class at the same time. A second teacher does not teach herself but manages a school with a total of 12 classes. In this school, half of the classes are taught by teachers with Montessori training, while the other half of the teachers have not had Montessori training [11].

Table 1 shows an overview of the participants and whether they were teaching an elementary class or managing a school and in which Austrian region they were working:

Due to the selection criteria, especially based on their training (elementary school teaching degree, Montessori training) of the teachers participating in the study, they can all be called experts. Thus, they can be recognized as having expert status in their field of activity [33, 35].

	Region	Teacher	Head of school
1.	Vorarlberg	X	
2.	Lower Austria	X	
3.	Styria	X	
4.	Lower Austria	X	
5.	Lower Austria	X	
6.	Vienna	X	
7.	Lower Austria	X	X
8.	Vienna		X
9.	Vorarlberg	X	

Table 1.
Overview of the participants.

3.1.2 Survey tool

The interviews with the experts were conducted as guided interviews. This method restricted maximum openness, allowing for all possibilities and expressions. The guide was designed to still allow for openness, but still be as structured as possible [35]. The guide consists of exploratory questions, guiding questions relevant to answering the research question, and the opportunity to ask ad hoc questions if the course of the interview necessitates it [34].

The questions about the subject area of mathematics arose from the research area presented [4, 12, 14, 26, 27, 30]. Teachers were asked about what materials they used prior to the conversion of instruction to distance learning and what role action orientation played in their mathematics instruction. Another question was related to the implementation of mathematics instruction during the distance learning phase. Finally, teachers were asked how they experienced the development of their students' mathematical performance and what conclusions they could draw from it [11].

3.1.3 Evaluation of the data

The interviews were conducted as video conferences via Zoom, recorded, and subsequently transcribed. The transcripts were subjected to a comprehensive data analysis and evaluated in a rule-based and theory-based manner. The qualitative content analysis according to Mayring [34] was applied. In order to be able to compare the statements of the interviews with each other, categories were deductively formed, which resulted from the questions [11].

3.2 Results of research

In order to trace the development of teaching from face-to-face teaching before the school closures through the different phases of distance learning, the teachers described their teaching, specifically including their mathematics teaching [11].

Prior to the school closures, all teachers interviewed facilitated open instruction, although to varying degrees. In no class, frontal teaching was mentioned as the primary method of instruction. All students of the interviewed teachers had many opportunities during open lessons to work on, practice, repeat, and consolidate mathematical lesson contents in an activity-oriented way with Montessori materials, but also with other didactic materials. Although the teaching of the interviewed teachers was very heterogeneous, child-centered, action-oriented, open teaching was important to all teachers [11].

3.2.1 Lesson organization during distance learning

During distance learning, all teachers provided workpacks of notebooks, books, and worksheets for the children, which parents picked up from school on Monday and returned on Friday for correction. The tasks contained in them served to repeat and practice already-known lesson content, especially in mathematics [11].

Over time, some teachers changed distance learning and included videoconferencing, explainer videos, or other digital offerings, and increased contact with children. The personal accompaniment of the children by the teachers was important to them, as the answer of one participant 8 shows: "We did not want to leave the children alone, we wanted to accompany them" [11]. The teachers showed great commitment in designing distance learning in general and tried out different methods and possibilities over time. Here, too, a very heterogeneous picture emerges, since due to the lack of guidelines from the Ministry, the teachers tried to design distance learning according to their own possibilities and the needs of the children [11].

3.2.2 Teaching mathematics during distance learning

A particular challenge for the teachers interviewed was teaching mathematics in distance learning. The greatest difficulty for them was the lack of material, which they used in face-to-face lessons, especially for the activity-oriented development of new content, and which the children did not have access to in distance learning. The teachers tried to deal with this problem in different ways [11].

One teacher provided Montessori materials for the students to use at home. The children could pick these up at school and share them. However, she found that the students could not handle them at home without an introduction by the teacher and thus did not benefit from the materials. In order to still be able to teach new learning content with the support of materials, she provided illustrations of materials and worked more with finger pictures. According to this teacher, the children not only lacked haptic experiences with the didactic materials, they were also unable to build up internal images. When the students were allowed to return to school after the distance learning phase, the teacher repeated the learning content again with the materials [11].

Another teacher also gave Montessori materials to take home but had a good experience with them. She combined the work with the plans in print format with video conferences in which a lot was discussed about mathematics, mathematical problems and possible solutions [11].

Another teacher made her own materials for the children to work on fractions in 4th grade and sent them home with the work packets. In this way, the students were able to use these materials to work on fractions and fractional arithmetic. She postponed topics that could not be worked on in an activity-based manner at home until after the distance learning phase and thus to the face-to-face classroom [11].

One teacher offered video conferences on mathematical topics. In these, she worked with a document camera and showed the work with the materials. The students were not able to work with the materials themselves, but they got an idea of how the material could be used. This teacher explicitly referred to the fact that the active work with the materials before the school closures was very valuable for the students. Students had developed internal images of the materials and what they could do with them that they could draw on during distance learning [11].

Other teachers emphasized the importance of inner images about actions with materials. During the videoconferences, the children referred to these actions that they performed in face-to-face lessons and were thus able to build a deeper understanding of mathematical actions. This was especially evident for those students who were not affected by school closures until the third or fourth grade and who were able to do a great deal of work with mathematical materials in their first years of school [11].

Several teachers reported that they worked on as little new content as possible during the distance learning period. They planned very carefully for what content could be worked on later during face-to-face periods when students had access to the materials again [11].

4. Conclusions

The study presented in this article explored the question of which features of activity-based, materials-supported mathematics instruction Montessori teachers judge as conducive to developing mathematical competencies during distance learning in times of the COVID-19 pandemic [11].

In summary, it can be seen that all Montessori teachers interviewed placed great emphasis on the use of didactic materials, especially Montessori materials, during face-to-face instruction. This led the teachers to find the transition to distance learning in mathematics particularly challenging. However, it was found that students who were able to learn and experience mathematics in an activity-based and material-supported way before the school closures were able to develop inner images of materials and the actions with them. They were able to use these internal images during distance learning to describe mathematical processes and communicate about mathematics. These results connect to previous research findings by Apfler [11] showing that an action-based engagement with didactic materials, especially Montessori materials in mathematics enable high levels of competence in the areas of working with numbers and communicating. Likewise, the connection with the current subject didactic discussion in mathematics, which thematize the importance of action orientation with suitable didactic materials for the construction of inner images as well as the detachment from the material [6, 8, 9, 36, 37].

There are currently no research findings on long-term consequences in the mathematical development of students in Austria.

The teachers emphasize the importance of Maria Montessori's concept also for children in the 21st century and underline the necessity to raise the status of Montessori education in Austria [11].

The limitations of the research presented arise from the small sample that resulted from the criteria. Due to the small number of study participants, no generally valid statements can be made, but only the results of the sample can be described. Furthermore, it should be critically noted that no well-founded statements can be made about the actual mathematical competencies of the students. The results are merely subjective assessments by the teachers. Standardized test procedures with pre- and posttest would have been necessary here, but this was not possible due to the unpredictability of the events [11].

Nevertheless, an interesting picture of the teachers' assessments emerges and the research question can thus be answered as follows: "The present study provides evidence that action-oriented, material-supported mathematics instruction favors the development of internal conceptual images and that these could be drawn upon during distance learning. This was helpful for the processing of mathematical tasks in distance learning as well as for the further development of the mathematical competencies of the students." [11].

These results provide relevant information for the training and further education of teachers, because the new curriculum for elementary school, which will come into force in 2023, also focuses on the active examination of didactic materials in mathematics [3]. It will be necessary to address this issue even more in initial and in-service training.

Author details

Sabine Apfler
University Teacher College Lower Austria, Baden, Austria

*Address all correspondence to: s.apfler@ph-noe.ac.at

References

[1] Sliwka A, Klopsch B. Disruptive Innovation! In: Fickermann D, Edelstein B, editors. Langsam vermisse ich die Schule. Münster, New York: Waxmann; 2020, 2020. pp. 216-229. DOI: 10.31244/9783830992318.14

[2] Bundesministerium für Bildung, Wissenschaft und Forschung. Lehrplan der Volksschule [Internet]. 2012. Available from: https://www.bmbwf.gv.at/dam/jcr:b89e56f6-7e9d-466d-9747-fa739d2d15e8/lp_vs_gesamt_14055.pdf [Accessed: July 08, 2023]

[3] Bundesministerium für Bildung, Wissenschaft und Forschung. Lehrplan der Volksschule, Pub. L. No. BGBI. II Nr. 1/2023, 130 (2023). [Internet]. 2023. Available from: https://www.ris.bka.gv.at/Dokumente/BgblAuth/BGBLA_2023_II_1/Anlagen_0001_CE7F0AA2_A925_4A4D_8C3C_355D12BD22D1.html [Accessed: July 08, 2023]

[4] Apfler S. Erwerb mathematischer Kompetenzen in der Regelvolksschule. Eine empirische Vergleichsstudie zu den Ergebnissen der „Informellen Kompetenzmessung" zwischen österreichischen Schülerinnen und Schülern mit und ohne Unterstützung der Montessori-Pädagogik [Internet]. 2018. Available from: https://opus4.kobv.de/opus4-uni-passau/files/559/apfler_sabine_Kompetenzen.pdf [Accessed: July 08, 2023]

[5] Schneider W, Küspert P, Krajewski K. Die Entwicklung mathematischer Kompetenzen. Paderborn: Schöningh; 2013

[6] Kuntze S. Vielfältige Darstellungen nutzen im Mathematikunterricht. In: Sprenger J, Wagner A, Zimmermann M, editors. Mathematik lernen, darstellen, deuten, verstehen. Didaktische Sichtweisen vom Kindergarten bis zur Hochschule. Wiesbaden: Springer Spektrum; 2013. pp. 17-33. DOI: 10.1007/978-3-658-01038-6

[7] Aebli H. Zwölf Grundformen des Lehrens. 6th ed. Stuttgart: Klett-Cotta; 1991

[8] Söbbeke E, Steenpaß A. Mathematische Deutungsprozesse zu Anschauungsmitteln unterstützen. In: Böttinger C, Bräuning K, Nührenbörger M, Schwarzkopf R, Söbbeke E, editors. Mathematik im Denken der Kinder. Anregungen zur mathematikdidaktischen Reflexion. Vol. 2010. Seelze: Kallmeyer; 2010. pp. 216-244

[9] Lorenz JH. Kinder begreifen Mathematik. In: Frühe mathematische Bildung und Förderung. 2nd ed. Stuttgart: Kohlhammer; 2016

[10] Apfler S, Summer A. Der mathematische Geist nach Maria Montessori. In: Hammerer F, Herbst T, Weinhäupl W, editors. Montessori-. Das Kind im MittelpunktPädagogik. Wien: Jugend & Volk; 2020. pp. 42-47

[11] Apfler S. Montessori-orientierter Unterricht und Distance Learning in der Volksschule. [Internet]. 2023. Available from: https://www.ph-noe.ac.at/fileadmin/root_phnoe/MitarbeiterInnen/Sabine_Apfler/Dissertation_APFLER_SABINE.pdf [Accessed: July 08, 2023]

[12] Summer A. Montessori versus "Regelschule." Evaluation der Mathematikleistungen von Schülerinnen und Schülern vierter Klassen in Montessorieinrichtungen und Regelvolksschulen. [Internet]. 2007.

Available from: https://www.imst.ac.at/imst-wiki/images/4/47/Dissertation_Summer_2007.pdf [Accessed: July 08, 2023]

[13] Luckmann M. Mathematisches Repertoire von Volksschulkindern in der dritten Schulstufe—Montessori- und Regelschulklassen im Vergleich. In: Eckert E, Hammerer F, editors. Untersuchungen und Ansätze zur Weiterentwicklung der Montessori-Pädagogik in Österreich. Innsbruck: StudienVerlag; 2009. pp. 96-124

[14] Hanewinkel N. Handlung sorientiertes Lernen mit dem Bruchrechenmaterial Maria Montessoris. In: Eine Analyse von Arbeitsweisen und mathematischen Verstehensprozessen bei Grundschulkindern. Berlin: LIT; 2007

[15] Weinhäupl W, Neuhauser M. Montessori einfach klar! Handreichung für die Arbeit mit Montessori-Materialien; Mathematik. Salzburg: Weinhäupl; 2016

[16] Burow O. Wie die digitale Revolution uns und die Schule verändert. In: Burow O, editor. Schule digital—Wie geht das? Wie die digitale Revolution uns und die Schule verändert. Weinheim, Basel: Beltz; 2019. pp. 12-60

[17] Tengler K, Schrammel N, Brandhofer G, Lernen trotz Corona. Chancen und Herausforderungen des Distance Learnings an österreichischen Schulen. Medienimpulse. 2020;**58**:2. DOI: 10.21243/mi-02-20-24

[18] Wacker A, Unger V, Rey T. Sind doch Corona-Ferien, oder nicht? In: Fickermann D, Edelstein B, editors. Langsam vermisse ich die Schule. Münster, New York: Waxmann; 2020. pp. 79-94. DOI: 10.31244/9783830992318.04

[19] Hummrich M. Schulbildung auf Distanz – "Beschulung Zuhause" in Zeiten von Corona. Stellungnahme des Vorstands der Sektion Schulpädagogik in der DGfE. [Internet]. 2020. Available from: https://www.dgfe.de/fileadmin/OrdnerRedakteure/Sektionen/Sek05_SchPaed/2020_Stellungnahme_Schulpaedagogik_Schule_Zuhause.pdf [Accessed: July 08, 2023]

[20] Janschitz G, Zehetner E, Fernandez K. Digitalisierung mit der Brechstange? Auswirkungen der Pandemie auf Digitalisierungsprozesse und digitale Ungleichheiten an österreichischen Schulen. 2022;**12**(2):387-406. DOI: 10.1007/s35834-022-00347-5

[21] Hattie J, Beywl W, Zierer K. Lernen sichtbar machen. In: Überarbeitete deutschsprachige Ausgabe von "Visible Learning". Hohengehren: Schneider; 2014

[22] Zierer K. Herausforderung Homeschooling—Theoretische Grundlagen und empirische Ergebnisse zum Fernunterricht. 2nd ed. Hohengehren: Schneider; 2021

[23] Holtgrewe U, Lindorfer M, Siller C, Vana I, Salamon N. Plötzlich bin ich in der Situation eines Youtubers—Lehrer_innen und ihre Erfahrungen mit Schüler_innen im Lockdown [Internet]. 2021. Available from: https://zenodo.org/record/4550640#.Y8kQKHbMLIU [Accessed: July 08, 2023]. 10.5281/zenodo.4550640

[24] Hoffmann I. Die Corona-Pandemie als Katalysator für Schulreformen? In: Fickermann D, Edelstein B, editors. Langsam vermisse ich die Schule. Münster, New York: Waxmann; 2020. pp. 95-101. DOI: 10.31244/9783830992318.05

[25] Irion T, Zylka J. Wie Grundschulkinder in Zeiten von Corona lernen können. Ein Interview über Mediennutzung

und Lernen an Grundschulen. Lehren und Lernen. Zeitschrift für Schule und Innovation aus Baden-Württemberg. 2020;**4**:11-13

[26] Engzell P, Frey A, Verhagen M Learning loss due to school closures during the COVID-19 pandemic [Internet]. 2021. Available from: https://osf.io/preprints/socarxiv/ve4z7/download [Accessed: July 08, 2023]

[27] Rollett W, Leitgeb T, Scharenberg K. Die COVID-19-Pandemie im Schuljahr 2020/21: Wie haben sich die kognitiven Grundfähigkeiten von Schulkindern der Primarstufe entwickelt? Zeitschrift für Bildungsforschung. 2022;**12**(2):353-368. DOI: 10.1007/s35834-022-00358-2

[28] Schreiner C, Kraler C, Berger F, Hagleitner W, Jesacher-Rößler L, Roßnagl S. Zunahme von Bildungsungerechtigkeit durch die COVID-19-Pandemie beim Übergang in die Sekundarstufe II: Leistungsschwache und wenig motivierte Schüler*innen als besonders vulnerable Gruppe. Zeitschrift für Bildungsforschung. 2022;**12**(2):369-386. DOI: 10.1007/s35834-022-00352-8

[29] Lillard A, Else-Quest N. Evaluating Montessori Education. [Internet]. 313: 1893-1894. 2006. Available from: https://www.montessori-science.org/Science_Evaluating_Montessori_Education_Lillard_.pdf [Accessed: July 08, 2023]

[30] Suffenplan W. Die Lernstandsergebnisse von VERA 2004 bei Montessori-Schulen und Montessori-Schulzweigen Nordrhein-Westfalens. Montessori. 2006;**44**(1/2):18-60

[31] Tuerr B. Die Umsetzung von Prinzipien der Montessori-Pädaogogik an österreichischen Volksschulen unter besonderer Berücksichtigung der Freiheit des Kindes. In: Eckert E, Hammerer F, editors. Untersuchungen und Ansätze zur Weiterentwicklung der Montessori-Pädagogik in Österreich. Innsbruck: Studien Verlag; 2009. pp. 55-74

[32] Zach R. Institutionelle und bildungstheoretische Bedingungen von Montessori-Volksschulklassen. In: Eckert E, Hammerer F, editors. Untersuchungen und Ansätze zur Weiterentwicklung der Montessori-Pädagogik in Österreich. Innsbruck: Studien Verlag; 2009. pp. 75-95

[33] Hug T, Poscheschnik G. Empirisch forschen. 3rd ed. München: UVK; 2020. DOI: 10.36198/9783838553030

[34] Mayring P. Qualitative Inhaltsanalyse. 13th ed. Weinheim, Basel: Beltz; 2022

[35] Helfferich C. Leitfaden- und Experteninterviews. In: Baur N, Blasius J, editors. Handbuch Methoden der empirischen Sozialforschung. 2nd ed. Wiesbaden: Springer; 2019. pp. 669-686. DOI: 10.1007/978-3-658-21308-4

[36] Schipper W, Wartha S. Diagnostik und Förderung von Kindern mit besonderen Schwierigkeiten beim Rechnenlernen. In: Fritz A, Schmidt S, Ricken G, editors. Handbuch Rechenschwäche. Lernwege, Schwierigkeiten und Hilfen bei Dyskalkulie. 3rd ed. Weinheim: Beltz; 2017. pp. 418-435

[37] Schulz A. Erfolgreich rechnen lernen. Prävention von Schwierigkeiten – Diagnose – Förderung [Internet]. 2020. Available from: https://bildungsserver.berlin-brandenburg.de/fileadmin/bbb/unterricht/faecher/naturwissenschaften/mathematik/Materialien_zur_Diagnose_und_Foerderung_im_Mathematikunterricht/Erfolgreich_rechnen_lernen_WEB_2019_12_20.pdf [Accessed: July 08, 2023]

Chapter 16

Innovative Methods in Computer Science Education

Kinga Pusztai

Abstract

With the appearance of rapid technological changes, the culture of receiving and communicating information, as well as attention, has also changed. Some of the pedagogical methods that were well-proven a few decades ago have now become obsolete. It is important to replace them with new methods that are related to the views of the Z and alpha generations. This chapter deals with the innovation of the ELTE IK Computer Science B.Sc. subject. My innovations are related to the method of digital storytelling and gamification. Besides a brief description of the methods, I will present the changes implemented in my courses, such as making videos or the use of edutainment applications. My innovations have been studied over five semesters, researching students' feedback on the changes and modifying some elements based on students' feedback. Furthermore, I was also curious about the changes' impact on students' results. From my research, it is clear that the introduced tools motivated the students to spend more time learning the subject, leading to deeper mastery.

Keywords: digital storytelling, making video, multimedia, gamification, edutainment, computational thinking

1. Introduction

Based on their relationship with the information society, we distinguish between generations X, Y, Z, and Alpha [1]. Members of Generation X, today's 36–50-year olds, are referred to as "digital immigrants." They are the ones who have come close to the Internet in their adulthood. Members of the Y generation were born in the 1980s and 1990s and are called "digital natives" because they were exposed to the Internet as young children. The members of Generation Z are mostly high school students, but their first representatives have already started their university studies. They are the "Facebook generation," who do not even know what life is like without the Internet anymore, their primary communication interface is no longer e-mail, but social networking. Lower school and younger children make up the Alpha generation. The use of personal communication devices is a matter of course for the members of the Alpha and Z generations, they wake up and go to bed with smartphones, they are always available and constantly connected to each other online. They can easily manage the rapid flow of information and often change their activities during multitasking. Thus, it is difficult to capture their attention with traditional, frontal tools. They prefer visual presentation over long, unstructured texts.

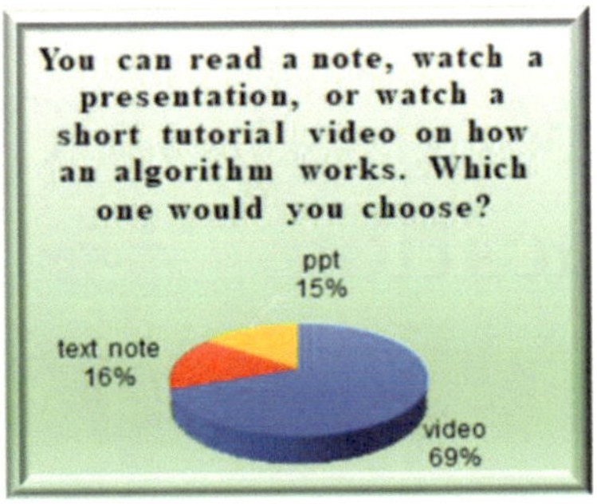

Figure 1.
Student feedback on the use of video.

Of course, there have also been many criticisms of generational theories (e.g., [2]). It is true that the differences between individuals are more diverse than the differences between generations, however, the learning habits of today's students have changed. In one of my surveys, I asked students about their learning habits using an online questionnaire. Over several semesters, I was able to interview a total of 128 students, 69% of whom preferred watching the tutorial videos, only 16% of whom chose the text notes, and 15% of whom chose the presentation (**Figure 1**).

All these results show that it would be important for us as teachers to introduce new methods in our classes that better support the communication and motivational structure of today's net generation.

A possible solution to this problem is the application of digital storytelling in education and gamification These methods also fit in with the philosophy of computational thinking, so their application supports the development of computational thinking skills.

2. Methods used in the courses

2.1 Digital storytelling

2.1.1 History of digital storytelling

The storytelling method was first complemented by visuality. In the 1970s and 1980s, several visual storytelling works were born in the United States that combined traditional, personal narration with visual elements Visual storytelling was able to reach a wider audience with its personal tone and the stories of everyday people, so it became popular.

The public could easily identify with the personal messages, "video postcards," that were introduced first in traveling theater performances, and later even on television. The stories made the audience recall their own experiences and the video postcards soon became very popular. At the same time, the method became more widely used in various fields of education [3].

With the explosive spread of digital technology, it has become easier to create personal messages and short films that were previously created only by mobilizing large technical equipment. The first workshop about "personal narrative short film" making was held in 1993 at the American Film Institute in San Francisco. The Center for Digital Storytelling (known as StoryCenter since 2015), led by Joe Lambert, was established at the same place a year later, and its members developed a digital storytelling methodology that combines classical storytelling with the multimedia possibilities of digital technology. The nonprofit organization's objectives included empowering people who are not professionals to display their life stories in a self-contained, individual character, cinematic way. They saw themselves as a movement, the focus of their workshops was on capturing individual life paths, they wanted to create a community with the participation of people in a similar situation. The StoryCenter has collaborated with nearly 100 US universities and colleges to incorporate digital storytelling into a range of courses [4].

In Europe, the method was formally introduced in 2003, when the regional service BBC Wales, based on the experiences of other digital storytelling courses held in various locations, started its own program and center, called Capture Wales, under the leadership of Daniel Meadows. At the same time, the Australian Centre for the Moving Image in Melbourne also began to work extensively with the Centre for Digital Storytelling.

BBC Wales reached a wide audience with its grand project where every night the audience could see a personal digital story related to the region. Due to the success of the show similar projects have been initiated in many educational, health development, and community areas. As a result, Wales had become the European Centre of digital storytelling.

Digital storytelling is interpreted and used differently in different countries and institutions. To present and discuss these, festivals are held annually in Wales and international conferences in different countries every few years.

2.1.2 Definitions of digital storytelling

There are many different definitions of "digital storytelling" (DST). Daniel Meadows defines digital stories as "short, personal multimedia tales told from the heart" or "multimedia sonnets from the people" [5].

Digital storytelling can be defined by Leslie Rule as "the modern expression of the ancient art of storytelling. Digital stories derive their power by weaving images, music, narrative, and voice together, thereby giving deep dimension and vivid colour to characters, situations, experiences, and insights" [6].

According to the most common definitions, DST is a new learning organization process that combines traditional storytelling with the use of digital tools [7]. The idea is that students do not use digital tools for their ends, but create unique narratives, specific multimedia creations that capture the attention and enthusiasm of their peers and generate communication on the topic within the learning community.

Some researchers (e.g., Lanszki [4] and Ohler [8]) move away from a tool-using approach to focus on the process, that is, the emphasis is not on digital but on storytelling. In their opinion, good digital storytelling is based on well-written text, and the use of technology is only for expressing content. "Medium and technology provide a supportive background for the creation of a fragmented world based on individual perception, summarized in a narrative framework, and thus based on causality" [9].

In the case of science subjects, it can be assumed that—due to the age of the students—few digital stories exploring individual life situations are produced. It is, therefore, necessary to interpret the definition of digital storytelling widely: not only individual life stories are included in the class of digital stories but also narrated audio-visual presentations created using the steps of the method. This is how DST has been interpreted by the professors at the University of Houston [10] and this is the interpretation I have used in my classes.

2.1.3 Introducing the digital storytelling method

A digital story is a 2–3-minute individual multimedia narrative, usually written in E/1, told in the student's own voice [4]. The narrator emphasizes the most important elements of the story in keeping with the tight time frame, and the logical priority of the events is increased [11]. Too long a film degrades the quality of the digital story.

The steps of the method are defined in the following five stages [12]:

1. Students first write the text that serves as the basis of the story. The ideal digital story is essential to the point and concise (Formally: approx. 250 words, which is an A/4 page).

2. In the second stage, students recite their written text, that is, create an audio file.

3. In the third stage, students design and create the visual elements needed for the text. It is important not to use more than 15–20 images.

4. The fourth stage is editing, where students use a video editing software of their choice (e.g., Microsoft Movie Maker, Sony Vegas, or online editing apps that can be used with smart devices such as WeVideo and Power Director) to create their digital story.

5. The final stage is to screen, discuss and evaluate the work.

Digital storytelling is also very popular with students because it uses activities that students already do in their peer groups: sharing pictures, videos, and stories. However, during the application of the method, these activities are supplemented by a phase of independent learning and thinking, as well as a critical and constructive interpretation of each other's work.

Using this method, learners are active participants in learning. They feel responsible for the quality of their work and therefore pay sustained attention to the content they are working on.

DST also promotes the development of internal motivation, as students create for the outcome and for the experience itself, while being pervaded with a sense of surpassing themselves and their abilities [13].

In addition to its positive effect on students [14] and performance [15], it improves students' problem-solving skills, the ability to learn independently, and the development of critical thinking.

The use of DST promotes students' active and creative participation in the teaching-learning process, in which independent and cooperative forms of.

One of the most important features of DST is that during the process, students go from the abstract to the concrete level step by step [16].

2.2 Gamification and edutainment

Nowadays, the most common definition comes from Deterding in 2011, which says: "Gamification is the use of game design elements in non-game contexts" [17]. The concept of gamification in the public mind appeared in the early 2000s and has become popular since 2010, but its origins can be traced far back.

Fuchs [18] found examples of the use of gamification in the army dating back to early Roman times.

Zichermann and Linder [19] give an example of how Napoleon used gamification in 1795 when he organized a nationwide puzzle with a reward of 12,000 francs to solve the problem of storing food supplies for war. Thanks to the game, even before Pasteur, master chef and pastry chef Nicolas Appert invented the pasteurization process, which allowed food to remain fresh for more than 4 months.

The ancestor of today's widespread loyalty programs is the Green Stamps scheme, implemented by S&H in 1896 [20], which allowed customers to collect green stamps and points in a point-collecting book. When the book was full, customers could redeem it for various interesting products.

Two very important concepts appear in the definitions of gamification: game elements and game mechanisms, which are often called game design techniques. Game elements refer to tools taken from traditional and video games and game mechanisms to the application of the operating principle.

Of course, tools will only work effectively if the mechanisms of a game are given: a game is voluntary, promising success, transparent, and properly delimited (providing proper time).

The "out-of-game context" term in the definition implies that the purpose of the game is different from that of gamification. The biggest difference between game and gamification is that a game is an activity providing entertainment or amusement whereas the purpose of gamification is to do something to achieve a predefined goal in real life.

One interesting subset of gamification is edutainment, which is "a portmanteau combining the words 'Education' and 'Entertainment'. It is delivering information and knowledge by means of entertainment for making it enjoyable" [21].

This term was first used by The Walt Disney Company in 1948 to talk about the "True Life Adventure" series [22].

Edutainment technology comes in many forms. A streaming video platform or a prepackaged learning product can be categorized as edutainment if it has both entertainment and educational value. Edutainment is very much an issue in developing modern digital and hybrid curricula for the classroom, and for supplementary educational use [23].

3. Application of the methods in my courses

"Algorithms and Data Structures" is a two-semester course that students can take in the second semester of the first year and the first semester of the second year. In both semesters, the courses consist of a two-credit lecture and a three-credit practice, which does not include a lab where they can program the learned algorithms. Students must take an exam on the lecture material and write mid-term and end-term tests on the practical material, which forms the basis of their practical grade. Many other courses are based on these courses, so it is important that students complete them on time.

Due to the theoretical nature and early appearance of the subject Algorithms and Data Structures, I found it necessary to use the possibilities of the above-mentioned innovative methods in the practical lessons. I also had the students comment on the changes with the student's anonymous questionnaires. in the form of an anonymous questionnaire.

As a first step, I introduced the point system in the semester [24], that is, it is not only the results of the tests that matter in the evaluation. Students must write two tests, each of 60 points. The compulsory minimum score for both papers is 20 points. (If someone does not reach the mandatory minimum on tests, they must retake them, that is, failed tests cannot be compensated for by other scores by any other scoring.) If the combined result of the 2 tests reaches 40 points, they can improve their practical grade by a further 20 points. I gave 7 different options for this, which are as follows:

- One of my biggest and most successful innovations was that I created a—usually playful—edutainment application for each class, with the deadline for submission at the end of the study period. If they solved the assignment completely well, they were given one point and initially 5 minutes of "opportunity." (I doubled that reward for a more difficult assignment.) The 5-minute "opportunity" can be used for class delays, missing a class in total, or to get more time on the test.

- "Opportunities" were later given for completing the quizzes correctly. Each practical lesson also included a quiz, which students had to complete with a 70%. I rewarded the solutions between 70% and 100% with "opportunities," which were measured in minutes.

- Students could also make tutorial videos, which I rewarded with a maximum of 20 points. This assignment could also be solved in groups.

- They could cut my recordings, thereby creating several shorter videos. For this, they could get a maximum of 10 points.

- Students are given homework in every class, which is not compulsory. They had to solve this task by the beginning of the next class at the latest. It was worth two points.

- They were given programming problems for some topics, which were accepted for submission until the end of the study period, and rewarded with a maximum of 10 points depending on the difficulty of the assignment.

- During the semester students were given the opportunity to carry out more serious research on a topic related to the course and to share it with us in some form for a maximum of 20 points

I also tried to change the course of the lessons, we started each lesson with a Kahoot, which was excellent for revising, activating the students, and creating a relaxed atmosphere.

3.1 Making videos

For my practical lessons, I have made available to the students supplementary materials on my website that explain or illustrate the operation of each algorithm and data structure. Some of these are YouTube videos, others are animations and visualizations from the website www.algoanim.ide.sk [25], and the most important elements are videos and presentations made by the students or by me.

When preparing the students' digital stories, I required some of the formal features of DST, but not all. For example, I did not ask for the text to be in E/1, but I drew their attention to the brevity and conciseness of the video. Furthermore, I did not accept the use of only still images, I required that all works should use animation. In terms of its form, I preferred the video, but I also accepted the presentation.

I used the method for five semesters. My goal was for the students to create a video of their own based on the shared resources—either as group work or independent creation. As a support, I provided consultation opportunities for the students, which were used by about a third of the students. The initial enthusiasm was great, several people even came to the consultation, but every semester some of the commitments were not realized. It also happened that they undertook a topic as a group, but in the end, only one student made the promised video. In total, 12 videos (8 of students and 4 of my own) and 4 presentations were completed, which can be available on my website.[1]

I graded the videos differently each semester. In general, the maximum number of points per video was 20. (That means, the maximum available extra point was obtained with this assignment.) If the students made a video together, they shared the 20 points. One semester, however, I offered only 10 points for making a video, but this proved to be insufficient. No video was made this semester, and the reason is that a demanding video takes a lot of time.

I preferred the group work method for making a video, as group work has many advantages. Initially, I did not even want to accept a video made by individual work, but due to the COVID-19 epidemic, personal classes were replaced by online classes, making collaborative work much more difficult. For the first few experimental semesters, only individual videos were made. Later, when education returned to traditional personal classes, more and more groups—usually consisting of two people—were formed. The videos created in groups were also worth 20 points, which the students divided among themselves. In addition, I asked the groups to keep a "project diary," which included who had completed each subtask. This showed that the students were excellent at the organizational assignments, generally undertaking and solving assignments in equal measure.

I was generally satisfied with the quality of the videos. Most of the videos were made with some kind of video editing software (**Figure 2a**), and the smaller part was a video version of an animated PowerPoint presentation (**Figure 2b**). I gave some aspects related to the content, which the students followed. I found a small mistake in only one video, which was later corrected.

I asked the students to make short videos of a maximum of 5 minutes. They tried to comply with this, only three videos exceeded the 5-minute limit, but they were very close to 10 minutes. Thus, the average time of the video was 4.95 minutes, the standard deviation was 3.22 and the median was 4.04. I think I will have to advertise this even more prominently in the future.

I was also curious about the students' feedbacks, which I investigated using online questionnaires.

(a) (b)

Figure 2.
Screenshot from a student video.

[1] https://people.inf.elte.hu/kinga/algoritmusok1/video.htm

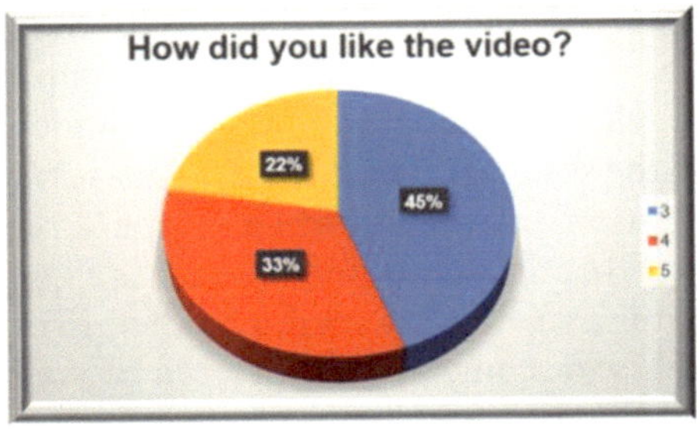

Figure 3.
Student feedback on the student videos.

In my first questionnaire, I researched their study habits. This was completed by 128 students. A five-point Likert scale was rated at an average of 4.77 that tutorial videos are available on my website, and students think it's a good idea to make a video for extra points. According to them, video production "involves the student in the learning, they feel it is important to prepare in detail, while making the video they understand perfectly how the algorithm works."

In another questionnaire, I was wondering how students rate their peers' videos. The questionnaire was completed by 22 students, with an average rating of 4.55 for all videos (**Figure 3**) and all but one student (93%) rated it as helpful to their learning.

After making the video, I asked the students about video making. At first, I was curious about their motivation. Many of them emphasized giving help over their peers, as they understood the material and found it easier to learn with the help of video. Some were inspired by the lack of easily understandable Hungarian literature for the course material. But there were those who chose this task to get extra points because they have been making videos for a long time and liked making a video.

Secondly, I was curious about the difference between design and implementation. Most of the students did not have unexpected difficulties or did not spend more time making the video because they felt free to implement it or had already planned the video in advance. However, there was a student who did not plan ahead and took more time to make the video. According to him, he had to start over several times, redoing it, because he came up with a new idea in the process, which would make the assignment more understandable or better illustrated.

Finally, I wondered whether it helped them to learn the course material, which they rated an average of four on a scale of five. They were also asked to write their personal opinions on the assignment, one of which was: "I think it is very useful because the person who does it is more immersed in the subject and someone who might be behind in the subject can understand it more easily than from literature."

This semester, I'm doing more research. It would be nice if a high-quality video could be made to all the material for the two-semester subject, but that would take a lot of time. Therefore, I thought of using the lessons I recorded during online teaching, cutting and possibly re-editing them to create short tutorial videos. These videos will not only help students to learn the material but may also be used by students to create their own videos. I think that by doing this, the time of video making will be reduced and the way it is done will possibly be simplified, which may help me to engage more students, so hopefully, more student videos will be made in a semester.

While cutting my recordings, another idea occurred to me: what if I gave this assignment to the students? Editing one lesson sounds easy, but it needs to be watched several times to do the assignment well. That way, the student who undertakes the assignment is unwittingly learning the material. Of course, the maximum score that can be obtained by doing this is only half of the score that can be obtained by making a real video.

My plan is working, the students are very happy to apply for the task, and I am sure that at least one video will be produced for each lesson of the semester. As a result of the edits, the videos are shorter and of better quality, so I think they are more useful than if I were to share the footage with the students.

I was also curious about the students' feedback, which I researched through an anonymous online questionnaire completed by 22 students. On a five-point Likert scale, the question "How much do you like the idea of make tutorial videos for each lesson by editing last year's recordings?" was rated 4.55 (with a standard deviation of 0.67 with five mode and median) 81.8% of students think that making shorter videos of the recordings, possibly more, is more useful than sharing uncut recordings, and only 18.2% of students chose uncut recordings (**Figure 4**). Students could also comment on my idea, one of them is: "It is easier to process the material in several smaller parts, videos are helpful."

3.2 Edutainment apps

During the two semesters, I created a total of 46 edutainment applications for a total of 34 topics. I used the following four frameworks to create these—usually playful—"apps": LearningApps, Kahoot, Quizlet, and HotPotatoes. These were either assigned as optional homework or were part of the class.

I have used the optional homework apps for several semesters and have modified them several times based on constructive feedback from students. This is one of the reasons why 38 apps were created for the 24 topics. Some of them are presented in Gamification in Higher Education [26], but all of them are available on my website.[2]

These tasks were characterized by high student participation, with a total of more than 4800 views of the apps, which means that on average an app had more than 141 views. (In Crops, the correct solution is less.) Student activity is also evidenced by the result that there was an optional task where 18 out of 20 students dealt with the task, which means 90% participation. Student feedback was surveyed using online anonymous questionnaires. A total of 343 students rated the apps (over five semesters) with an average rating of 4.56 (with a standard deviation of 0.56) and 96.2%

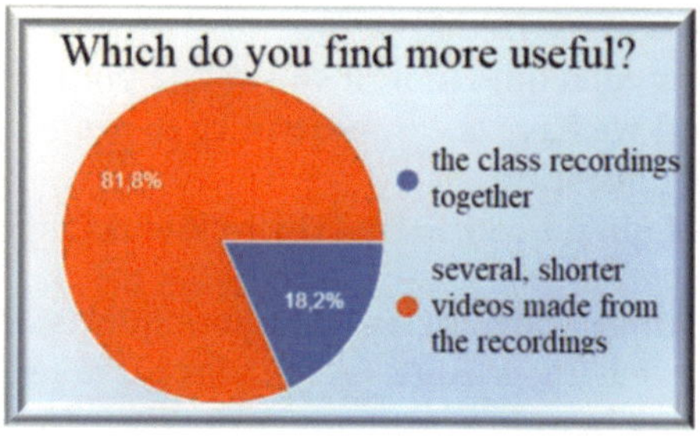

Figure 4.
Feedback from students on which type of video is more useful.

[2] https://people.inf.elte.hu/kinga/algoritmusok2/seged_kitoltott.htm
https://people.inf.elte.hu/kinga/algoritmusok1/seged_kitöltött.htm

said that the app helped them with their learning. Although there was no obligation to write a review of the apps, there were more than I expected (90). Some of these were "Useful and easy to learn with," "Very good, it made me realize that I had some shortcomings," "It summarizes what you need to know in a concise way, I find it very useful and cool," "It's imaginative and helps a lot to put into practice the knowledge you've acquired," "I found the apps very useful, they added a lot to writing successful tests," "They helped me a lot in understanding algorithms/data structures as I prefer visual learning. It is a playful way to understand the material, reducing the energy involved," "I think these assignments have really, really boosted my performance in the subject. Playful and interactive learning can move university students in the same way. Especially in the midst of such a dry online education, these easier, yet very helpful and enjoyable tasks were refreshing. It was nice to be able to do them until they were 100%, it was also positive feedback and we always had to guide ourselves to the right solutions. The extra credit was also a great motivation, but anyone who did it just for that reason must have learned along the way, wittingly or unwittingly. I think it's a very good idea and I'm grateful that we had the opportunity to do it from this subject."

In the second semester, I started the lessons with Kahoot quizzes. I find it very useful because it is easy to use, it does not take too much time, but the students really like it, it brightens up the course of the lessons, it briefly summarizes the knowledge of the previous lesson, it confronts the students with their shortcomings and after using it, the students pay more attention. In addition, it also sends me feedback on the success of the solutions to each question and the student's performances.

To use a Kahoot quiz in class, it is not enough to have a teacher's computer and a projector or a smart board, each student needs to have a phone or laptop. To start, I launch the app on the teacher's computer from Kahoot.com, which projects a game pin. Students join the game by entering the projected pin and a nickname on Kahoot.it. After that, the game starts. Each question is projected to the audience (**Figure 5a**), and they answer using their phones, where they see only the colors and shapes of the answers (**Figure 5b**). A given amount of time (usually 20 seconds) is allowed for each question. The time can be varied; I increased it to 30 seconds for several questions. There are several types of questions to choose from, the best known being simple quizzes with up to four choices, then true/false questions, type answer questions, and puzzles. Questions can have multiple correct answers. Students are awarded points for correct answers, which also take into account speed. After each question, it shows the podium positions (1–3) based on the points scored, and we have the opportunity to project the image again and discuss the answer to the question.

In compiling a Kahoot quiz, I considered the following aspects.

- First of all, I consider it important that a Kahoot quiz is not too long and does not want to repeat too much knowledge. (A quiz for me consists of 5–7 questions.) A Kahoot that is too long becomes boring takes a lot of time out of the lesson and contains more information than the students can remember, so it is not effective.

- It is important to pay attention to the time limit. If the question is too long, then more time should be allocated, but if you overestimate the question and allocate more time than necessary, then the competitive spirit is reduced, and thus the students' motivation.

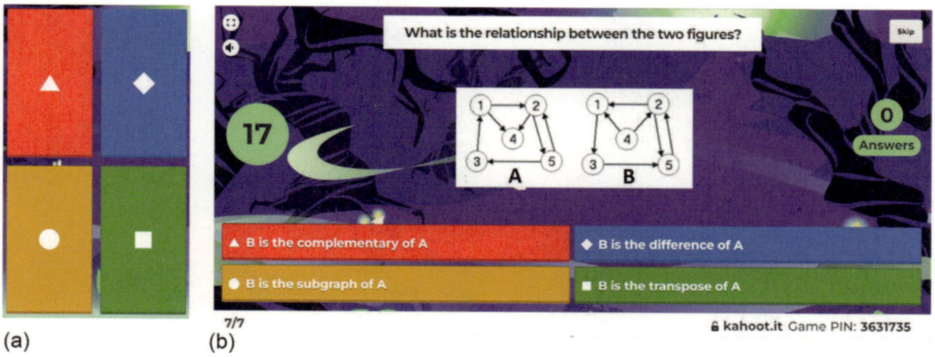

Figure 5.
(a) Kahoot question, (b) Kahoot question on student's phone.

- One of Kahoot's drawbacks is that it shows the images in a small size. An image is at its largest when it is lying flat and its aspect ratio is closest to 350:220.

I researched the students' feedback using an anonymous questionnaire, which was filled out by 21 students. All of them liked the fact that we repeat Kahoot at the beginning of the lesson, and on a five-point scale, they rated its usefulness with an average of 4.65 (standard deviation of 0.81, a mode and median of 5). Some opinions about it: "It creates a good mood for the entire lesson", "Absolutely good, playful but also useful," "Useful for interactive revision and checking existing knowledge."

3.3 Canvas quizzes

The success of edutainment applications gave the next idea on how to better engage students in online lessons. For each class of the two semesters, my colleagues and I, at my suggestion, prepared a 10-point canvas quiz that students had to complete with a minimum of 7 points. I rewarded results above 7 points with "opportunities," which allowed them to get more time on a test or miss one more class. The quizzes were to be completed as homework and the essence of which is to deepen the knowledge learned in class.

When editing the quizzes, we paid attention to the following principles:

- It should be solved in about 10 minutes.
- Include a theoretical question, but most of it should be a practical task to deepen the knowledge of the algorithms and data structures learned in class.
- The quiz should be able to be completed several times.
- The correct solutions should be visible after a certain time.

A total of 54 students filled out the questionnaire about the quizzes. The reliability index of the questionnaire is 0.88.

On a scale of one to five, the question "How much did you like the quizzes used in the practice?" question was rated 4.19 (with a standard deviation of 0.89), and the

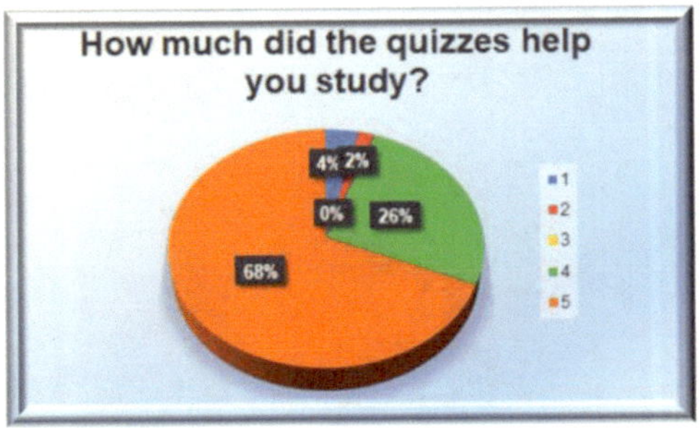

Figure 6.
Student feedback on the use of quizzes.

question "How much did the quizzes help you learn?" was rated 4.54 (with a standard deviation of 0.91) (**Figure 6**). While the most common rating for the first question was a 4, for the second question more than half of the answers were 5. Students were also allowed to comment on the quizzes, which was not compulsory, but still received more comments than expected. Some of them were "The quizzes help a lot in preparation, I think it's an excellent idea," "They are very funny and creative. I like that they playfully provide knowledge. It worked for me," "I find both the Canvas quizzes and the apps very useful. Many times, the student may think he understands the material, but during the quizzes he encounters different examples than in class and it can be easy to find out that he hasn't understood everything well," "They were very motivating." "The quizzes helped me a lot in learning, I especially liked that it was not necessary to solve it the first time, but several attempts were possible." "Sometimes you have to 'hunt' for the information to fill it in correctly, but they are absolutely good for practice. Some specific cases did not occur in practice, so it absolutely supported the understanding of the material."

4. Impact of using innovative methods

4.1 Comparison of practical grades

In the above chapters, I have presented the changes made in Algorithms and Data Structures I and II and the positive feedback of the students, but I think it is more important that the positive impact of the changes is reflected in the student's grades, so I have researched this as well.

In my research, I compared the results of my experimental groups with the results of groups whose courses did not include the innovative elements I introduced. Completing the course is a two-step process, students must first obtain a practical grade and then take an exam. Although the primary objective of the course I taught was to obtain a practical grade, I made a comparison of the results of both grades.

For the practical grade, I used my previous groups as a control group, whose students attended the course in a traditional way. During the semester, students are required to write two tests. The tests of the "Test" groups consisted of five tasks, for which they were supposed to be given 1.5 hours, but I allowed an extension of time for those who requested. The "Innovative" groups consisted of 6 tasks per test, also given 1.5 hours, but the extension of time was not automatic, but conditional. (Students can get "opportunities" during the lessons, which can also be used to

Code	Mid-term		End-term		Total score		Grades			
	Average	STD[a]	Average	STD[a]	Average	STD[a]	Average	STD[a]	Mode	Median
Results of innovative groups										
19_4	42.2	8.23	47.43	7.38	89.63	16.47	4.54	0.94	3	3
19_11	35.41	7.71	38.44	10.79	79.5	21.66	3.6	0.84	3	3
20_6	42.47	9.99	43.79	10.98	94.98	24.79	4.24	1.03	5	5
20_13	39	9.93	39.76	10.36	86.32	24.03	3.9	1.04	4	4
Average	**39.77**	**8965**	**42.355**	**98.775**	**87.6075**	**21.7375**	**4.07**	**0.9625**	**3.75**	**3.75**
Results of test groups										
15_12	29.27	10.74	24.33	8.24	56.8	20.2	2.67	1.1	3	2
15_13	24.82	9.27	21.27	6.53	47.65	22.68	2.08	1.12	1	2
16_13	25.55	12.8	32.19	5.9	58.62	21.59	2.8	1	2	3
18_8	38.52	13.23	33.81	12.53	75.73	25.88	3.38	0.95	2	3
18_11	38.37	8.85	39.55	9.68	82.78	20.38	3.73	1	3	3
Average	**31.306**	**10.978**	**30.23**	**8576**	**64.316**	**22.146**	**2932**	**1034**	**2.2**	**2.6**
Difference	8464	−2013	12.125	13.015	23.2915	−0.4085	1138	−0.0715	1.55	1.15

[a]*Standard deviation.*

Table 1.
Comparison of practical grades.

extend time.) Five tasks of the test were similar to the "Test" group, and task 6 was to write an algorithm using a learned technique. (The students found this task the most difficult.) It can therefore be said that the "Innovative" groups had to solve a more difficult test in the same amount of time compared to the "Test" groups. **Table 1** shows the results of my "Test" and "Innovative" groups. My "Innovative" group scored 8.464 (39 and 31.306) better on average on the mid-term test and 12.125 (42.355 and 30.23) better on the end-term test, for a total of 23.29 (87.6075 and 64.316) more points for the semester, for an average grade improvement of 1.55 (3.75 and 2.2). The changes in the standard deviation values of my "Innovative" groups (2.013 lower than the mid-term paper, 1.3 higher than the end-term paper, 0.4 lower for the total score, and 0.07 lower for the marks) indicate, that the students performed more consistently. The mode of grades of the "Innovative" groups varied between 3 and 5 (average 3.75), while this value of the "Test" groups ranged between 1 and 3 (average 2.2). In the "Innovative" groups, the median of the grade also ranged between 3 and 5 (average 3.75), while in the "Test" groups this value ranged between 2 and 3 (average 2.6).

The results also show that the students took advantage of the opportunities to get extra points, so they invested more work in the course during the semester.

4.2 Comparison of exam grades

I chose a different control group when examining the exam marks. I thought that my survey would better reflect reality if I compared the performance of the examined groups on the same assignments, so I took all students from the entire year as the control group. Furthermore, by choosing the control group in this way, I was able to more easily solve the problem of comparing the results of the students who took the online exam with the results of their fellow students who took the online exam, while I was able to compare the results of the students who participated in the face-to-face exam with the results of the students who participated in the face-to-face exam. This allowed me to conduct my observations over several semesters, which led to more reliable results.

Figure 7 shows the results of my "Test" and "Innovative" groups. Similar to the practical grades, the improvements were observed here as well (0.63, 0.44, 0.24, and 0.32), with an average improvement of 0.4 compared to their year-group peers.

Based on my statistical tests, it can be said that my innovations positively impacted not only the students' motivation but also their performance.

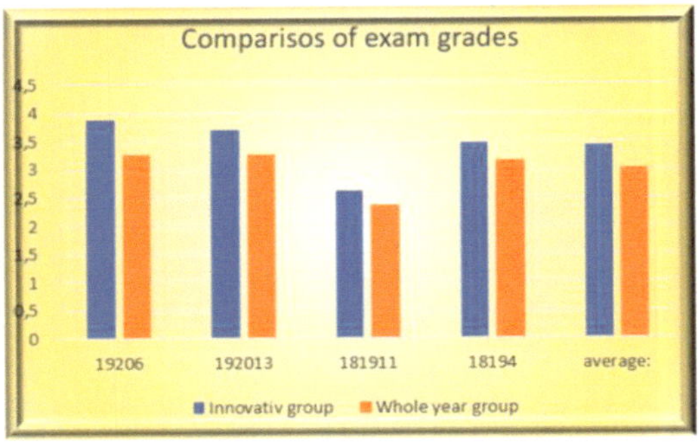

Figure 7.
Comparison of exam grades.

5. Conclusion

In a rapidly changing world, education needs to keep pace with developments. As today's growing generation is born into an online world, education needs to open up to smart devices. The application of digital storytelling or gamification in education shows us a possible path to this.

My article deals with the innovation of a specific university course and its effects. First, I briefly introduced the methods I used, then I showed how I innovated in my Algorithms and Data Structures I and II courses, and finally, I examined how my changes impacted my students. In addition to the student's feedback, I also investigated the impact on their performance.

In summary, the high participation of students in the surveys, the content of their answers, and their achievement clearly proved that the use of these computational thinking teaching activities increases the efficiency and quality of the knowledge transfer process, helps in understanding and deepening the course material, and adds color to the lesson. In all the surveys, students clearly expressed their satisfaction and became more interested and motivated.

Based on all of this, I recommend using the innovative methods presented in the article in higher education, which can be used to improve computational thinking skills even more.

Author details

Kinga Pusztai
ELTE IK, Budapest, Hungary

*Address all correspondence to: kinga@inf.elte.hu

References

[1] Nagyés Á, Kölcsey A. Mit takar az alfa-generáció? Metszetek. 2017;**6**(3):20-30

[2] J. Ollé, A spekulatív jellegű digitális nemzedékelméletek kritikai áttekintése, 2014. Available from: https://www.slideshare.net/ollejanos/a-spekulativ-jellegu-digitalis-nemzedekelmeletek-kritikai-attekintese [Accessed: October 10, 2021]

[3] E. Anthropolis, A digitális történetmesélés története, 2012. Available from: https://digitalistortenetmeseles.hu/tortenete/ [Accessed: April 3, 2022]

[4] Lanszki A. Digitális történetmesélés és tanulói tartalom(re)konstrukció. Új Pedagógiai Szemle. 2016;**66**(3-4):82-88

[5] EduTech Wiki, Digital storytelling, 2019. Available from: http://edutechwiki.unige.ch/en/Digital_storytelling. [Accessed: April 11, 2022]

[6] Sadik A. Digital storytelling: a meaningful technology-integrated approach for engaged student learning. Educational Technology Research and Development. 2008;**56**(4):487-506

[7] Michalski P, Hodgesés D, Banister S. Digital storytelling in the middle childhood, special education classroom: A teacher's story of adaptations. Teaching Exceptional Children Plus. 2015;**8**(4):3

[8] Ohler J. Digital Storytelling in the Classroom, New Media Pathways to Literacy, Learning and Creativity. Thousand Oaks, CA: Corwin Press; 2013

[9] A. Lanszki, A tanulói aktivitás szerepe a digitális történetmesélésben, in Interaktív oktatásinformatika, Lévai Papp-Danka, Szerk., Budapest: ELTE Eötvös Kiadó, 2015, pp. 79-91

[10] University of Houston, What is digital storytelling?, 2021. [Online]. Available from: https://digitalstorytelling.coe.uh.edu/page.cfm?id=27&cid=27&sublinkid=29 [Accessed: October 26, 2021]

[11] Meadows D. Digital storytelling: Research-based practice in new media. Visual Communication. 2003;**2**(2):189-193

[12] H. C. Barrett, How to create simple digital stories, 2009. [Online]. Available from: http://electronicportfolios.com/digistory/howto.html [Accessed: April 02, 2021]

[13] A. Lanszki, A digitális történetmesélés mint komplex tanulásszervezési eljárás, in Digitális történetmesélés a nevelési-oktatási folyamatban, A. Lanszki, Szerk., Eger, Líceum Kiadó, 2017, pp. 22-43

[14] Abdolmanafi-Rokniés SJ, Qarajeh M. Digital storytelling in EFL classrooms: The effect on the oral performance. International Journal of Language and Linguistics. 2014;**4**:252-257

[15] Smeda N, Dakichés E, Sharda N. The effectiveness of digital storytelling in the classrooms: A comprehensive study. Smart Learning Environments. 2014;**1**(6):21

[16] A. Lanszki, A digitális történetmesélés relevanciája a tanárképzésben, in Felkészülés a pályára, felkszülés az életre, F. Iván, Szerk., Eger, EKF Líceum Kiadó, 2015, p. 89–97

[17] S. Deterding, M. Sicart, L. Nacke, K. O'Haraés, D. Dixon, Gamification: Using

game-design elements in non-gaming contexts, Proceedings of the 2011 Annual Conference Extended Abstracts on Human Factors in Computing Systems, pp. 2425-2428, 2011

[18] Németh T. English Knight: Gamifying the EFL Classroom. Piliscsaba: Pázmány Péter Katolikus Egyetem Bölcsészet- és Társadalomtudományi Kar; 2015

[19] Zichermannés G, Linder J. The Gamification Revolution: How Leaders Leverage Game Mechanics to Crush the Competition. New York: McGraw-Hill; 2013

[20] Nagy JR. A gamification megjelenése a magyar középiskolákban. Budapest: BMGE; 2017

[21] Sidrah S. This is how edutainment can be a creative solution to all your problems!, [Online]. Available from: https://leverageedu.com/blog/edutainment/ [Accessed: January 15, 2021]

[22] A. Belyh, Edutainment, [Online]. Available from: https://www.cleverism.com/lexicon/edutainment/ [Accessed: November 1, 2022]

[23] Technopedia, Edutainment. [Online]. Available from: https://www.techopedia.com/definition/5506/edutainment. [Accessed: February 23, 2021]

[24] Nádoriés G, Prievara T. IKT módszertan: Kézikönyv az info-kommunikációs eszközök tanórai használatához. Budapest: Tanárblog.hu; 2012

[25] Veghés L, Udvaros J. Possibilities of creating interactive 2D animations for education using HTML5 canvas javascript libraries. eLearning and Software for Education. 2020;**2**:269-274

[26] Kovácsné Pusztai K. Gamification in higher education. Teaching Mathematics and Computer Science. 2021;**18**(2):87-106

Chapter 17

Integration BIM and Emerging Technologies in Architectural Academic Programs

Nada Abdelhai

Abstract

Building information modelling (BIM) provides major potential to improve project performance through its integration with emerging technologies, which are proven benefits in the architecture, engineering and construction (AEC) industry, such as better data sharing, increased productivity, safety improvement and enhanced visualization. As the evolution of BIM integration with the emerging technologies has progressively increased its influence on the (AEC) sector, an architectural academic programme must introduce BIM through its integration with the new technologies practices into architectural academic programs, in order to fulfil the requirements of this sector. The chapter's objective is to provide students, academic researchers and practitioners with a comprehensive updated review of the importance of adopting BIM integrated with emerging technologies in the architectural academic programme. The methodology comprises literature review to investigate the current developments in this subject area by reviewing and explaining the use of BIM integration with emerging technologies in construction project phases and presents practical examples of architectural projects that used these technologies. The chapter concluded that architectural education that supports emerging technologies has great impact on AEC. The chapter suggested considerable guidelines to facilitate introducing BIM integration with emerging technologies in the architectural academic programme and to find the possibility of applying the guidelines in architectural academic programs.

Keywords: BIM, Building Information modelling, emerging technologies, integration, architectural, academic programs

1. Introduction

World technological trends with great speed are involved in architectural and construction practice, also the building industry is rapidly evolving. The technology of BIM is integrated with these innovations and enhances in this direction. The companies are looking to benefit from computer technology to improve design and construction processes. At the same time when new emerging technologies sound to be created every day, understanding how to use these technologies appropriately to improve design and construction is very important [1]. The emerging technologies refer to information and communication technologies that facilitate the development, storage and handling of information and promote the various forms of communication between human beings and electronic systems and between electronic systems in digital binary computing systems [2].

The rapid development of BIM integration with emerging technologies has forced architects to make continuous advances in architectural design methodologies. The emergence of building information mdelling (BIM) has influenced the development of the architecture, engineering and construction (AEC) industry. These changes have created challenges for educators and students, where BIM integration with emerging technologies is not systematically taught in universities in practice. These new challenges need educational standards and new architectural programs to meet the new requirements, however, it is very important to find the approach to introduce these technologies into the architectural education programme to form specialists with high-level skills. (There is a great demand to ensure that graduates are able to use these technologies integrated with BIM) [3, 4].

The chapter explores a comprehensive review of BIM integration with emerging technologies including Artificial Intelligence (AI), Cloud technologies, internet of things (IoT), Virtual and augmented reality (VR/AR), Laser Scanning, 3D printing and Drone Technology, and identify the types of tools and methods of implementation for these technologies that provide the greatest benefit to the architectural academic program and the AEC industry. A systematic literature review was carried out on this subject area.

2. The concept and adoption of BIM

1. Building information modelling is a BUSINESS PROCESS for generating and leveraging building data to design, construct and operate the building during its life cycle. BIM allows all stakeholders to have access to the same information at the same time through interoperability between technology platforms.

2. Building information modelling is the digital representation of the physical and functional characteristics of a facility. As such it serves as a shared knowledge resource for information about a facility, forming a reliable basis for decisions during its life cycle from inception onwards.

3. Building information modelling is the ORGANIZATION & CONTROL of the business process by utilizing the information in the digital prototype to affect the sharing of information over the entire lifecycle of an asset. The benefits include centralised and visual communication, early exploration of options, sustainability, efficient design, integration of disciplines, site control, as-built documentation, etc., effectively developing an asset lifecycle process and model from conception to final retirement [5].

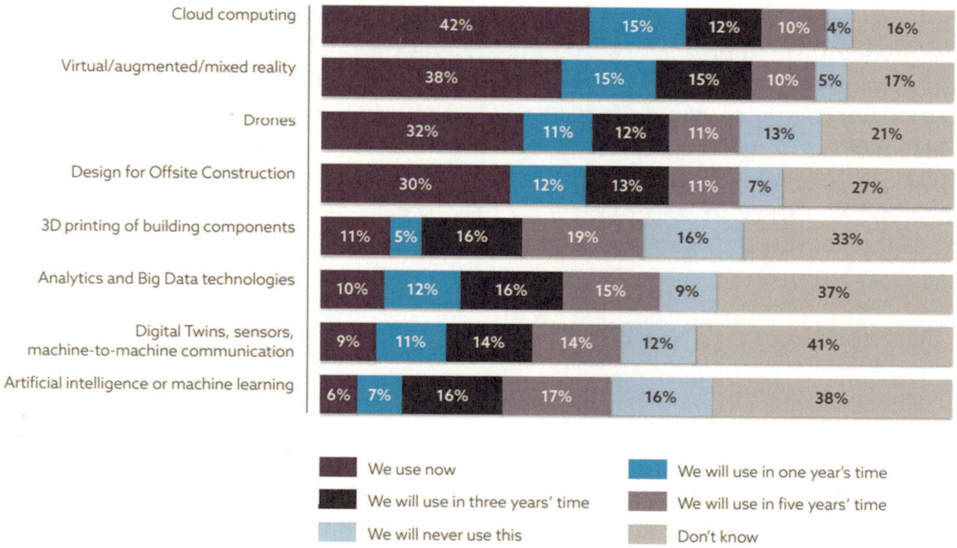

Figure 1.
BIM adoption between the years 2011 and 2020.

BIM changes the way architects think and design. In the last several years the use of this technology has grown in the architecture, engineering and construction (AEC) industry and sustainability.

Figure 1 shows the variation in the use of BIM by designers and other construction experts between the years 2011 and 2020, where the percentage of users in 2011 was (13%) and in 2020 (73%). It is noted that the percentage of BIM users, who are designers and other construction experts, has increased over time [6].

3. BIM and global education in architectural academic programs

BIM education and BIM awareness/uptake are still at different levels of implementation across the globe. Most countries are reporting BIM education being provided to architecture engineering and construction (AEC) students by their higher education and technical training institutions. The number of courses being offered is not significantly increasing year on year, with countries/regions such as Australia, Canada, Chile and Finland perhaps being the exception. This may indicate that the number of courses offering some form of BIM education may be reaching its saturation point in many countries/regions. However, many countries and regions are reporting that the content of such courses is now being expanded to include more sophisticated elements of BIM [7].

It was realized that BIM education in university curricula is an important demand for fulfilling educational adaptation in real work in the architecture industry. It is believed that BIM in academia is the fastest way to fill the gap between the education and real work and accelerates BIM knowledge and skills. It is a general believe that the fundamental education has to cover the lack of BIM-skilled professionals to fill the gap between BIM and the architecture industry.

Schools of architecture are adopting BIM in different courses in the curriculum: Digital graphic representation (DGR), workshop, design studio, BIM course, construction technology, construction management, research project and training. The predominant approach of schools is to adopt BIM in Design Studio, however, there are cases where design studios are integrated with other courses. Another approach is to teach tools and BIM in a specific BIM course. There are also cases in which tools are taught on courses of short duration (one week) non as workshops, which are generally design studios. Similarly, the workshops can be isolated or integrated with some courses. A BIM course may be elective or integrate the curriculum, in isolation or integrated with another course [8].

4. Methodology

The methodology comprises literature review to investigate the current developments in this subject area. The literature review included:

1. A review of emerging technologies and the importance of BIM integration with emerging technologies in an architectural academic programme.

2. A review and explanation of the use of BIM integration with emerging technologies in construction project phases.

3. Practical examples of using BIM integration with emerging technologies in architectural projects.

4. Guidelines for introducing BIM integration with emerging technologies in an architectural academic programme.

5. BIM integration with emerging technologies

BIM provides the ability to simulate a construction project in a simulated environment. A building information model is created digitally by using BIM technology. Upon completion, BIM will include detailed engineering and related data to support the design, procurement, manufacturing and construction activities vital to construct the building. Because of the greater potential in the AEC industry, many researchers have used BIM integration with other emerging technologies, although all emerging technologies have their own advantages. For example, BIM integration with Artificial Intelligence (AI), BIM integration with Cloud technologies, BIM integration with internet of things (IoT), BIM integration with Virtual and augmented reality (VR/AR), BIM integration with laser scanning, BIM integration with 3D printing and BIM integration with Drone Technology. All these technologies are considered promising emerging technologies in the AEC industry. The benefits of using these emerging technologies in the AEC industry are known to be enormous [9].

BIM and perfect information management are the bases for introducing new digital ways and tools of working. These technologies go parallel with BIM to enhance

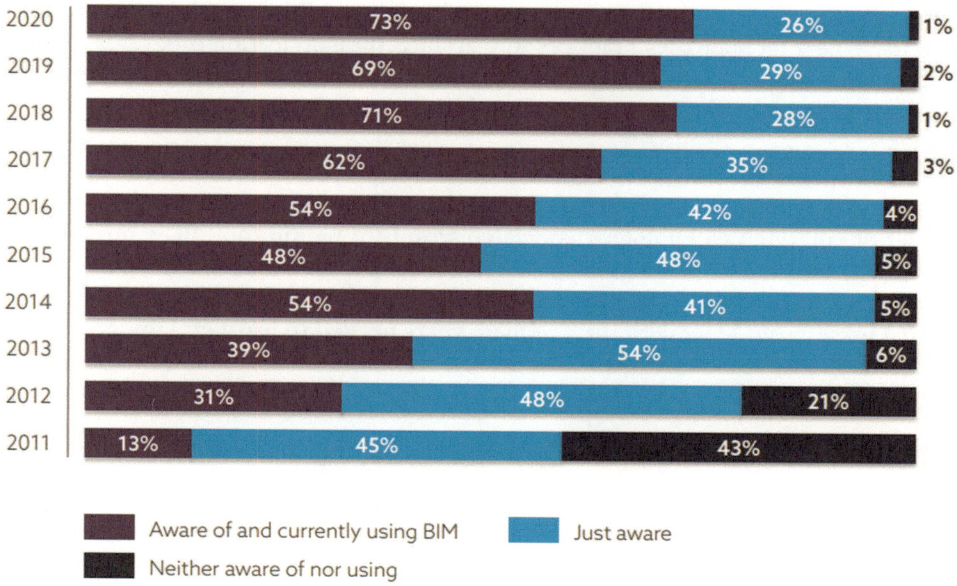

Figure 2.
The use of the emerging technologies in 2020.

the visualization of designs for clients, link different types of data across a project, and share technical data with other project team members [6]. Since the beginning of the AEC industry, it has followed ways to minimize costs, enhance visualization, improve efficiency and data sharing, increase productivity, reduce building waste, improve safety and boost quality and sustainable performance, all while seeking to minimize delivery time. The AEC industry is still dependent on traditional architectural drawings and procedures in order to lead a business. At the same time, the AEC professionals realize the significance of the emerging technologies in obtaining more accurate and smart modelling. Since the AEC industry is highly competitive, the AEC firms that successfully perform the most recent technologies will be capable to outperform the competition due to their ability to adopt them. In spite of massive automation opportunities, the AEC sector has only lately started the transformation from tradition to automation [9].

Figure 2 shows the result of a survey conducted in 2020, in which designers and other construction experts were asked about BIM and the expected use of the emerging technologies. It was noted that the most used technologies are in the following arrangement: Cloud computing, virtual/ augmented/ mixed reality, drones, design for offsite construction, 3D printing of building components, analytical and big data technologies, digital twins, sensors, machine to machine communication, artificial intelligence or machine learning [6].

6. The importance of introducing BIM integrated with emerging technologies in architectural academic programs

As the evolution of BIM integration with emerging technologies has progressively increased its influence on the architecture, engineering and construction (AEC) sector, architectural academic program must introduce BIM through its integration with these new technologies practices into architectural education programs, In order to fulfil the requirements of this sector. The importance of introducing BIM integrated with emerging technologies in architectural academic programs is as follows:

1. The ability to graduate from the university with significant knowledge of the latest emerging technologies associated with BIM within the AEC industries.

2. To provide students with the adequate years needed to learn emerging technologies associated with BIM experience at the time of their graduation.

3. To ensure a systematic understanding of the key aspects of the system of new technologies and their role in architecture.

4. To provide the student with the basic managerial skills in the digital age [10].

7. A review of emerging technologies and the importance of BIM integration with emerging technologies in the architectural academic program

These emerging technologies and the importance of BIM integration with emerging technologies in architectural academic program will be reviewed as follow:

7.1 Artificial intelligence (AI)

Artificial intelligence is the ability of a machine to learn in a similar way to a human being, to absorb new information and use it to develop its own intelligence system. Artificial intelligence (AI) is a multidisciplinary technology with a lot of branches, but machine learning is currently the most prevalent sector of artificial intelligence.

The term 'machine learning' refers to the ability of a machine to take in patterns in data and use them to think independently and produce their own solutions. The more data is entered into the machine, the better it can understand and provide insights. This is key in construction as much of the current processes are dependent on human capability. AI allows us to simplify these processes, reduce cost, risk and time and improve the efficiency and output of the project.

7.1.1 The importance of BIM integration with artificial intelligence (AI)

BIM software can use machine learning to understand data, discover patterns, and from this, and make independent decisions about how to automate and develop the model-building process. BIM software gathers a lot of data, which (AI) operates to discover the possibilities of every aspect of a construction project and find the preferable solution faster than the human brain can. This speeds up operations and, in addition, reduces the risk of human mistakes, which can improve safety at sites. AI has shown that now there is ability to lead BIM to the next level, to boost advance the industry. Productivity has always been a problem in construction and as a result, the industry has developed at a much slower pace than any other industry. The AI could be the solution to many problems that happened in the past as follows:

- Building design: AI allows the user to enter design standards or a set of 'rules' into the system so that the machine can generate the most viable output based on your needs. In terms of BIM, this can be used to form site footprints, floor plan designs and more. All of these plans relate to each other as well, meaning that if you change measurements in a site footprint through the process, for instance, your machine will have information to make the needful modifications in all aspects of the design to ensure the perfect accuracy all over the project.

- Safety and risk reduction: BIM has improved safety on site, making it easier to put in place additional safety measures before a project is implemented. BIM models with the help of AI can carry this to the next level, predicting incidents on site before they happen. Through machine learning, BIM software now has the power to analyse construction projects from the image alone and recognize hazards such as those for workers at heights, slips and falls.

- Productivity improvement: More investment in building technology in recent years has led to the development of BIM with the help of AI, which has made operations in all areas more efficient. This new technology helped eliminate inefficiencies that were delaying things down, reducing errors and accelerating the project completion.

- Continuous update: Systems using AI always learn from past and continuing projects. This means that they are capable to update on a nearly everyday basis, providing the most efficient and effective information to construction employees as soon as possible. This will assist in the improvement and evolution of the industry, assist to discover novel design solutions faster and permit their participation in all areas [11].

7.2 Cloud technologies

Cloud computing is a prompt emerging technology that can be used by communication devices such as computers, tablets and smartphones [12]. Cloud computing is a model for enabling ubiquitous, convenient and on-demand network access to a common set of configurable computing resources (such as networks, servers, stores, applications and services) that can be provisioned and released quickly with minimal management effort or service provider interaction.

The architecture of cloud computing includes four layers: application, platforms, infrastructure and hardware (from top to bottom). The hardware layer consists of the cloud's physical resources (i.e. computer equipment such as servers). The infrastructure layer is also known as the 'virtualization layer' because it uses virtualization technologies to manage computing resources by partitioning physical resources. The platform layer consists of operating systems and application frameworks that reduce the burden on the virtual machine, and thus acts as a kind of virtualized server. The top layer is the application layer or the actual cloud application. These layers are loosely coupled so that each layer can evolve separately [13]. There are several methods of grouping cloud computing technology. Cloud computing can be split into four major groups:

- A private cloud: It is used exclusively by a single organization including many consumers.
- A community cloud: It is used exclusively by a particular community of consumers from organizations with participated interests.
- A public cloud: It is an open access cloud that the general public can access.
- A hybrid cloud: It is an infrastructure that comprises two or more cloud deployment models (e.g. private, community and public). These models provide various business chances for many organizations as they can be integrated in different ways. Organizations can select the right collection that meets their needs when moving to the cloud environment [14, 15].

7.2.1 The importance of BIM integration with cloud technologies

The integrated Cloud-BIM model, considered the second generation of BIM, allows higher levels of collaboration and a more effective real-time communication platform for project team members. The Advantages of BIM integration with Cloud technologies are:

- Ease of access, storage manageability and high-performance computing abilities are the Advantages of using cloud BIM technology for managing and storing BIM data.
- Regardless of geographic location, Cloud BIM technology would provide real-time tracking of construction, collaboration, clash monitoring and sharing of information among members of the construction team.
- The integration of cloud computing in the BIM process could reinforce efficient planning through the life span of a project.
- Cloud BIM can help with sharing as-built drawings for remodelling of projects.
- The use of cloud BIM has led to collaboration in other areas such as the city information modelling (CIM) with the notion of significantly improving urban development and urban management, It supports the exchange of knowledge, multiservice and multi-field cooperation to create a digital city.
- The advantages of cloud computing are accessibility from any place, scalability and, most importantly, cost-effectiveness.
- Integrating mobile devices such as PCs, tablets, smartphones and personal digital assistants with Cloud BIM will enhance the ability to track real-time location data.

- Cloud technologies are used for the web interface, which can use a server to handle large BIM information. Cloud BIM can provide multiple users with real-time services over the Internet, at the same time, it allows for distributed storage of huge data using multiple servers, performs parallel computing and analysis and displays all online BIM models in a three-dimensional (3D) way to browse the web standard.

- Cloud BIM is also used for cost estimating, real-time quality checks and for site memos. BIM Field can be effectively used for on-site project management [16].

7.3 Internet of things (IoT)

IoT technology sensors are actually simple devices, can be applied to infrastructures and buildings and can create very high innovation potential in terms of managing the technological and financial aspects of their optimal and efficient operations. When connected to special devices, they can measure light intensity, humidity levels, temperatures or any other types of mechanical or physical information.

7.3.1 The importance of BIM integration with IoT

BIM allows the forming of a digital model that represents the digital twin of the work undertaken or to be implemented. The real model, or building asset, can be managed over the digital twin (a model of the asset in a cloud-based service) in highly advanced ways that also avail integration with IoT sensors. By integrating these results with BIM technologies, the resulting response is real-time communication between the detected data and the 3D model (also used for visual purposes) that opens up an infinite number of opportunities in terms of effectiveness and project management.

For instance, the potential of updating the project documents based on the actual progress of the work, accessing management matters and criticalities and the possibility to optimize the whole process on the foundation of such data and all in real time. The IoT sensors store a regular connection between the physical environment and the digital model, opening up new possibilities in the field of project management. This large amount of data represents a support foundation for each decision-making and management stage of the construction process, from the construction site to management and maintenance over time.

For example, during the construction phase, sensors allow you to obtain all relevant data relating to the existence of staff on the construction site, the workflows, peak presences and any overlapping tasks. They can gather real-time information considering the use of materials, equipment and machinery and get a constantly updated status on the progress of the work and supplies.

Thanks to cloud platforms, this large database is naturally available to all project participants, following an approach fully consistent with the BIM philosophy [17].

There are other advantages when integrating BIM with the (IoT) are:

- IoT is a key component of the smart city concept, as it brings together 'the physical, digital and human systems' in the built environment to provide a sustainable, prosperous and inclusive future for its citizens [18].

- IoT enables a real-time understanding of every aspect and component of a building and its operation, increasing the accuracy and availability of information.
- IoT supports employees in the AEC sector to overcome one of the limitations of BIM, for example, static build information and increase the efficiency of services to a higher level [19, 20].

7.4 Virtual and augmented reality (VR/AR)

Other visualization technologies that have shown major potential for use in collaborative operations are virtual reality (VR) and augmented reality (AR). Virtual reality has a major benefit to the AEC industry. Architects can represent and evaluate their design on a scale of 1:1 and walk through their project as if it had been built; by this technique, clients can understand the design and modify the design to suit their requirements. As a medium, VR has three specific characteristics. It is interactive (users can interact with models), spatial (models are represented in three spatial dimensions) and real-time (feedback from actions is given without noticeable pauses). With the ability to exploit and reuse information directly from models, current multidisciplinary collaboration can evolve towards integrated multidisciplinary collaboration on models.

7.4.1 The importance of BIM integration with Virtual and augmented reality (VR/AR)

- VR can become a highly effective instrument for exploration of digitally modelled architecture by providing a spatio-visual representation of the design object. Also, VR allows stereovision and so a depth perception in digital environments by using stereoscopic head-mounted displays (HMDs). The degree of immersion is directly related to image quality and the reaction rate of the HMD. There is a case in construction projects is that some stakeholders are not from the AEC sector and have no familiarity with architectural drawings and construction documents. A common problem is that information and design concepts are not presented in a way that all stakeholders can understand well. In this context, real-time visualizations and virtual reality were demonstrated to offer an effective communication platform.
- VR allows stakeholders and employees in the AES sector to experience and discuss something that does not yet exist from a common perspective. Rather than speaking in imagination, virtual reality provides us a more realistic frame of reference. As a result, it narrows the deficiency of understanding between customers and architects, and between visible and invisible thinkers.
- The use of VR during various stages of project design development and construction is an attraction for the relevant stakeholders. Many participation or project presentation sessions can be stressful and boring for participants. The act of wearing HMDs and being separated from the real world can have something interesting to people, similar to the allure of playing with arcades or other gaming devices [21].

7.5 3D laser scanning

The main advantages of using 3D laser scanning are, for example, scanning a building or a road and obtaining measurements and information about the objects or building. This information can later be used for structural redesign, architectural redesign and dimensions. There are many applications for this technology that allows improving accuracy, time consumption and safety when it comes to scanning. It is not easy to draw a line between surveying, construction, engineering and architecture when discussing laser scanning.

7.5.1 The importance of BIM integration with 3D laser scanning

The expansion in the refurbishment market has resulted in greater coverage of Scan-to-BIM in the industry. As a result of market needs, nearly all CAD developers started working on supporting the point cloud data with their platform, and now almost all products in the market offer this feature. Traditional measurement techniques have been taking more time and human effort to measure the building and then create drawings/sketches to acquire BIM models for the building. Other methods were standard data collection techniques like using a total station and then using AutoCAD to produce drawings. After that, taking these files in Revit, Archicad or such architectural modelling software serves as a reference for preparing a 3D model of the building in BIM tools. Although traditional techniques lack completeness and time efficiency compared to 'scan to BIM' in most cases, the issue of cost depends on many other factors like the level of detail, usage, space and location of the project. The combination of modern technologies, such as laser scanning, with BIM can be a powerful tool for process improvement, providing accurate data in minimal time. The benefits of using BIM in architectural engineering are related to alternative design and computation analytics, materials analytics, data management, operational plans and information exchange between different entities [22].

7.6 3D printing

3D printing is an additive manufacturing process for making 3D solid objects from a digital file. The materials used to produce the object are various types of concrete, hard plastic, carbon fibre, metal, etc. The use of 3D printing has developed in the latest years with technological advances in the engineering sector [23]. 3D printing engineering models are becoming an interesting alternative. Architectural models are commonly made of wood, cardboard, plastic or other mouldable materials. Architects need models to study aspects of their design. It is all the time, modified to obtain an accurate concept of their idea in architectural design. 3D printing at a construction scale will have a wide diversity of applications within the AEC sector. The major advantages of these technologies include speeding up the construction process, reducing labour cost, increasing accuracy, decreasing complexity, greater job integration and reducing waste produced. A variety of different approaches have been demonstrated yet, which include on-site and off-site manufacturing of buildings and building components, using industrial robots, tethered autonomous vehicles and gantry systems [24].

Demonstrations of construction 3D printing technologies to date have included the fabrication of housing, building construction components (cladding, structural panels, and columns), bridges, civil infrastructure, artificial reefs, follies and sculptures. It may also allow construction to be carried out in harsh or hazardous environments that are unsuitable for human workforce such as places with special conditions [25].

7.6.1 The importance of BIM integration with 3D printing

The effective combination of BIM technology and 3D printing technology, each of which takes advantage of the advantages of information technology, is a unique new theory and method that has been formed in the process of adapting to the future development of the construction industry. The key to implementing BIM + 3D technology is the need to break the barrier between BIM technology and 3D printing technology. BIM provides building model information, and 3D printers follow the corresponding path to print. In order to enable BIM data to be input to a 3D printer, it is necessary to transform BIM data into 3D printer data code to achieve the industrial production process of prefabricated buildings. The corresponding component information will also be changed in the data processing system. After the path planning is completed, the 3D model composed of all the graphic elements can be visually displayed in BIM software to timely optimize the design plan and modify and monitor the printing process. The 3D printing device is driven by the control system, which generates the printing program code to drive the control system after the path planning is completed. The specific process is as follows: The pumping device with building materials (concrete mixture) is driven by the nozzle through the operation of the robot arm, and the BIM model instructions are printed layer by layer according to the input printing code, forming prefabricated components or directly generating the actual construction. Real-time data during the process of the control system will also be transmitted to the BIM management system to achieve dynamic and visual management of safety, cost, quality and progress during the production of 3D-printed assembled buildings [26]. 3D printing technology can print complex components, it has advantages such as formed the building without a mould in the actual construction, shortening the product manufacturing cycle, reducing costs, energy saving, improving production efficiency, and protect environment [27].

7.7 Drone technology

The drone is a small aircraft that comes with a compact design in different sizes. The drones are operated remotely with the help of a controller. Project managers and AEC specialists use drones to record and report the progress of a construction project. Drones can capture precise project data over a large area in a short period. These data can be used by BIM professionals to revise their model and make the BIM modelling process more reliable.

7.7.1 The importance of BIM integration with drone technology

Drones are changing the way buildings are designed. One useful benefit of using drones in architecture is their application in surveying. The drone has a small size and its ease of manoeuvrability make it an ideal tool for getting into hard-to-reach places. The aerial photography provided by the drones allows the engineer and client to visualize what the final product will look like. Aerial footage allows an architect to design precise plans. Drones are efficient and can get mapping projects done in a fraction of the time a project would normally take. Mapping is also more accurate, eliminating errors and costly revisions. The data collected by the drone could eliminate the need to hire land surveyors for topographic surveys.

The AEC sector benefits greatly from the use of drones. Drones save construction projects time and money. With their ability to collect real-time data on the job site, drones increase workflow and efficiency [28].

8. A review and explaining the use of BIM integration with emerging technologies in construction project phases

The chapter will review and explain the use of BIM integration with emerging technologies in construction project phases. The project phases are Project Initiating, project planning, project executing, project controlling, project closing and project operation and maintenance.

8.1 Project initiating phase

The Project Initiating Phase is defined as the formal recognition of the beginning of a project. This involves identifying stakeholders and developing a project charter [29]. None of the emerging technologies was reviewed in this phase, as no literature was found showing the effect of emerging technologies in this phase.

8.2 Project planning phase

The Project Planning Phase is defined as creating and maintaining a workable plan to accomplish the purpose of the project. This involves developing a project management plan which encompasses all of the project knowledge areas [29].

8.2.1 BIM integration with emerging technologies used in the project planning phase

- BIM can facilitate many aspects of building design including 3-D visualization, Design coordination and clash detection, Value engineering, Constructability review and analysis and Building performance analysis. BIM can be used to provide project teams with an early-warning system for cost and schedule overruns. In this case, a prototype is produced – combining BIM with earned-value analysis – to assist in this area by using a real-life project as a sample [30]. BIM can assist project teams to establish sustainable procurement practices [31].

- AR can help with design management. This was tested using a panoramic view of a renovated construction site and superimposing BIM data on that view. Then, users with construction experience - and without experience with such type of technology - were able to successfully examine the design in a semi-augmented reality experience. Quantitative results received by the moderator showed a positive response to semi-augmented reality conditions [32]. AR is also useful for moving away from 2D paper plan documents for decision making, to further automated task systems [33].

- AR, it was found that the use of artificial intelligence improved the level of accuracy in Bill of Quantity, it also has been proven to be accurate price of a project. This helps contractors estimate projects more accurately [34]. AI can be used to ensure structural steel adheres to guidelines set by standards and building codes for fire resistance and safety. Providing more precision in performance on steel structures against fire. In addition, AI can be used to understand what will happen to various types of materials like concrete, masonry, steel and wood—under maximum temperature conditions [35].

- BIM and AR, with computer-based simulation software, can be used to effectively plan occupational health and safety (OHS) for labourers on construction sites, by simulating labour-based scenarios of emergency evacuation that can occur under various conditions during the project execution phase. This simulation can also be created to depict different construction methods. The result given is the approximate total evacuation times per floor for all labourers and shows the areas of overcrowding [36]. BIM supports occupational health and safety, it provides a visual understanding of the construction site before the project execution phase starts. So, BIM can help in planning for possible dangers in the future [37].

- Drones can influence the range by scanning the ground to determine the boundaries of the site. This technology replaces the need for bulky tools and equipment that may be necessary to carry out the task. This gives an opportunity for faster services at a lower cost.

- 3D printing provides the opportunity to minimize the amount of time spent on a project. Specifically, for 3D printing metal structural components, the total time to produce a conventional wall was reduced by 35 per cent. 3DP, partly because of its time-saving effect, provides the opportunity to save on project costs. This is due to the reduced labour required due to the raised automation of production [38].

8.3 Project executing phase

The project executing phase is defined as carrying out the project plan. This involves directing and managing project work, with a focus on quality, communications, risk, procurement and stakeholder engagement [29].

8.3.1 BIM integration with emerging technologies used in executing phase

- Cloud technology enables construction practitioners to have faster access to a range of cloud computing, software, networking, servers, database, storage and analytics services for effective project management. Cloud computing provides enhanced access to project data and site images just by connecting remote mobile devices to a dedicated remote server. Designers, engineers and contractors can take advantage of cloud computing to collaborate on BIM projects in real time even though they are in different geographic locations [39].

- Drones and related technologies have an impact on the field of (OHS) in construction. This is done by gathering drones and transmitting real-time video to construction managers, for site safety check purposes. Drones have been shown to help manage the supply chain at construction sites. They help manage this by moving goods from one point to another on the site. Once supplies are dropped on site, they are detected, identified and tracked using the Global Positioning System (GPS). After that, they are automatically taken to their appropriate location [40]. Drones weighing just 12 Kg, it can raise concrete blocks and hold them in place, without any human interaction, using a laser-based positioning system. This technique shows the promise of building walls without using scaffolding [41].

- AR has been identified as useful in occupational health and safety scenarios, specifically danger recognition and avoidance. With a wearable device, a labourer can receive information about dangerous conditions without having to analyse the data they receive on their devices. There have been positive results with the on-Site hazard identification system, which uses an image capture device to gather images from the construction site. This information is provided to the wearable and reported to the operator [42].

- AI has a different way of making an impact on occupational health and safety, specifically in tunnelling, using a tunnel boring machine (TBM). Since the TBM is very sensitive to geological changes, and because it is hard to monitor the state of equipment and rock mass in real time to make critical decisions, artificial intelligence plays a vital role in this area. This is done using machine learning to extract the big data collected by TBM to create new rock mass prediction models, along with using artificial neural networks to define the quality of the rock mass. These innovations can assist prohibit accidents that can lead to real-time changes to a boring block of rock, which in turn can save lives [43]. AI is known to help manage contracts in construction, especially in managing conflicts. AI has been used to understand data and predict clashes, AI can successfully mine data on historical projects to identify relevant projects and understand the types of clashes that may appear. This information will better prepare contract managers for the implementation of the project, as they can prepare their forecasts accordingly, and reduce risks [44].

- IoT can help fill in the gaps needed to provide up-to-date, real-time information on the construction supply chain. Mobile internet is a group of devices connected with internet capability and all connected to one system for a complete view [45].

- 3D printing has a significant impact on material supply. A particle layer as a base on the ground is needed before this as a first step in the concrete construction process. Using 3DP technology, particles of different sizes, materials and overall design can be generated for different scenarios, creating a new opportunity for structural design and construction [46]. Cement printing offers the advantages of being cost-effective, highly efficient, flexible in design and environmentally friendly. Reducing labour would be a side effect of performing this technology [47]. Similarly, this technique was also used in the construction of walls.

8.4 Project controlling phase

The Project Controlling Phase is defined as monitoring progress and taking corrective action when necessary. This involves monitoring and controlling project work while managing change, with a focus on scope, schedule, quality, resources, monitoring communications, risks and stakeholder engagement and controlling procurements [29].

8.4.1 BIM integration with emerging technologies used in the project controlling phase

- AR is known to provide a sense of realism to projects. An intuitive experience is created where 4D scheduling information is attached to AR objects and displayed to the user. This helps control the schedule, all in a realistic simulation type. This can be especially useful for long-distance managers who need to be refreshed on schedules. This is an improvement for completing the same type of task through physical fieldwork [48]. AR was sought and identified as an influence on site and building inspections through an intuitive interactive approach. Data can be captured immediately, which will lastly minimize response time to resolve inspection issues that require quick action. This is possible due to the integration between BIM and AR platforms [49]. AR can be used by site inspectors for inspections of tunnel construction projects. AR aids are in progress. The application uses BIM as a basis to get a realistic view of the project in real time. Then, teams can control the project and provide real-time updates of actual progress against planned progress using on-site augmented reality in real time. Generally, this system provides a more effective way to control projects [50].

- Drones also influence when it comes to quality control. Drones have been sought as a more cost-effective and effective means of building inspection, providing high accuracy in selecting images and video for the teams [51].

8.5 Project closing (and Handover) phase

The project closing phase is defined as bringing the project and paperwork to one hundred per cent completion and entering the history in a database [52]. Once this occurs, a handover will happen to the operations management team.

8.5.1 BIM and emerging technologies used in project closing phase

- BIM has been checked to improve facility management by providing monitoring, documentation and sharing of information from previous project phases into the maintenance and operations phase. This is particularly useful for large projects in mechanical, electrical and plumbing applications. These applications are

No	Project	BIM integration with emerging technologies	The impact of technology used on the project
1.	The Curzon Building – Birmingham City University	3D Scanning	1. Reduced errors 2. Clearer views of designs 3. Centralised project information 4. Accurate as-built models/drawings 5. All model objects in the model

No	Project	BIM integration with emerging technologies	The impact of technology used on the project
2.	Silber Boulevard, MK – All Saint Development	Cloud computing	Improved communication over drawings
3.	Walthamstow Hospital	3D Scanning	1. Reduced errors 2. Model objects in the model
4.	Herdwick Gate, Shipston-on Stour	Artificial Intelligence	Cut and fill was completed by remote control plant/machinery
5.	Pre-Boarding Zone	BIM/VR	1. Client engagement with VR 2. Enhanced design and construction 3. Better speed of construction
6.	Grow on Space Business Development Space, Market Harborough	VR	1. Client has good understanding of the design 2. Client also has a better notion of the size of spaces in their asset, something which rendered images cannot convey 3. Positive reaction from this and other customers led to own investment in more VR tech.
7.	Fairfax Street Student Rooms, Coventry	Drones	1. Remote monitoring 2. Enhanced quality check
8.	Student Halls	Laser scanning	1. Speed up the design process 2. Verification of on-site data
9.	M62 SMP	Cloud	We will be able to embed the information from site easily into the cloud to be saved in CDE for handover documentation
10.	Office accommodation	Cloud	1. Security of data 2. Minimize manual backups 3. Data collection and duplication
11.	Bishopsgate	VR	1. Safer, more efficient logistics arrangements 2. Project management ream improved planning 3. Impact of design changes visualized to show option 1 vs option 2 for contractor-client dialogue
12.	New Museum of London	Drones	1. Use of drones to 3D scan inaccessible/unsafe roof 2. Design team had real scanned data to earlier in project 3. Less change in future
13.	Camden Lock	Drones	Quick, easy-to-understand progress updates for teams and clients
14.	Data centres	VR	1. Collaboratively solving problems in multidisciplinary teams 2. Supporting ‘right first time’ delivery 3. De-risking the construction sequence
15.	Salesforce Tower – San Francisco	IoT	The IoT-enabled smart grid within the buildings rapidly transfers energy from a surplus node to a deficit node or saves the surplus energy to use later. It makes it possible to use renewable energy sources.

Table 1.
BIM integration with emerging technologies used in some projects in the UK, and the impact of its use on the project.

envisioned not only for the construction team but for the operations team to advance the building lifecycle [52].

8.6 Project operation and maintenance

Activities that take place during the phase comprise replacement of materials, equipment maintenance and equipment requiring replacement and minor renovations to allow for a review of facility usage. This stage is sometimes referred to as facility management (FM) [53].

8.6.1 BIM integration with emerging technologies used in Project operation and maintenance phase

- IoT application is more in post-construction as compared to design and construction phases and is seamlessly integrated into the information network [39].

9. Practical examples of using BIM integration with emerging technologies in architectural projects

Table 1 shows BIM integration with emerging technologies used in some projects in the UK, and the impact of its use on the project [39].

10. Guidelines for introducing BIM integration with emerging technologies in an architectural academic programme

BIM integration with emerging technologies are proven helpful in the architecture, engineering and construction (AEC) industry due to their various advantages to projects and stakeholders, so the chapter suggested considerable guidelines to facilitate introducing these technologies in architectural academic programme as shown in the following:

10.1 Develop a new education curriculum module

It is expected that there will be more demand for the emerging technologies associated with BIM such as AI, Cloud technologies, (IoT), (VR, AR), 3D laser and 3D printing and drone technology. These technologies will affect architectural education and training systems, which create a need to develop new skills. Education courses could be divided along the semesters.

10.2 Establish an interdisciplinary curriculum system across integration

There is difficulty for a single discipline in architecture to meet the development needs of all emerging technologies integrated with BIM, such as (AI),cloud technologies ,(IoT), (VR, AR), 3D Laser and 3D printing Scanning and drone technology, so architectural curricula must be developed to include different disciplines in this field.

10.3 Create relationship between the AEC sector and emerging technologies associated with BIM education in an architectural academic programme

Close collaboration between the AEC sector and academia is fundamental to the two-way exchange between training and reality. The experience of cooperative professionals from the AEC sector with teaching qualifications is fundamental to prompt students and develop their personal skills in relation to the AEC sector requirements. Also, it is preferable to engage stakeholders and professionals in the case studies of the projects to provide advice on how to improve the AEC sector with emerging technologies associated with BIM. Close collaboration between employers and educators in the AEC sector will enable the creation of a stronger shared vision in the training of BIM integration with emerging technologies. Connecting the AEC sector and architectural academic programme will improve traditional teaching.

10.4 Strengthening the qualifications of teaching staff

It is very important to build a team of teachers with the ability to solve complex engineering problems. Teaching staff must strengthen their learning and practice in order to meet the needs of new knowledge, theories and technologies. In particular, they must learn BIM integration with emerging technologies. Also, teachers who truly understand the innovation of technology should be hired, to reflect these qualifications on the development of student's skills and their innovative ability. Faculty members should have: Share recent ideas that reinforce student success, explore new tools and techniques for teaching, interact with innovative and emerging technologies and Acquire knowledge of how to grasp and apply technology for learning.

10.5 Provide teaching resources

Teaching BIM integration with emerging technologies needs to provide textbooks, tutorials, Computer labs, strong internet and other electronic and technological equipment. BIM typically requires regularly new and upgraded hardware to run the processing-intensive software.

11. Conclusion

This chapter was designed to provide students, academic researchers and practitioners with a comprehensive updated review of the importance of adopting BIM integrated with emerging technologies in architectural academic programme, and reviewed these technologies such as artificial intelligence (AI), cloud technologies, internet of things, virtual and augmented reality (VR/AR), laser scanning, 3D printing and Drone Technology.

The chapter reviews the importance of BIM integration with emerging technologies in the architectural academic program, a review and explanation of the use of BIM integration with emerging technologies in a construction project phases, and presented practical examples of using BIM integration with emerging technologies in architectural projects.

The chapter concluded that there is increasing adoption of BIM integrated with emerging technologies in the field of AEC and the increasing demand for professionals in this field. The architectural education that supports emerging technologies has great impact on architectural practice and AEC. The presented projects where BIM integration with emerging technologies was used have been characterized by reducing cost and time and improving quality. BIM integrated with emerging technologies provide new ways of learning, while at the same time requiring educational institutions' faculty to collaborate and reorganize architectural programs, teaching materials and study plan.

The chapter suggested considerable guidelines to facilitate introducing BIM integration with these technologies in architectural academic programmes such as, developing new education curriculum module, establishing an interdisciplinary curriculum system across integration, creating relationship between AEC sector and emerging technologies associated with BIM Education in architectural academic programme, strengthening the qualifications of teaching staff and providing teaching resources. The recommendation is to study each emerging technology independently and find the possibility of applying the guidelines presented in this chapter for BIM integration with emerging technologies in architectural academic programs.

Conflict of interest

The author declares no conflict of interest.

Author details

Nada Abdelhai[1,2]

1 Taif University, Taif, Saudi Arabia

2 Sharq Elniel College, School of Architecture and Environmental Planning, Khartoum, Sudan

*Address all correspondence to: nmabdelhai@tu.edu.sa

References

[1] Emerging Technologies BIM Research Group, Arizona State University [Internet]. Available from: https://etbimlab.com/ [Accessed: May 16, 2022]

[2] Digital Technologies in the Architecture, Engineering and Construction (AEC) Industry—A Bibliometric—Qualitative Literature Review of Research Activities [Internet]. 2021. Available from: https://www.researchgate.net/publication/352189856_Digital_Technologies_in_the_Architecture_Enginee [Accessed: May 16, 2022]

[3] Tsai MH, Chen KL and Chang YL. Development of a Project-Based Online Course for BIM Learning [Internet]. 2019. Available from: http://www.mdpi.com/2071-1050/11/20/5772/htm.Pdf [Accessed: May 21, 2022]

[4] Krivonogov A, Zakharova G, Kruglikov S, Plotnikov S. Implementation of BIM-Technologies in the Educational Program of the Architectural University [Internet]. 2018. Available from: http://www.matecconferences.org/articles/matecconf/pdf/2018/05/matecconf_bd2018_01001.pdf [Accessed: May 22, 2022]

[5] National BIM Standard- United States- Version 3 [Internet]. 2015. Available from: http://www.buildingsmart.org/resources/terms-and-definitions [Accessed: May 17, 2022]

[6] 10TH BIM Report 2020 [Internet]. 2020. Available from: https://architecturaltechnology.com/static/3f388415-32f9-408d-85cc2c1adf13d012/TheNBSBIMReport2020.pdf [Accessed: June 8, 2022]

[7] BIM Education – Global 2021 Update Report [Internet]. 2021. Available from: https://www.icis.org/.../BIM-Education-Global-2021-Update-Report-V8.pdf [Accessed: May 8, 2022]

[8] Ramadan N. An Approach to Introduce BIM into schools of architecture curriculum in Greater Khartoum. In: Proceedings of the ICCCEEE International Conference on Computer, Control, Electrical, and Electronics Engineering (ICCCEEE); 27-29 February 2021. Khartoum: ICCCEEE; 2021

[9] Digital Technologies in the Architecture, Engineering and Construction (AEC) Industry—A Bibliometric—Qualitative Literature Review of Research Activities [Internet]. 2021. Available from: www.researchgate.net/publication/352189856_Digital_Technologies_in_the_Architecture_Engine [Accessed: June 3, 2022]

[10] Rachid A. Development of a Framework to Support Embedding BIM within Undergraduate Architectural Programmes in Lebanon. Salford, Manchester: Salford University; 2021

[11] How Artificial Intelligence is Improving the Efficiency of BIM [Internet]. 2020. Available from: https://aitopics.org/doc/news:5DF44A07 [Accessed: May 26, 2022]

[12] Cloud Computing - ISSUES, Research and Implementations [Internet]. 2009. Available from: www.academia.edu/2374327/Cloud_Computing__Issues_Research_and_Implementations [Accessed: June 2, 2022]

[13] Cloud Computing: State-of-the-Art and Research Challenges [Internet]. 2010. Available from: www.semanticscholar.org/paper/Cloud-computing%

3A-state-of-the-art-and-research-Zha ng-Chen [Accessed: June 2, 2022].

[14] The NIST Definition of Cloud Computing [Internet]. 2011. Available from: www.nist.gov/publications/nist-definition-cloud-computing [Accessed: June 3, 2022]

[15] A Review of Cloud- Based BIM Technology in the Construction Sector [Internet]. 2014. Available from: https://www.researchgate.net/publication/305585997_A_review_of_cloud-based_BIM_technology_in_the [Accessed: April 24, 2022]

[16] Cloud BIM Technology as a Means of Collaboration and Project Integration in Smart Cities [Internet]. 2021. Available from: https://www.academia.edu/39556826/SMART_CITIES_and_BIM_Technologies_Effects_of_BIM_on_proj [Accessed: June 1, 2022]

[17] IoT in BIM [Internet]. 2021. Available from: https://biblus.accasoftwa re.com/en/iot-in-bim [Accessed: April 26, 2022]

[18] Integration of BIM, Web Maps and IoT for Supporting Comfort Analysis [Internet]. 2017. Available from: https://doi.org/10.5194/isprs-annals-IV-4-W4-221-2017 [Accessed: July 1, 2022]

[19] History of 3D Printing: It's Older than You Think [Internet]. (No date) Available from: https://www.autodesk.com/redshift/history-of-3d-printing/ [Accessed: June 30, 2022]

[20] BIM and IoT for the AEC Industry: A systematic Literature Mapping [Internet]. 2020. Available from: www.proceedings.blucher.com.br/article-deta ils/bim-and-iot-for-the-aec-industry-a-systematic-literat [Accessed: April 26, 2022]

[21] Hossein M. BIM Implementation in Architectural Practices: Towards Advanced Collaborative Approaches Based on Digital Technologies. Catalonia: Polytechnic University; 2019

[22] Raza M. BIM for Existing Buildings-A Study of Terrestrial Laser Scanning and Conventional Measurement Technique. Berlin: University of Applied Science; 2017

[23] A review of 3D concrete printing systems and materials properties: current status and future research prospects [Internet]. 2018. Available from: https://www.researchgate.net/pub lication/324507765_A_review_of_3D [Accessed: August 12, 2022]

[24] Nithesh N. Development of Concrete 3D P. Aalto University: Helsinki; 2018

[25] 3D Printing in Architecture, Engineering and Construction- Concrete 3D Printing [Internet]. 2019. Available from: www.researchgate.net/publica tion/348826088_3D_Printing_in_Arch itecture_Engineering_and_Co [Accessed: April 26, 2022]

[26] Research on the Application of 3D Printing and BIM Technology in the Direction of Bridges [Internet]. 2021. Available from: www.e3sconferences.org/articles/e3sconf/abs/2021/33/e3sconf_aesee2021_02051/e3sconf_a [Accessed: May 29, 2022]

[27] BIM for 3D Printing in Construction-Technology Foundations and Industry Practice [Internet]. 2018. Available from: www.researchgate.net/publication/327757885_BIM_for_3D_Printing_in_Construction_Technology [Accessed: May 1, 2022]

[28] The Use of Drone Technology In Architecture, Engineering, And

Construction [Internet]. 2020. Available from: www.sistaffing.com/2020/01/30/the-use-of-drone-technology-in-architecture-engineering-and-constr [Accessed: May 1, 2022]

[29] A Guide to the Project Management Body of Knowledge [Internet]. 2017. Available from: https://faspa.ir/wp-content/uploads/2017/09/PMBOK6-2017.pdf [Accessed: June 30, 2022]

[30] Study on BIM-based construction project cost and schedule risk early warning [Internet]. 2015. Available from: https://www.researchgate.net/publication/282900205_Study_on_BIM based_construction_project_cost_an [Accessed: August 12, 2022]

[31] A Mixed Review of the Adoption of Building Information Modelling (BIM) for Sustainability [Internet]. 2017. DOI: 10.1016/j.jclepro.2016.09.222 [Accessed: June 30, 2022]

[32] Integrating BIM and Panorama to Create a Semi-Augmented-Reality Experience of a Construction Site [Internet]. 2016. DOI: 10.1080/15578771.2016.1240117 [Accessed: June 30, 2022]

[33] Exploring the Performance of an Augmented Reality Application for Construction Layout Tasks [Internet]. 2019. DOI: 10.1007/s11042-019-08063-5 [Accessed: June 30, 2022]

[34] Optimized Artificial Intelligence Models for Predicting Project Award Price. Automation in Construction [Internet]. 2015. DOI: 10.1016/j.autcon.2015.02.006 [Accessed: June 30, 2022]

[35] Properties and Material Models for Common Construction Materials at Elevated Temperatures [Internet]. 2019. DOI: 10.1016/j.conbuildmat.2019.04.182 [Accessed: July 1. 2022]

[36] Planning labor Evacuation for Construction Sites using BIM and Agent-Based Simulation [Internet]. 2018. DOI: 10.1016/j.ssci [Accessed: July 1, 2022]

[37] Building Information Modeling and Safety Management: A Systematic Review [Internet]. 2018. DOI: 10.1016/j.ssci.2017.08.015 [Accessed: July 1, 2022]

[38] Metal 3D printing in construction: A review of methods, research, applications, opportunities and challenges [Internet]. 2019. Available from: https://www.researchgate.net/publication/330073297_Metal_3D_printing_in_construction_A_review_of_me [Accessed: August 12, 2022]

[39] Ireti N. Application and Effects of Emerging Technologies on Variation Minimization in the UK Construction Projects. London: Coventry University; 2020

[40] Applications of Multirotor Drone Technologies in Construction Management [Internet]. 2018. Available from: https://www.researchgate.net/publication/324365476_Applications_of_multirotor_drone_technologies_ [Accessed: August 12, 2022]

[41] Feasibility Study for Drone-Based Masonry Construction of Real-Scale Structures [Internet]. 2018. DOI: 10.1016/j.autcon.2018.06.015 [Accessed: July 1, 2022]

[42] Image-Based Construction Hazard Avoidance System using Augmented Reality in Wearable Device [Internet]. 2017. DOI: 10.1016/j.autcon.2017.06.014 [Accessed: July 1, 2022]

[43] Applications of Multirotor Drone Technologies in Construction

Management [Internet]. 2019. DOI: 10.1080/15623599.2018.1452101 [Accessed: July 3, 2022]

[44] Classifying Influential Information to Discover Rule Sets for Project Disputes and Possible Resolutions [Internet]. 2016. DOI: 10.1016/j.ijproman.2016.10.001 [Accessed: July 2, 2022]

[45] Mobile Internet based Construction Supply Chain Management: A critical review. Automation in Construction [Internet]. 2016. DOI: 10.1016/j.autcon.2016.08.020. [Accessed: July 2, 2022]

[46] Particle-bed 3D printing in concrete construction – Possibilities and challenges. Cement and Concrete Research [Internet]. 2018. DOI: 10.1016/j.cemconres.2018.05.018 [Accessed: July 2, 2022]

[47] State-of-the-art of 3D Printing Technology of Cementitious Material—an Emerging Technique for Construction [Internet]. 2018. DOI: 10.1007/s11431-016-9077-7 [Accessed: July 2, 2022]

[48] Improvement of Realism of 4D Objects Using Augmented Reality Objects and Actual Images of a Construction Site [Internet]. 2018. DOI: 10.1007/s12205-017-0734-3 [Accessed: July 2, 2022]

[49] Conceptual framework for integrating BIM and Augmented Reality in Construction Management [Internet]. 2020. DOI: 10.3846/jcem.2020.11803 [Accessed: July 3, 2022]

[50] Mobile Augmented Reality Applications for Construction Projects [Internet]. 2018. DOI: 10.1108/CI-02-2017-0013 [Accessed: July 3, 2022]

[51] Automated Content-Based Filtering for Enhanced Vision-Based Documentation in Construction Toward Exploiting Big Visual Data from Drones. Automation in Construction [Internet]. 2019. DOI: 10.1016/j.autcon.2019.102831 [Accessed: July 3, 2022]

[52] Construction and Facility Management of Large MEP Projects Using a Multi-Scale Building Information Model [Internet]. 2016. Available from: https://www.researchgate.net/publication/305825542_Construction_and_facility_management_of_large_ [Accessed: August 12, 2022]

[53] The life cycle of a building project [Internet]. No date. Available from: https://psu.pb.unizin.org/introductiontothebuildingindustry/chapter/lifecycle/ [Accessed: April 7, 2022]

Chapter 18

Current and Emerging Trends within Higher Education That Enhance Diversity and Inclusivity in Undergraduate STEM Provision

Marie Bassford

Abstract

Diversity and inclusivity rest at the heart of the current focus within Higher Education to not only attract students from all backgrounds into studying Science, Technology, Engineering and Mathematics (STEM) subjects at University, but create a welcoming and supportive learning environment that retains STEM students from underrepresented groups. A range of initiatives are highlighted within this chapter that explore opportunities to create a more diverse and inclusive STEM Higher Education community, and include the following: the development of outreach programmes and targeted recruitment efforts to attract students from underrepresented groups, scholarships and bursaries for students from disadvantaged backgrounds, mentoring and support programmes for minority students and the creation of dedicated learning environments, that provide additional support and resources, such as mentoring, tutoring and counselling services. The chapter also considers the diversification of STEM courses that focus on the social and cultural aspects of STEM and efforts to increase under-representation of minority groups in STEM faculty within Higher Education. The initiatives explored within this chapter not only create a more diverse and inclusive STEM Higher Education community benefitting the students but also help to address the wider societal challenges faced by STEM fields.

Keywords: diversity, inclusivity, higher education, science, technology, engineering, mathematics

1. Introduction

Diversity and inclusivity are becoming increasingly important issues within science, technology, engineering and mathematics - STEM for short – in Higher Education (HE) as universities strive to create a more welcoming and supportive learning environment for students from all backgrounds, that retain students from underrepresented groups. This chapter explores opportunities to create a more diverse and inclusive STEM HE community, which will not only benefit the students but also help address the wider societal challenges faced by STEM fields.

The UK STEM provision landscape is evolving with individual universities placing significant focus on creating their own specific initiatives to enhance diversity and inclusivity within their offering of undergraduate programmes. Even in 2023, the picture shows room for much improvement with statistics related to under-represented groups in STEM shining a spotlight on the areas that require continued efforts. To set the scene, let us begin by considering that in the 2019/2020 academic year, there were approximately 563,835 undergraduate students studying STEM subjects in UK universities. In 2019/2020 the most popular STEM subjects among undergraduates were Engineering and Technology, Biological Sciences, and Physical Sciences [1].

In the last decade there has been an unprecedented growth in students opting to take STEM subjects, suggesting the impact early STEM uptake in schools is having on higher education. Acceptances to engineering courses, for example, are up 21% from 25,995 in 2011 to 31,545 in 2020. Artificial Intelligence courses at university have seen an even greater increase, with acceptances onto computer science courses rising by almost 50%, from 20,420 in 2011 to 30,090 in 2020; a rise that is anticipated to continue given recent advancements in AI [2].

It is noteworthy that, historically access to STEM subjects have been harder for young people, particularly among students from disadvantaged backgrounds and females. However, in just a decade, between 2010 and 2020, there was a 49% increase in the number of women accepted into full-time STEM undergraduate courses. The number of UK 18-year-olds from disadvantaged backgrounds in this same 1-year span accepted into full time undergraduate STEM courses increased by 79%. This suggests that the increased support in STEM subjects is benefitting young people, with a gradual lift on potential barriers to access based on gender, race or class.

There have been a number of recent measures undertaken by the Department for Education to support STEM teaching. These include 'Isaac Physics', an online platform with resources designed to support students transition from GCSE through sixth form to university [3], a range of support for teacher CPD (e.g., Science Learning Partnerships, Stimulating Physics Network [4]), the Teaching for Mastery programme in mathematics [5] and an £84 million investment programme to improve computing teaching. Given that STEM subjects have a positive impact on the economy and society, the measures taken indicate a positive push in terms of applications to study STEM subjects at UK University, which complements the drive by universities to implement widening participation initiatives to increase access to higher education, focused particularly in STEM subjects.

Let us now consider some of the under-represented groups within STEM in Higher Education. In the 2020/2021 academic year, women accounted for 40.5% of undergraduate students enrolled in STEM subjects in the UK, with the gender balance varying among different STEM disciplines. There was a small increase in the percentage of females studying science subjects between 2019/20 and 2020/21, which remains lower than the overall percentage studying science subjects (45%). Notably, women make up a larger proportion in Biological Sciences (58.3%) compared to Engineering and Technology (17.8%) [1]. Furthermore, the proportion of female engineering professionals in the UK is among the lowest in Europe.

Ethnic diversity representation is still a challenge in some STEM fields. Black and Minority Ethnic (BME) students accounted for 24.7% of undergraduate STEM students in the UK in the 2019/2020 academic year, highlighting the need for continued efforts to increase representation and support the success of BME students in STEM fields. The proportion of black first year undergraduate enrolments in science subjects increased by 4 percentage points in 2020/2021.

The proportion of STEM students with known disabilities has more than doubled, from 7.5% (12,585 students) in 2007/2008 to 15.5% (33,530 students) in 2018/2019 [6]. On the surface this might suggest that, during that ten-year period, STEM subjects were attracting students with disabilities. However, a more complex picture is revealed upon breakdown of the statistics. From 2007/8 to 2018/19, the number of first-degree STEM students reporting mental health conditions rose from 675 to 8695 – a 1184% increase, suggesting the overall increase is being driven by an exponential rise in the proportion of students *declaring* learning differences and mental health conditions. In the same period, the number of first-degree STEM students reporting a learning difference increased from 6500 to 12,000 – an 84.7% increase. Interestingly, non-STEM subjects also experienced this upward trend which likely reflects greater awareness of, and improved support for, students with learning differences and mental health issues at universities.

Despite improvements in assistive technology and student support, students with disabilities who wish to study in STEM still encounter considerable barriers, with the proportion of students with physical disabilities hardly changing over the same ten-year period. For example, STEM students who are deaf or have a serious hearing impairment made up 0.3% of STEM students in 2007/08 (470 students) and 0.3% in 2018/19 (590 students). In some instances, the proportion has even decreased slightly, for example, the number and proportion of STEM students who are blind or have a serious visual impairment has decreased from 0.2% (285 students) in 2007/08 to 0.1% (270 students) in 2018/19 [7].

In the 2019/2020 academic year, 24.9% of undergraduate students in STEM subjects were from POLAR4 (the measure of what proportion of young people in a given area participate in higher education) quintiles 1 and 2, indicating lower participation neighbourhoods. These statistics provide a snapshot of the current undergraduate STEM provision, inclusivity, and diversity data within the UK in the last decade or so. It is important to note that the landscape continues to evolve, and individual universities have their own specific initiatives and data. This chapter aims to provide an overview of some of the current and emerging initiatives within HE, collated from a variety of online data sources, that highlight trending areas of good practice for enhancing diversity and inclusivity in undergraduate STEM provision.

2. Outreach programmes

The development of outreach programmes targeting underrepresented groups in UK STEM universities has been a crucial and ongoing endeavour. Recognising the need for diversity and inclusivity in STEM fields, institutions have made significant efforts to address the barriers faced by underrepresented groups and provide them with equal opportunities. Outreach programmes play a vital role in engaging underrepresented communities, particularly at an early stage, to spark interest and create pathways into STEM education. These programmes often involve collaborations between universities, local schools, and community organisations. They aim to inspire and motivate students from diverse backgrounds, showcasing the exciting possibilities and real-world applications of science, technology, engineering, and mathematics. Outreach activities may include science fairs, hands-on workshops, career talks, and interactive demonstrations, among others. By taking STEM education beyond the confines of university campuses, these programmes reach students who may not have considered pursuing STEM disciplines otherwise.

UK universities are well practised with devising and running successful outreach programmes, with dedicated teams focussing on such activities, whilst a number of institutions are specifically targeting underrepresented groups in STEM fields. The University of Southampton hosts the Girls into Engineering and Science (GIES) programme [8], which aims to inspire and encourage girls to pursue engineering and science careers. The programme included interactive workshops, industry visits, mentoring, and residential programmes to promote gender diversity in STEM fields. Another popular initiative run in partnership with an outreach team at Southampton University is known as 'Dragonfly Days', where Year 9 girls from local schools have an opportunity to meet and work with female engineers from the Faculty of Engineering and the Environment. As well as doing practical investigations and experiments, there are opportunities to quiz researchers about their path to university and to explore careers in STEM in general [9].

The University of Strathclyde's Engineering the Future for Girls initiative [10] offers hands-on engineering experiences and workshops specifically designed to engage and inspire girls in STEM. The programme provides an opportunity for girls to interact with engineering professionals and explore various STEM disciplines. After two years of running events virtually, in 2023 participants spent four days on campus hearing from university staff and industry partners who provided an insight into studying and working in the field of engineering.

The University of Manchester's Manchester Access Programme (MAP) includes outreach activities targeting underrepresented groups in STEM subjects [11]. The programme provides support, guidance, and subject-specific workshops to inspire and encourage students to pursue STEM degrees at university, making sure that talented, capable students can benefit from a university education, regardless of their background and identity.

The above represent just a small sample of university-led outreach initiatives targeting STEM programmes in some of the under-represented groups. There are also independent initiatives that aim to make a positive difference to under-represented groups working in, entering or studying STEM subjects. The University of Edinburgh partners with EQUATE Scotland, an organisation that promotes gender equality in STEM fields [12]. They offer various initiatives and programmes to encourage and support women pursuing STEM careers, including mentoring, networking events, and professional development opportunities.

STEM Learning [13] is committed to STEM education from primary, secondary to post-16 and employment, by delivering teacher continuing professional development in STEM subjects, bringing STEM role models into schools as part of the STEM Ambassador Programme or providing bespoke, long-term support for groups of schools in collaboration with companies through partnerships. In collaboration with the UK Government, employers, organisations and educational establishments (both schools and universities) deliver positive STEM interactions for teachers, young people and beyond. Volunteers from academia and industry offer their time, passion and energy to visit schools, colleges, and youth group organisations who aspire to infuse the real world of work into STEM subjects.

Since 2005, the Primary Engineer® initiative [14] have developed an inclusive engineering curriculum that spans Early Years, through Primary, Secondary, Further and Higher Education institutions. The development of young people through engagement with engineering and the promotion of engineering careers through inspiring programmes and competitions sit at the heart of its aims. They have developed a project-based learning approach to education which enables young people to engage with practical mathematics and science alongside creative problem solving. The integrated curriculum nature of the programmes develops resilience and curiosity via strong links to engineers in academia and the industry which provides an opportunity for prospective students to expand their knowledge of engineering careers, career paths and opportunities, with numerous Universities throughout the UK now involved. Whilst not specifically targeting under-represented group, Primary Engineer aims to improve the development of engineering skills for teachers and practitioners addressing the inequalities in engineering.

The Smallpeice Trust [15] provide opportunities for young people to fuel their passion for engineering, supported by charities, parents, teachers, universities, national and international companies such a Jaguar Land Rover, Lloyd's Register Foundation, Lockheed Martin, National Cyber Security Centre, National Grid, National Nuclear Laboratory, RAF, Shell, Siemens, Tomorrow's Engineers and the UK Electronics Skills Foundation. In 1991 the Smallpeice Trust launched a Scholarship programme to encourage and stimulate high-ability 15–16 year olds to consider Engineering and Technical Design as a career. In 2023, Loughborough University hosted a course for 12 to 14-year-old female students that let them experience life away from home and combined fun, action packed days with a programme of evening social activities.

Whether university-led or driven by initiatives, such as those highlighted above, the first step in enhancing diversity and inclusivity in STEM subjects within Higher Education lies clearly with outreach and providing a range of opportunities for potential students from any background or demographic to access STEM and spark that initial interest, whilst demonstrating the exciting opportunities that are available to them and removing any perceived barriers to applying for STEM-related courses. The earlier the spark is ignited, the greater the likelihood that potential students will select any necessary pre-University qualifications, such as A-levels in Mathematics or Physics, for example, or BTECs, that enable them to meet the application requirements for an undergraduate programme in a STEM subject. Sadly, for some students, by the time they are eligible to apply for a university programme, options for STEM subjects are removed by qualification pre-requisites they simply do not have, emphasising the importance of STEM-related outreach activities from an early age as well as targeting recruitment to STEM courses at University at post-16.

3. Targeted recruitment efforts

UK universities are increasingly focusing on targeted recruitment efforts to attract students from underrepresented groups. This includes attending recruitment fairs and events in areas with high levels of diversity, working with local schools and colleges to encourage students to consider STEM courses, and providing information and support to students and parents about the benefits of studying STEM.

In 2022 the University of Liverpool and University of South Wales became alliance members of Black British in STEM (BBSTEM) [16], an initiative which aims to support and champion Black individuals across science, technology, engineering and maths. With recruitment and retention rates particularly low among Black students in STEM subjects was established as a non-profit organisation in 2017. BBSTEM aims to increase the representation of Black scientists in academia and in industry whilst inspirinh young Black people to pursue STEM subjects in university and beyond. The University's membership has a number of benefits for students including employer led insight days, access to industry and academia opportunities from employers actively welcoming diverse talent, 1–1 mentoring from a BBSTEM professional member, opportunities to build professional networks and events and programmes.

Some universities, such as the University of Warwick use contextual offers to take into account an applicant's individual circumstances and background. This approach allows for a more holistic assessment of students and recognises the potential and achievements of applicants from diverse backgrounds, including those who may have faced educational disadvantages [17].

These examples demonstrate the efforts of UK universities to actively recruit and support students from underrepresented groups in STEM subjects. By providing targeted outreach, mentoring, and support, universities are working to address the underrepresentation of certain groups and create a more diverse and inclusive STEM community.

4. Scholarships and bursaries

Scholarships and bursaries have been instrumental in addressing the financial barriers faced by students from disadvantaged backgrounds. Access to quality education should not be limited by economic constraints, and universities have recognised the importance of providing financial aid to ensure that talented individuals have an equal chance to pursue STEM degrees. Scholarships and bursaries may cover tuition fees, living expenses, and other educational costs, alleviating the financial burden on students and allowing them to focus on their studies.

Numerous UK universities offer STEM scholarships and bursaries specifically for students from disadvantaged backgrounds, such as:

1. The University of Birmingham Access to Birmingham (A2B) Scholarship [18] provides financial support to students from low-income backgrounds. The scholarship offers a tuition fee waiver and a maintenance grant, making higher education more accessible for students pursuing STEM degrees.

2. The University of Edinburgh Undergraduate Mathematics Scholarships offers Mathematics Scholarships for undergraduate students from underrepresented backgrounds [19]. These scholarships provide financial assistance to cover tuition fees and living costs, enabling students to pursue their studies in mathematics.

3. The University of Manchester offers the Manchester Bursary [20], which provides financial support to students from low-income households. The bursary is available for undergraduate students studying STEM subjects and helps cover tuition fees and living expenses.

4. Imperial College London provides the President's Undergraduate Scholarships [21], which are awarded to students from underrepresented backgrounds who demonstrate academic excellence. The scholarships cover full tuition fees and provide a living cost allowance for undergraduate studies in STEM fields.

5. The University of Bristol offers Access to Bristol Scholarships for students from underrepresented backgrounds [22]. These scholarships provide financial support to cover tuition fees and living costs, making higher education more accessible for students pursuing STEM degrees.

6. University of Warwick: The University of Warwick's Warwick Scholars Program provides scholarships and bursaries to students from underrepresented backgrounds [23]. These financial awards aim to support students studying STEM subjects and alleviate the financial burden of tuition fees and living costs.

5. Mentoring and support

Mentoring and support programs are instrumental in supporting minority students throughout their academic journeys on STEM courses. These programmes provide guidance, mentorship, and a support network for students from underrepresented backgrounds, helping them navigate the challenges they may face and providing them with the necessary tools to succeed. Target Oxbridge [24], is a mentoring programme that specifically supports black African and Caribbean students in their applications to Oxford and Cambridge. The programme provides one-to-one mentoring, advice on the application process, interview preparation, and academic support to increase the representation of black students in STEM fields.

Once enrolled, mentors, who are often experienced faculty members or professionals in STEM fields, offer advice, encouragement, and academic support to mentees. These programmes also foster a sense of belonging, creating a safe space for minority students to share their experiences, seek guidance, and develop their skills and confidence. Mentoring and support programmes typically include one-on-one mentorship, group mentoring sessions, networking events, and skill-building workshops. The University of Leeds provides a comprehensive personal development programme called Leeds for Life [25], which includes mentoring and networking opportunities for students. The programme connects students with alumni mentors who offer guidance, support, and industry insights, creating a supportive network for students from diverse backgrounds.

The University of Birmingham also operates a Student Mentor Scheme [26] that pairs new students with trained mentors who can provide guidance and support throughout their studies. The scheme aims to create a supportive environment where students can seek advice, build connections, and foster a sense of belonging.

Universities support undergraduate Black Asian and Minority Ethnic students in a variety of ways, often through dedicated programmes, such as the Be More Empowered (BME) for Success Programme at the University of Bristol [27] and the University of Manchester's BME Success Programme [28] who support undergraduate BAME students by improving the experience and celebrate the success of students from these backgrounds.

UK universities have made significant progress in developing programmes to address the underrepresentation of minority students in STEM fields. They have established dedicated offices, departments, or centres that focus on diversity and inclusion initiatives, working collaboratively with students, faculty, and staff. These institutions often conduct research to identify barriers and challenges faced by underrepresented groups, allowing for evidence-based programme development and continuous improvement. The Engineering Science Learning Centre is a programme at De Montfort University built around optional support sessions created by and run by student for students, as part of the Royal Academy of Engineering Diversity Impact Programme [29]. The student-oriented centre provides a range of activities that help to close the existing, as well as perceived, skills gap between students from underrepresented groups and their peers.

Peer Assisted Study Support (PASS) schemes exist in many universities, albeit with different variants of the acronym, but all who offer academic support to students through peer-led group study sessions. The schemes provide opportunities for students from diverse backgrounds to receive additional help, build study skills, and connect with peers in a supportive learning environment.

6. Diversification of STEM courses; decolonising the curriculum

UK universities are also diversifying their STEM courses to make them more inclusive and welcoming to students from diverse backgrounds. For example, UCL Mathematics Bridging Course, specifically designed to support students from diverse backgrounds who may not have had the opportunity to study certain topics in depth [30]. The course helps bridge any gaps in mathematical knowledge and builds a strong foundation for success in STEM degrees.

Universities are also creating courses that focus on the social and cultural aspects of STEM, as well as courses that are more interdisciplinary and include elements of other disciplines, such as the arts and humanities. All universities are committed to developing inclusive curriculum that represents a diverse range of perspectives and experiences. This includes incorporating diverse case studies, diverse authors and researchers, and engaging with a broad range of cultural contexts to provide a more inclusive and enriching educational experience for all students. Most institutions are actively working on decolonizing the curriculum across various disciplines, including STEM subjects. This involves critically examining and addressing Eurocentric biases and incorporating a more diverse range of perspectives, theories, and contributions from scholars and researchers from different cultural backgrounds.

7. Creation of supportive and inclusive learning environments

UK universities are also working to create supportive learning environments that are welcoming to students from diverse backgrounds. This includes providing additional support and resources to students from underrepresented groups, such as mentoring, tutoring, and counselling services, as well as creating spaces for students to meet and network with each other. In STEM subjects, with often a strong emphasis on practical projects and laboratory work it is important that inclusive laboratory environments are created by providing accessible facilities and adapting experimental procedures to accommodate students with disabilities. This ensures that all students, regardless of their physical and sensory differences, can fully participate and engage in practical STEM activities [31]. Furthermore, many facets of differences between atypical and neurodivergent students can be recognised by carefully designing assessments that recognise important STEM skills, such as problem-finding [32].

8. Student wellbeing and counselling services

Student wellbeing and counselling support services are generally well-established in both physical on-campus and virtual form offering support services to help students address personal and emotional challenges. These services create a supportive environment where students from diverse backgrounds can access professional help and support their well-being.

Disability support within Higher Education is now well established, with the typical offering including assistance with applying for the Disabled Students' Allowances funding, referral for an assessment with an Educational Psychologist, referral for Autistic Spectrum Disorder assessment, dentification of individual reasonable adjustments, arrangements for study needs assessments, liaison with academic and other university departments and Exam support. Emerging technology now share a space for supporting students, for example Brain in Hand [33] is a digital self-management support system for people who need support in a variety of ways including help in remembering things, managing anxiety as well as planning and making decisions. The technology is not condition-specific, but is often used by people who are autistic or managing anxiety-related mental health challenges. Combining practical human support and digital self-management technology, Brain in Hand helps people live more independently. Many Higher Education students are funded by the Disabled Students' Allowance (DSA), a government grant which helps disabled students overcome barriers to learning.

9. Student societies and networks

Student Societies and Networks should not be underestimated in terms of the added value to student experience, beyond the obvious social advantages. All UK universities host a plethora of student societies and networks that cater to diverse interests, cultures, and identities. These societies provide spaces for students to meet, network, and support each other, fostering a sense of community and belonging. A good example of a successful, STEM-related society is Formula Student (FS), Europe's most established educational engineering competition [34]. The FS competition aims to develop enterprising and innovative young engineers and encourages more young people to take up a career in engineering, with over 100 university teams taking part every year. The competition typically forms part of a degree-level project and is viewed by the motorsport industry as the standard for engineering graduates to meet, transitioning them from university to the workplace. FS is a kite-mark for real-world engineering experience, combining practical engineering experience with soft skills including business planning and project management, celebrating its 25th anniversary in 2023.

Multifaith Chaplaincy, such as the University of Bristol's Multifaith Chaplaincy provides a welcoming and inclusive space for students of all faiths and beliefs. The chaplaincy offers pastoral care, counselling, and a range of religious and spiritual activities, fostering a supportive community and promoting interfaith dialogue and understanding.

10. Increased diversity in STEM staff representation

Finally, there is a growing recognition of the need to increase diversity among STEM academic staff. This includes efforts to recruit more women and minority staff members, as well as providing support and training to existing staff members to help them better understand and support the needs of students from diverse backgrounds.

In all universities, you will find an Equality and Diversity Unit that works to promote diversity and inclusion across the institution, including in STEM disciplines.

These units support initiatives to recruit and retain a diverse faculty and provides training and resources to faculty members to create inclusive learning environments. Additionally, universities have established Equality, Diversity, and Inclusion Committees (EDIC) that oversee initiatives to improve diversity and inclusion among staff members. These committees work to address barriers, supports recruitment efforts, and provide training and resources to promote inclusive teaching and mentoring practices.

Specific examples of successful initiatives include the University of Cambridge's Women in Science, Engineering, and Technology initiative (WiSETI) [35] which aims to promote gender diversity in STEM disciplines and provides support, mentoring, and training to female faculty members, as well as hosting networking events and recognising and celebrating the achievements of women in STEM.

Efforts are being made to increase gender diversity in STEM across the UK. For instance, the Women in Science and Engineering (WISE) Campaign [36] aims to achieve gender balance in the UK's STEM workforce. In addition to running events and conferences that aim to enable and promote the participation, contribution and success of women in the UK STEM workforce many institutions are engaging with WISE via membership which gives access to numerous additional benefits such as discounts and newsletters.

Advance HE is a British charity and professional membership scheme promoting excellence in higher education with some notable initiatives worth including here. Aurora is their leadership development initiative for women, run as a unique partnership bringing together leadership experts, higher education providers and research institutes to take positive action to address the under-representation of women in leadership positions in the sector [37]. Since its launch in 2013 more than 10,000 women from nearly 200 different institutions across the UK and the Republic of Ireland have participated in Aurora. The programme provides leadership training, networking opportunities, and mentoring to female staff members, enabling them to progress in their careers and contribute to a more diverse faculty.

Advance HE's Race Equality Charter helps institutions to identify and address the barriers facing Black, Asian and minority ethnic staff and students, while also providing a framework for action and improvement. Institutions can apply for a bronze or silver award, depending on their progress [38]. In April 2023, De Montfort University (DMU) became the first higher education institution in the UK to win a Silver Award under Advance HE's Race Equality Charter, demonstrating its commitment to addressing racial inequalities in higher education.

The Athena Swan Charter [39] is a framework which is used across the globe to support and transform gender equality within higher education and research. It was established in 2005 to encourage and recognise commitment to advancing the careers of women in STEM as well as medicine (STEMM) employment. The Charter is now used to address gender inequality more broadly, and not just barriers to progression that affect women.

These charters promote diversity among staff, for example, through targeted recruitment efforts, unconscious bias training, flexible working policies, mentoring programmes, and support networks for minority staff members. They highlight the growing recognition among UK universities of the importance of increasing diversity among STEM faculty. Through recruitment efforts, training programmes, and support networks, universities are working to attract and retain faculty members from underrepresented groups, promote inclusive teaching practices, and create a more diverse and supportive academic environment. These efforts are aimed at creating a more diverse and inclusive STEM community, which will not only benefit the students but also help address the wider societal challenges faced by STEM fields.

The Higher Education Statistics Agency (HESA), experts in UK higher education data and analysis, provides data that shows that the proportion of STEM academic staff declaring disabilities has increased over the time period (from 2% in 2007/08 to 3.8% in 2018/19), which is a much smaller percentage of STEM staff with disabilities compared to students. Furthermore, their data highlights decreasing representation at each major STEM career transition stage. The percentage of people with a known disability in the general population increases with age, but academic staff with a disability are less represented at each level of seniority. The data does not provide an explanation for the differences in the proportion of disabled staff and students, however an extensive review of available literature, analysis of HESA data for staff, and a series of interviews were carried out with disabled STEM staff to investigate the barriers to scientists declaring their disability and, importantly, what could be done to overcome them [40]. Various barriers were identified to declaring disability for STEM staff, including fear of stigma or discrimination, lack of clarity about the process for obtaining reasonable adjustments, and few role models with disabilities for early career scientists. Their conclusion was that several factors appeared to have a positive impact on declaration rates, including institutions providing clear definitions of disability as well as detailed information about available adjustments.

11. Conclusion

These examples demonstrate the efforts of UK universities to create supportive learning environments that cater to the diverse needs of students and highlight the efforts of UK universities to diversify their STEM courses and create a more inclusive and welcoming environment for students from diverse backgrounds. By adopting inclusive curriculum practices, considering individual circumstances in admissions processes, and implementing targeted support programmes, universities are actively working to address barriers and promote equal opportunities in STEM education. By providing mentoring, tutoring, counselling services, and spaces for networking and socialising, these universities aim to ensure that students from diverse backgrounds feel welcomed, supported, and connected throughout their academic journey.

If we stand back and consider the prospects for students beyond the UK, data suggests that graduates in the fields of STEM are sought after globally, but are often in short supply [41], with many countries actively encouraging enrolment in STEM to aid important growth industries like computer sciences, medtech, mobility or digital services. The success has been variable among different countries; UNESCO Institute for Statistics [41] reported that tertiary students in Malaysia and Tunisia are among the most likely to graduate in a STEM field, with between 43.5 and almost 40 percent of students receiving a respective degree (out of all countries where recent data was available). India, with a sizeable share of 34 percent of students picking STEM, [42] produced the most graduates in total in the field, due to its population of around 1.4 billion people.

Data for China was not available from UNESCO, however, in 2016, the World Economic Forum reported that China would exceed India's number producing 4.7 million STEM graduates a year [43]. Yet, according to the National Science Foundation, China classifies engineering and science fields quite broadly, leading to a lack of comparability in the data. Other countries with strong cohorts of more than 30 percent STEM graduates include United Arab Emirates, Germany, Belarus and South

Korea. In general, countries that have managed to produce a higher share of STEM graduates than elsewhere are more likely to be found in the Arab world, in Eastern Europe and also in East Asia. After Tunisia, the share of STEM degree recipients is greater than 29% in Algeria, Mauretania and Morocco, due to the prevalence of computer engineering [44] in the region. The Arab Gulf - a place that has recently been pushing to innovate its economies - is producing an above-average number of STEM graduates, including the UAE and Oman. With the exception of Germany, Western Europe is not a STEM axis, with only 26 percent of UK graduates coming from STEM courses, as do 25 percent in France and 23 percent in Spain. Even fewer graduate in the Americas, with shares of 19.6 percent and 17.5 percent in the U.S. and Brazil, respectively [45].

In recent years, the impact of these efforts by UK universities has been noticeable. More students from under-represented backgrounds are enrolling onto STEM HE courses, and the number of graduates from diverse communities is increasing. By fostering a diverse and inclusive environment, universities are benefiting from the unique perspectives, experiences, and talents that these students bring, enhancing innovation, creativity, and problem-solving within STEM disciplines. However, there is still work to be done. Continuous evaluation and improvement of outreach, scholarship, and mentoring programmes are essential to ensure their effectiveness. Collaboration among universities, industry partners, and government organisations is crucial to sustaining and expanding these initiatives. It is important to address systemic barriers, such as unconscious biases in recruitment and admissions processes, and to promote inclusivity at every level of STEM education and research. The initiatives described in this chapter aim to break down barriers, provide equal opportunities, and create an inclusive environment that allows all students to thrive and contribute to the scientific and technological advancements of the future. By investing in these initiatives, UK institutions are nurturing a new generation of STEM professionals who reflect the diversity of our society, breaking down barriers and promoting inclusivity, innovation, and creativity in the fields of science, technology, engineering, and mathematics. This commitment to embracing and supporting students from diverse backgrounds ensures that the future workforce is more representative, equitable, and better equipped to address the complex challenges facing our world.

Conflict of interest

The author declares no conflict of interest.

Author details

Marie Bassford
De Montfort University, Leicester, United Kingdom

*Address all correspondence to: mbassford@dmu.ac.uk

References

[1] Higher Education Statistics Agency, HESA [Internet]. 2023. Available from: https://www.hesa.ac.uk/ [Accessed: August 18, 2023]

[2] More Young People are Taking STEM Subjects than Ever Before, the Education Hub blog, Gov.UK [Internet]. 2021. Available from: https://educationhub.blog.gov.uk/2021/02/09/more-young-people-are-taking-stem-subjects-than-ever-before/ [Accessed: August 18, 2023]

[3] Isaac Physics [Internet]. 2023. Available from: https://isaacphysics.org/ [Accessed: August 18, 2023]

[4] Stimulating Physics Network, STEM Learning [Internet]. 2023. Available from: https://www.stem.org.uk/secondary/cpd/stimulating-physics-network [Accessed: August 18, 2023]

[5] Teaching for Mastery, National Centre for Excellence in the Teaching of Mathematics [Internet]. 2023. Available from: https://www.ncetm.org.uk/teaching-for-mastery/ [Accessed: August 18, 2023]

[6] Disability in STEM – Research Commissioned by the Royal Society, The Royal Society [Internet]. 2021. Available from: https://royalsociety.org/topics-policy/diversity-in-science/disability-reports/ [Accessed: August 18, 2023]

[7] AdvanceHE [Internet]. 2023. Available from: https://www.advance-he.ac.uk/ [Accessed: August 18, 2023]

[8] Girls into Engineering and Science (GIES) [Internet]. 2023. Available from: https://www.smallpeicetrust.org.uk/course-page/54f6bbba-b870-ed11-81ac-0022481b5db9 [Accessed: August 18, 2023]

[9] Dragonfly Day, University of Southampton [Internet]. 2023. Available from: https://www.southampton.ac.uk/engineering/outreach/dragonfly-day.page [Accessed: August 18, 2023]

[10] Engineering the Future for Girls, Strathclyde University [Internet]. 2023. Available from: https://www.strath.ac.uk/engineering/outreach/engineeringthefutureforgirls/ [Accessed: August 18, 2023]

[11] Manchester Access Programme, The University of Manchester [Internet]. 2023. Available from: https://www.manchester.ac.uk/study/undergraduate/contextual-admissions/map/ [Accessed: August 18, 2023]

[12] EQUATE Scotland [Internet]. 2023. Available from: https://equatescotland.org.uk/ [Accessed: August 18, 2023]

[13] STEM Learning [Internet]. 2023. Available from: https://www.stem.org.uk/ [Accessed: August 18, 2023]

[14] Primary Engineer [Internet]. 2023. Available from: https://www.primaryengineer.com/ [Accessed: August 18, 2023]

[15] Smallpeice Trust [Internet]. 2023. Available from: https://www.smallpeicetrust.org.uk/ [Accessed: August 18, 2023]

[16] BBSTEM, Black British Professionals in STEM [Internet]. 2023. Available from: https://bbstem.co.uk/ [Accessed: August 18, 2023]

[17] Contextual Offers, University of Warwick [Internet]. 2023. Available from: https://warwick.ac.uk/study/undergraduate/apply/

contextual-offers/#:~:text=The%20Contextual%20Offer%20will%20be,requirement%20for%20a%20Mathematics%20offer [Accessed: August 18, 2023]

[18] Access to Birmingham (A2B) [Internet]. 2023. Available from: https://www.birmingham.ac.uk/teachers/year-12/pathways-to-birmingham/a2b/index.aspx [Accessed: August 18, 2023]

[19] Scholarships and Student Funding, The University of Edinburgh [Internet]. 2023. Available from: https://www.ed.ac.uk/student-funding/undergraduate/international/mathematics [Accessed: August 18, 2023]

[20] Bursaries and Scholarships. The University of Manchester [Internet]. 2023. Available from: https://www.studentsupport.manchester.ac.uk/finances/funding-opportunities/all/bursaries-and-scholarships/ [Accessed: August 18, 2023]

[21] Undergraduate Scholarships, Imperial College London [Internet]. 2023. Available from: https://www.imperial.ac.uk/study/fees-and-funding/scholarships-search/undergraduate/ [Accessed: August 18, 2023]

[22] Access to Bristol Bursary, University of Bristol [Internet]. 2023. Available from: https://www.bristol.ac.uk/students/support/finances/scholarships/access/ [Accessed: August 18, 2023]

[23] Women in Engineering Scholars programme, University of Warwick [Internet]. 2023. Available from: https://warwick.ac.uk/fac/sci/eng/undergraduate/scholarships/womeninengineering [Accessed: August 18, 2023]

[24] Target Oxbridge [Internet]. 2023. Available from: https://targetoxbridge.co.uk/ [Accessed: August 18, 2023]

[25] Leeds for Life, University of Leeds [Internet]. 2023. Available from: https://students.leeds.ac.uk/info/1000058/leeds_for_life [Accessed: August 18, 2023]

[26] Student Mentor Scheme, University of Birmingham [Internet]. 2023. Available from: https://www.birmingham.ac.uk/study/student-experience/advice-support/student-mentor [Accessed: August 18, 2023]

[27] Be More Empowered (BME) for Success Programme, the University of Bristol [Internet]. 2023. Available from: https://www.bristol.ac.uk/directory/student-inclusion-services/student-inclusion-team/bame/ [Accessed: August 18, 2023]

[28] BME Success Programme, The University of Manchester [Internet]. 2023. Available from: https://www.manchester.ac.uk/connect/teachers/students/widening-participation/bame-programme/ [Accessed: August 18, 2023]

[29] Engineering Sciences Learning Centre and Employability Programme, Royal Academy of Engineering (RAEng), De Montfort University [Internet]. 2022. Available from: https://www.dmu.ac.uk/about-dmu/news/2022/april/dmu-wins-national-grant-to-champion-underrepresented-students.aspx [Accessed: August 18, 2023]

[30] Bridging Course, University College London [Internet]. 2023. Available from: https://www.ucl.ac.uk/maths/study/bscmsci-mathematics/bridging-course [Accessed: August 18, 2023]

[31] Bassford M. Sensory sensitivities and the student learning experience in Higher Education. In: AdvanceHE ANTF Symposium. United Kingdom: De Montfort University; 2023

[32] Bassford M, Lilliott E. Neurodiverse students and problem finding; creating

opportunities to incorporate special interests and harness the power of hyperfocus. In: AdvanceHE Teaching and Learning Conference. United Kingdom: Keele University; 2023

[33] Brain in hand [Internet]. 2023. Available from: https://braininhand.co.uk/

[34] Formula Student, Institution of Mechanical Engineers [Internet]. 2023. Available from: https://www.imeche.org/events/formula-student [Accessed: August 18, 2023]

[35] Women in Science, Engineering, and Technology (WiSETI) Initiative, University of Cambridge [Internet]. 2023. Available from: https://www.equality.admin.cam.ac.uk/projects/wiseti [Accessed: August 18, 2023]

[36] Women in Science and Engineering (WISE) Campaign [Internet]. 2023. Available from: https://www.wisecampaign.org.uk/about-us/ [Accessed: August 18, 2023]

[37] Aurora. AdvanceHE Leadership Development Initiative for Women Aurora [Internet]. 2023. Available from: https://www.advance-he.ac.uk/programme-events/developing-leadership/aurora [Accessed: August 18, 2023]

[38] Race Equality Charter, AdvanceHE [Internet]. 2023. Available from: https://www.advance-he.ac.uk/equality-charters/race-equality-charter [Accessed: August 18, 2023]

[39] Athena Swan Charter, AdvanceHE [Internet]. 2023. Available from: https://www.advance-he.ac.uk/equality-charters/transformed-uk-athena-swan-charter [Accessed: August 18, 2023]

[40] Addressing barriers for STEM students and staff with disabilities [Internet]. 2021. Available from: https://www.advance-he.ac.uk/news-and-views/addressing-barriers-stem-students-and-staff-disabilities#:~:text=The%20proportion%20of%20STEM%20students,shows%20a%20more%20complex%20picture [Accessed: August 18, 2023]

[41] UNESCO Institute for Statistics. [Internet]. 2023. Available from: http://data.uis.unesco.org/Index.aspx [Accessed: August 18, 2023]

[42] Tertiary Education in the Asia-Pacific Region – Statistics & Facts, Statista [Internet]. 2022. Available from: https://www.statista.com/topics/5974/tertiary-education-in-asia-pacific/#topicOverview [Accessed: August 18, 2023]

[43] The Countries with the Most STEM Graduates, Statista [Internet]. 2017. Available from: https://www.statista.com/chart/7913/the-countries-with-the-most-stem-graduates/ [Accessed: August 18, 2023]

[44] Best Universities for Computer Engineering in Africa, Edurank [Internet]. 2023. Available from: https://edurank.org/cs/computer-engineering/af/ [Accessed: August 18, 2023]

[45] Where Students Choose STEM Degrees, Statista [Internet]. 2023. Available from: https://www.statista.com/chart/22927/share-and-total-number-of-stem-graduates-by-country/#:~:text=Only%2026%20percent%20of%20UK,and%2023%20percent%20in%20Spain [Accessed: August 18, 2023]

Chapter 19

Educational Innovation in Higher Education with the Use of Management Simulators for the Development of Competencies

Alba Guzmán-Duque and Luisa Chalarca-Guzmán

Abstract

The teaching-learning processes in higher education require innovative tools that allow competency-based training. However, knowledge is measured numerically, ignoring disciplinary performance. Science, Technology, Engineering and Mathematics (STEM) are methodologies that provide spaces for learning practical with the use of technology and tools in the formation of competencies because the student to assume roles in decision making in simulated companies. This research is quantitative and correlational and presents the perception of 290 university students on the development of managerial competencies—individual and group—with the use of simulators. Individually, 55.9% indicate scenario planning, and in the group 47.9% are oriented toward the search for efficiency in the organizational environment. The ANOVA technique showed differences according to gender, where women create personal relationships with the members of their teams to make business decisions ($p < 0.000$), and men create competitive teams to obtain results ($p < 0.000$). The importance of the use of simulators for the development of managerial competencies is highlighted, and the need to adapt spaces that facilitate the formation of capacities and skills to use these tools efficiently is indicated so that Higher Education Institutions use innovative strategies to improve the competencies of their university students through STEM methodologies.

Keywords: management simulators, managerial competencies, university students, Higher Education Institutions, education innovation

1. Introduction

Technology forces Higher Education Institutions (HEI) to include innovation in educational processes for teaching-learning through scenarios that are conducive to the acquisition of disciplinary knowledge [1], with quality academic programs in the academy and in the organization [2], where professionals are trained by competencies to promote the successful performance of their graduates in the real world [3]. However, in any university career, difficulties are faced to apply knowledge in companies, because organizations prevent access to professional practices in disciplinary environments to make decisions [4]. This is justified because companies fear the consequences of inexperienced decisions, which can be costly [5] and directly influence financial, commercial, or organizational stability. Thus, professionals in their last semesters and recent graduates are hired for jobs that have little to do with their area of training because they have theoretical knowledge but little or no practice in the real sector [6]. As a consequence, the fear of companies to hire university students for decision making in real environments is reiterated, and as mentioned, a decision with unintentional errors has a negative impact on the performance of an organization.

Guzmán Duque and del Moral Pérez [7] affirm that there are three ways to apply the knowledge acquired: case studies that are presented in classes for students to propose alternatives to solve business situations; however, this proposal remains in speculation because there is no way to check the success of such decisions, and sometimes. Resolutions of other companies that were successful are applied, ignoring the student's personal idea; internships in companies, where the professional should make decisions in the managerial field but may face consequences because there are financial, organizational, and administrative risks, so the company prefers to place the professional in activities that are not related to his discipline because he fears for organizational, commercial, financial, and so on losses. The student takes on the role of managing a simulated company, applying strategies and having the possibility of observing the consequences of their actions, thus avoiding catastrophic mistakes in reality when working in sensitive areas. Precisely, the teaching of Science, Technology, Engineering and Mathematics (STEM) and its advantages in different disciplines allow the student to analyze different facts and use them in cognitive processes that favor the practical application of theories [8]. On the other hand, young people and digital natives were impacted by technologies that were a solution to their educational problems but negatively influenced their academic progress, challenging educational models and giving way to methodologies that integrate technological devices [9].

In this sense, HEIs are responsible because the educational models are oriented to measure the knowledge of their students through numerical assessments, ignoring the competencies required by the professional [10]. Although the education models in Latin America are based on competencies, the assessment is still numerical, that is, based on "knowing" and not on the "doing" of the professional; therefore, the training of the "being" as an actor is relegated to the background [11]. This shows a serious problem, because the teaching-learning processes are oriented to the retention of knowledge, ignoring the training environments by competencies, when in the digital era, data are available to the public, and it is the professional who performs an analysis of different variables to make a value judgment; this is the aspect that distances us from artificial intelligence [10]. Research prior to this publication focuses on the development of digital competencies for the professionalization of their students [12], the use of strategies that facilitate the formation of generic competencies [13], and the orientation toward competencies required in companies to perform in management positions [14], but the review of the processes to develop disciplinary competencies individually and in groups with the use of STEM methodologies is lacking.

In higher education, technology allows bringing educational communities closer in teaching-learning [15], although there is a difficulty for the institutions due to its rapid advancement, which implies inconsistencies in its rapid implementation [16] because it is required to face challenges of connectivity, its adoption, the formation of digital competencies, pedagogical aspects, and the fight against change [17]. The importance of the use of technology implies the need to access different ways to apply knowledge and achieve success in the student's academic results [18]. There are pedagogical strategies for use in the Classroom, such as intelligent tutoring systems, MOOCs, online games, collaborative platforms, chatbots [19], development of technological capabilities for leveraging these processes, to create joint knowledge and facilitate learning, etc. [20], virtual learning environments, virtual reality, videoconferencing, social networks and mobile learning [16].

Precisely, simulators allow an approach with reality for the application of disciplinary knowledge with the use of gamification strategies that facilitate teaching to become a learning habit [21]. In fact, in HEIs, these tools the creation of academic spaces, where students face real problems and integrate case studies with business practices to approach the real sector according to the degree [22] and to improve the quality of higher education based on the development of competencies of their students to keep the community satisfied and loyal to the institution [23].

This chapter shows how the use of managerial simulators, considered elements of STEM methodologies for the formation of managerial competencies—individual and group—in Business Administration undergraduates, can be used to improve their performance when they graduate and are immersed in the labor market. The differences that exist in the competencies according to the gender of the university student are highlighted and the abilities and skills required by the professional to maximize the use of management simulators are presented. At the end, it is specified how these tools are differentiating elements of STEM as facilitators in the development of competencies and application of knowledge in the disciplinary field.

2. Innovation in higher education

Currently, HEIs need to include technological tools in their teaching-learning processes, allowing them to innovate in order to improve the quality of education and maintain the attention of students [18], who require spaces that favor the formation of skills and competencies for the use and adoption of technologies in higher education [12, 17]. Sometimes, these important tools are underutilized, taking away the importance they offer and their contribution in the educational field to facilitate the grouping of content through Wikipedia, for example, and to organize classrooms to maintain academic interaction in Moodle [20]. If we add to this aspect the rapid advancement of technological tools and the low use and assimilation of educational communities, we can deduce the need to develop digital competence to maximize the use of technology and, thus, offer online spaces for the formation of disciplinary competencies [16]. The above implies the preparation of environments that implementation of technological tools in educational processes in higher education, which, although costly, can be used to improve the quality in academic careers [23].

A study by Chugh et al. [16] shows that the trend in the use of technology in higher education is present in different areas of knowledge, highlighting the social sciences and engineering with the use of social networks, videos, and websites with scientific content, while ignoring other tools such as adaptive learning, augmented reality, artificial intelligence, mobile learning, QR codes, synchronous tech, gamification, learning management systems, and rapid response. On the other hand, the training of professionals and the inclusion of technology are required to propose innovative solutions to social problems [11], where its appropriation is sought in the academy; for example, artificial intelligence promotes the development of skills related to critical analysis [10], the use of technologies inside and outside the classroom improves the acquisition of knowledge [19], simulators allow an increase in the quality of the profile of their graduates because they establish environments for making decisions that are conducive to competency-based training [24], and the use of STEM favors spaces for efficiency in teaching-learning [25].

3. STEM methodologies in management disciplines

Educational processes involve technology as mechanisms for teaching-learning for the development of skills that facilitate university students to perform in the workplace [1]. It is evident that it is immersed in young people and digital natives, because they find in their tools ways to solve any kind of problem in the educational field, which implies addiction to these devices [9]. The methodologies for teaching STEM, allow the student to analyze the different facts and use them according to cognitive processes that favor the practical application of theories [8], with which university students can train by competencies to perform in their managerial disciplinary field [14].

These methods are oriented toward courses related to basic sciences and can be used in any area of knowledge; for example, in socioeconomic contexts, university students can acquire skills and competencies necessary for their performance in the managerial field in terms of information analysis and problem solving [8]. Strohmaier, Ehmke, Härtig and Leiss [26] show that for the efficient use of these methodologies, STEM-based courses should be oriented with a syntax for student understanding without the pre-knowledge being an inconvenience for student participation, and in this way, the university student manages the technological tools in the teaching-learning process.

These technologies require adjustments for their optimal application in the classroom in HEIs, as indicated by different authors and as shown in **Figure 1**. In organizational processes, it is necessary to urge teachers to demand more in STEM methodologies activities and the use of technology, avoiding paternalism and being flexible in qualifications [27]; to adapt it in education so that teachers use it flexibly with students [26]; to provide spaces that allow the development of digital competencies of teachers for an efficient use of technology in the classroom [9]; and to promote the participation and interaction of teachers and students to take advantage of these technologies [1]. And, in the educational processes, teachers must improve the understanding of the methodology so that they are used naturally in their courses and so that students are satisfied [8]; adapt the documents used within the classroom to STEM methodologies according to the educational level and the inclusion of technologies to form the competencies required by students [26]; establish parameters for universal understanding that includes any student, without taking into account the pre-knowledge of the use of the methodology but considering their knowledge about the topics [14]; encourage motivation, interest, and the increase of previous knowledge for an efficient use in teaching-learning; and promote the participation of students in an active and voluntary way to have more satisfactory results, where they are reflected in their personal results [25].

The above scenario identifies those challenges that are required from HEIs to strengthen the application of STEM methodologies and the commitment of teachers

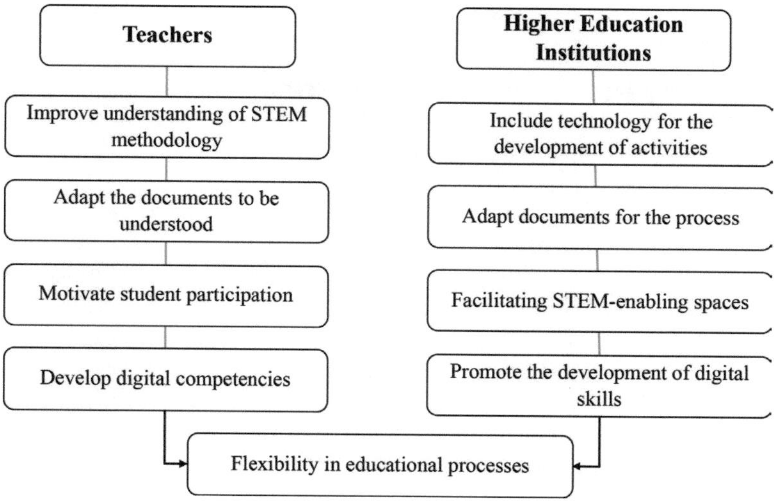

Figure 1.
STEM challenges in higher education.

to use them in a natural way to include students in the teaching-learning process in a flexible manner. It should be considered that most of the university students who use technology in the classroom are from generation Z, which implies the need to offer new methodologies that guarantee them autonomous learning, outside the classroom and according to their expectations so that they participate voluntarily and obtain their own satisfactions with the learning of their knowledge [9]. In fact, the negative influence of technology addiction in the development of young people's competencies and in their academic performance implies that HEI educational models are required to use methodologies that integrate technological devices, as is the case of STEM to satisfy them and keep them attentive in teaching-learning [9]. An experiment by Flegr et al. [25] used video as a tool in distance learning, in times of COVID, when it was necessary to devise alternatives to make practical experiments for the understanding of a subject. On the other hand, Minichiello et al. [27] evidenced the use of digital tools to make courses more flexible in times of pandemic, where participation and practical learning of different topics were encouraged. In this way and following the use of these technological tools to improve educational processes, the use of simulators allows to students of any discipline to apply knowledge and use simulated face-to-face and virtual environments that favor training by competencies [1] and the appropriation of methodologies such as STEM in the field of learning.

4. Management simulators for higher education

Simulators in the university environment facilitate the application of knowledge of their students in scenarios that show the results of their decisions in real time [4], where HEIs reduce costs because they save the implementation of on-site laboratories, which require large spaces for their processes, while the virtual simulator provides a suitable environment without requiring a physical space [6], which is a perfect scenario for students of administrative careers to assume managerial roles in decision-making. Simulators are digital environments for the application of knowledge in scenarios and in real time [22]; they contribute to the development of disciplinary competencies [28] and to decision making to learn from mistakes and correct actions, with the analysis of the results of simulated companies [29]. These tools are used in all disciplines because they are flexible and involve their actors in education to keep the attention of students and facilitate their approach to the business environment without having experience [16].

The contribution of simulators for learning involves a context and a methodology that motivates students to participate in educational processes, because they are accompanied by their teachers as advisors, rather than tutors, to promote teaching-learning [28, 30]. In fact, problem-based learning [31] and gamification are used, because they foster environments that promote competition and force the student to take a role in front of his situation and face the simulated reality to perform in his disciplinary field [16, 30]. The use of gamification strategies in higher education allows the student to feel satisfied, participate, and turn learning into a habit in a given discipline [21]. Simulators offer advantages that guide the student to assume a managerial role, through the person, the process, and the company, as shown in **Figure 2**.

Studies by Navarria et al. [6]; Guzmán Duque and del Moral Pérez [7]; Schmeller et al. [24]; and Cristofaro et al. [30] show that simulators provide feedback within the disciplinary field from the decisions made by students, encourage participation to make group decisions, and motivate them to achieve success because a simulated competitive environment is generated, without organizational risks through a holistic view of the company without causing collateral effects [4, 5] and for the integration of the areas of the company because it favors the overall view of the organization [32].

As previously mentioned, a major difficulty for professionals is to integrate theory with practice without presenting risks for organizations. Guzmán Duque and del Moral Pérez [7] in **Figure 3** show the three ways to do it, case studies, business practices, and simulators, where the importance for professionals to confront theoretical knowledge with business practices is highlighted, the latter being an impediment to the recruitment of university students, because organizations do not want to risk that an inexperienced subject makes managerial decisions that can have costly or perhaps irreversible consequences.

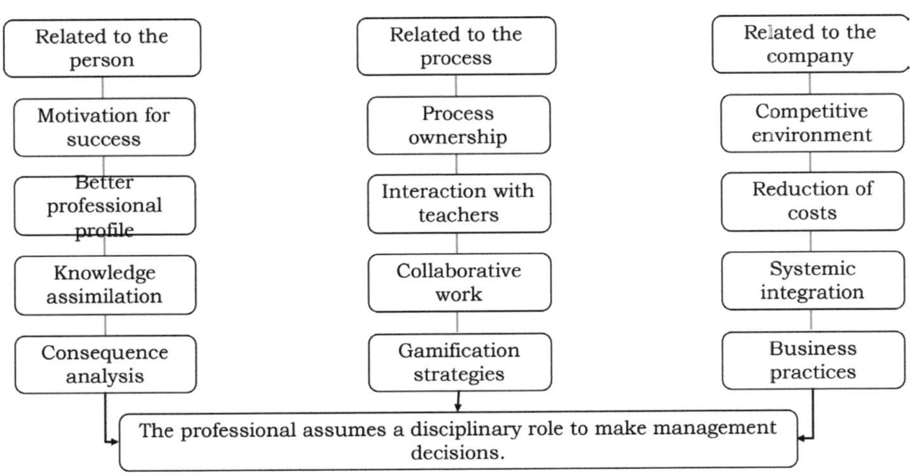

Figure 2.
Benefits of management simulators.

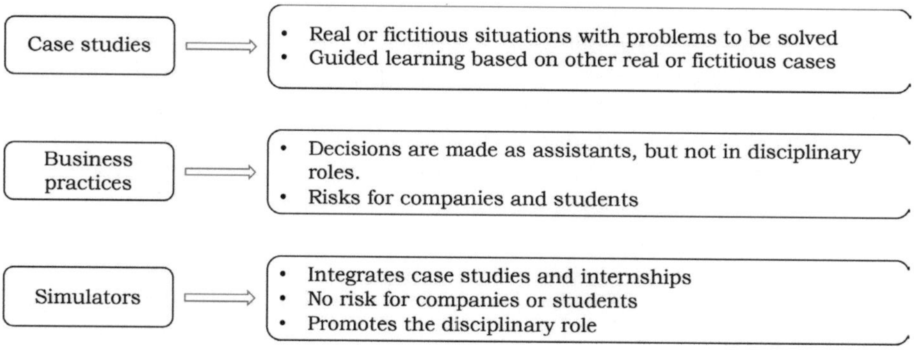

Figure 3.
Integration of knowledge through application of theories into practice.

Finally, students through the efficient use of simulators in the classroom develop competencies because it allows them to improve their teamwork skills [7]; make decisions related to their discipline, generating confidence in the process [29, 33]; and favors the application of their knowledge in practice [32].

5. Competency-based learning

5.1 Methodological context

In order to present the managerial competencies acquired with the use of simulators, a research—quantitative and correlational—was conducted and had the response of 290 Business Administration undergraduates to analyze their perception on the development of individual and group managerial competencies with the use of simulators and the importance of these tools as STEM methodologies in the teaching-learning process. This research is based on a study conducted in 2018 where they asked what these competencies were and from that space to determine if the simulator favors individual and/or group competencies for professional performance.

The participants were university students from the Business Administration program (N = 2700 approximately), of which 91% were employed in the disciplinary field at the time of the measurement (March–April 2023). The sample was calculated by finite sampling with a 95% confidence level and a 5.5% error and was composed of 290 professionals with a gender distribution of 66% women and 34% men; this was generated because in the socioeconomic sciences, the enrollment in women is higher than that of men. 52% are "between 18 and 25 years old," 41% are "between 26 and 35 years old," and 7% are older than 36.

Data collection was carried out with an instrument composed of four dimensions related to the students' perception of the acquisition of managerial competencies with the use of the management simulator. The variables were measured with Likert-type questions from 1 to 4 in order of importance (1 = none; 2 = somewhat; 3 = quite a lot; 4 = a lot). The dimensions addressed were generic competencies considering the capacities and skills developed and disciplinary competencies between the individual and group.

The capacities were measured through: (1) analysis and synthesis of results; (2) adaptation to change; (3) analysis of quantitative information; (4) critical reading; (5) comprehensive handling of English; (6) identification, approach, and resolution of problems; (7) application of knowledge in practice; (8) group work; (9) creativity for problem solving; (10) organization and time planning. And the skills considered: (11) use of IT and office automation; (12) search and analysis of information; (13) independent work; (14) oral and written communication; (15) commitment and responsibility to society; (16) conservation and preservation of the environment; (17) leadership in processes; (18) problem solving; (19) ethics, commitment, honesty, and integrity; and (20) interpersonal skills.

Individual managerial competencies addressed: (21) decision-making under pressure; (22) strategy selection and implementation; (23) acquiring the managerial role; (24) positioning oneself at the top; (25) autonomous learning; (26) teamwork; (27) leadership; (28) knowledge application; (29) financial analysis; and (30) management scenario planning. Meanwhile, the group managerial competencies addressed: (31) participation in the process; (32) analysis of administrative scenarios; (33) creation of relationships with the team; (34) creation of competitive teams; (35) communication to analyze results; (36) search for efficiency; (37) collaborative learning; (38) knowledge of competitors; (39) leadership in processes; and (40) leadership in the distribution of roles.

The instrument was validated with Cronbach's alpha statistic with a result of 0.87, which indicates internal consistency of the dimensions and reliability in its use [34]. Additionally, with the factor analysis, the relationship between each component was measured, determining the composition of each dimension, as follows: abilities (KMO = 0.919; Chi-square 2229.928 with 28 degrees of freedom; VE = 73.8%), skills (KMO = 0.935; Chi-square 2530.293 with 28 degrees of freedom; VE = 77.2%), individual disciplinary competencies (KMO = 0.924; Chi-square 3674.035 with 36 degrees of freedom; VE = 81.1%), and group disciplinary competencies (KMO = 0.942; Chi-square 4016.266 with 45 degrees of freedom; VE = 80.1%). Subsequently, the instrument was sent to university students, who answered the form voluntarily to determine their perception of the development of their managerial competencies with the use of simulators in the month of March–April 2023. The data analysis used descriptive statistical techniques, and the ANOVA technique was used to determine the difference according to gender; the calculations were made in the R software.

5.2 Capacities and skills of business managers

To maximize the advantages of the simulator, it is necessary for university students to improve their transversal competencies, classified between abilities to perform an administrative function and abilities to perform in a social environment. In this sense, **Table 1** shows the capabilities of business managers. In the level of great importance, 50% think that the most relevant ability is the analysis of quantitative information, 47.9% creativity for problem solving, 44.5% organization and time planning, 44.1% analysis and synthesis of results, 41.4% application of knowledge in practice, and 40.3% critical reading.

The skills allow a university student to develop in academic and social environments to solve situations in any area and favor integration with the use of technology. They are transversal because they are developed in the educational environment, and the integration with the skills to form a professional unites the "knowing" with the "doing" and later in the competencies with the "being." Precisely, the skills that university students stated as the most important to improve their performance in the

Capabilities	Not important	Slightly important	Quite important	Very important	Mean	Standard deviation
1. Analysis and synthesis of results	0.0	15.5	40.3	44.1	3.29	0.719
2. Adaptation to change	2.1	12.4	46.2	39.3	3.23	0.742
3. Analysis of quantitative information	0.0	12.4	37.6	50.0	3.38	0.696
4. Critical reading	3.8	17.6	38.3	40.3	3.15	0.843
5. Comprehensive handling of English	5.2	42.8	39.7	12.4	2.59	0.771
6. Identification, approach, and resolution of problems	1.7	13.1	46.6	38.6	3.22	0.734
7. Application of knowledge in practice	1.4	11.4	45.9	41.4	3.27	0.714
8. Group work	5.2	11.7	43.4	39.7	3.18	0.832
9. Creativity for problem solving	2.4	7.9	41.7	47.9	3.35	0.73
10. Organization and time planning	3.1	7.9	44.5	44.5	3.30	0.747

Table 1.
Business managers' capabilities.

simulator and that are observed in **Table 2** are: oral and written communication by 49.7%, interpersonal skills by 46.9%, and problem solving by 44.5%.

5.3 Individual and group management competencies of business managers

Simulators are powerful tools for management competency training. In a study by Guzmán Duque and del Moral Pérez [7], management skills and competencies perceived by students were measured as shown in **Figure 4**. In addition to the above, there is research that has measured the competencies of university students, where the development of interpersonal skills [29, 33] negotiation skills [32], improvement of professional performance and financial analysis in the management of the company to achieve results [24], and improvement in their graduation profiles [7] were detected.

It is observed that the simulators allow the university student to apply his knowledge, and when he assumes the managerial role, he is involved in the decision-making process to manage his simulated company, seeking results to be competitive in the market. The above shows the importance of the managerial competencies that are promoted in these academic spaces for university students, preparing them for their working life and allowing them to develop skills and abilities that strengthen their personality in the role that corresponds to them. The importance of training by

Skills	Not important	Slightly important	Quite important	Very important	Mean	Standard deviation
1. Use of IT and office automation	3.4	12.8	45.5	38.3	3.19	0.785
2. Search and analysis of information	3.4	12.8	45.5	38.3	3.21	0.763
3. Independent work	2.8	12.4	45.9	39	3.39	0.673
4. Oral and written communication	0.0	10.7	39.7	49.7	3.26	0.735
5. Commitment and responsibility to society	1.4	13.1	43.4	42.1	3.27	0.741
6. Conservation and preservation of the environment	1.7	12.4	43.4	42.4	3.21	0.788
7. Leadership in processes	2.8	14.5	42.1	40.7	3.32	0.688
8. Problem solving	0.0	12.8	42.8	44.5	3.30	0.707
9. Ethics, commitment, honesty, and integrity	0.7	10	45.9	43.4	3.32	0.679
10. Interpersonal skills	0.0	11	42.1	46.9	3.36	0.673

Table 2.
Skills of the business administrator.

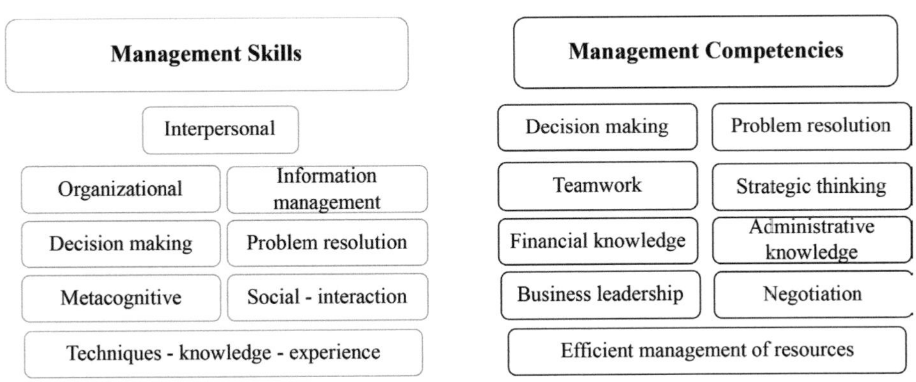

Figure 4.
Skills and competencies of business managers using management simulators.

disciplinary competencies in the area of study lies in the ease with which graduates can enter the labor market. It is evident that professionals have the option of being businessmen or entrepreneurs; however, some may be oriented toward the labor field. In this sense, the literature evidences the measurement of disciplinary competencies in the academic processes with the use of simulators, but it has failed to consider two areas, the personal and the group, which are shown in **Tables 3** and **4**. 9% emphasize scenario planning, 54.5% the application of knowledge acquired in the career and autonomous learning to improve knowledge, 53.8% the acquisition of the managerial role, 53.4% the selection and application of strategies to solve situations, 52.4% the use of financial analysis techniques, and 51.4% indicate that it is leadership in the management of a team. All the above show strengths for the Business Administrator, which are acquired in the teaching-learning process and are necessary for performance in the disciplinary field.

Table 4 shows the group management competencies that are considered to have improved with the use of the simulators, where 47.9% of the university students emphasize the search for efficiency in the organizational environment, 44.1% leadership in decision-making processes, 45.5% the creation of competitive teams that work for results, 44.1% leadership in the distribution of activities, 43.4% communication among participants to analyze results, 43.1% collaborative learning through teamwork, and 42.1% knowledge of competitors. The results show that collaborative work increases in the different scenarios, allowing to strengthen decision making in the simulation.

Individual competencies	Not important	Slightly important	Quite important	Very important	Mean	Standard deviation
1. Decision-making under pressure	3.1	4.8	42.4	49.7	3.39	0.722
2. Strategy selection and implementation	2.4	5.9	38.3	53.4	3.43	0.713
3. Acquiring the managerial role	3.1	4.5	38.6	53.8	3.43	0.723
4. Positioning oneself at the top	3.8	7.9	41.4	46.9	3.31	0.777
5. Autonomous learning	3.1	6.6	35.9	54.5	3.42	0.75
6. Teamwork	7.9	12.8	36.2	43.1	3.14	0.926
7. Leadership	2.4	5.9	40.3	51.4	3.41	0.711
8. Knowledge application	3.1	3.4	38.3	55.2	3.46	0.711
9. Financial analysis	1.4	8.6	37.6	52.4	3.41	0.706
10. Management scenario planning	2.1	5.9	36.2	55.9	3.46	0.701

Table 3.
Individual managerial competencies of business managers.

Group competencies	Not important	Slightly important	Quite important	Very important	Mean	Standard deviation
1. Participation in the process	4.8	12.1	43.8	39.3	3.18	0.824
2. Analysis of administrative scenarios	3.4	10.7	47.2	38.6	3.21	0.767
3. Creation of relationships with the team	5.9	11	44.8	38.3	3.16	0.84
4. Creation of competitive teams	4.8	10.7	39	45.5	3.25	0.833
5. Communication to analyze results	4.8	10.3	41.4	43.4	3.23	0.824
6. Search for efficiency	3.8	5.2	43.1	47.9	3.35	0.749
7. Collaborative learning	3.1	9.7	44.1	43.1	3.27	0.761
8. Knowledge of competitors	2.4	7.2	48.3	42.1	3.30	0.708
9. Leadership in processes	4.1	6.9	42.1	46.9	3.32	0.778
10. Leadership in the distribution of roles	4.8	5.5	45.5	44.1	3.29	0.78

Table 4.
Group management competencies developed with the use of simulators.

Subsequently, significant differences were detected between genders and disciplinary abilities, skills, and competencies after using the ANOVA method. Among the abilities, women indicate that the most relevant is to analyze quantitative information ($p < 0.000$), while men bet on group work ($p < 0.000$), and among the skills, women indicate that it is the search for and analysis of information ($p < 0.000$), and men indicate problem solving ($p < 0.000$). Among individual competencies, the university women say that leadership is more important in the management of the work team ($p < 0.003$), while the university men consider that they maintain autonomous learning to improve their knowledge ($p < 0.012$); and among group competencies, women create personal relationships with the members of their teams to make business decisions ($p < 0.000$), and men create competitive teams that are oriented to obtain results ($p < 0.000$).

6. Conclusions

Technology promotes access to different tools that facilitate the teaching-learning processes to favor the quality of education, and STEM methodologies promote favorable scenarios for university students' learning. Precisely, management simulators allow university students to have a holistic view of the company because all areas are integrated to solve problems, without causing collateral effects, and understand the effects of making decisions in real environments. In fact, gamification strategies used in teaching-learning are innovative in higher education because they generate competitive environments that force the university students to assume a managerial role to act as if they were in a real business practice and participate, as specified by Guzmán Duque and del Moral Pérez [7] and Cristofaro et al. [30] in their research. We agree with Dinata et al. [5] and Schmeller et al. [24] in the ease for the university student to integrate the company in a systemic way, and as evidenced in the research, it offers them an advantage to perform in managerial areas because it will be beneficial to have a disciplinary experience before entering the workplace. For this reason, the graduate profile is improved because competencies are developed for subsequent hiring, coinciding with Navarria et al. [6] because managerial practices are facilitated and with Craig et al. [8] in the application of knowledge, with skill being an advantage of STEM methodologies. From the research, the detection of those capabilities and skills that are required for the maximum use of the simulator in the teaching-learning environment to have more efficient academic processes with the use of technology is contributed.

It was found that students who use virtual simulators continue with the tendency to apply knowledge, improve their competencies, and, coinciding with Guzmán Duque and del Moral Pérez [7], favor competitive environments for men and collaborative environments for university women. In addition, they allow those competencies that were developed in learning environments to become strengths for graduates when they perform in their work environment, coinciding with García et al. [4] on the need for professionals to perform practices that facilitate their performance in the managerial field when exercising their professions and with Minichiello et al. [27] in seeking scenarios that facilitate the understanding of knowledge in a practical way, following STEM methodologies.

On the other hand, the university students indicated that the performance of their functions in their simulated jobs was improved because in the simulator, they applied their knowledge, improved the financial indicators of the simulated companies, and took decisions in environments under pressure, through the analysis of situations and teamwork for the continuous improvement of their companies. This complements the research of García et al. [4]; Dinata et al. [5]; Guzmán Duque and del Moral Pérez [7]; Meissner et al. [22]; and Cristofaro et al. [30], who indicate the need to develop managerial competencies that favor the results of organizations through the performance of professionals, and those of Heitzmann et al. [1]; Craig et al. [8]; and Flegr et al. [26], who affirm the need to use technology and STEM methodologies to improve the training of professionals.

After the results presented, it is necessary to indicate that it is necessary to include environments for training by competencies in ethics, innovation, and empathy in professionals with the use of strategies that promote the improvement of intra and interpersonal competencies. In this way, students are committed from the beginning of the simulations to participate in healthy competitions without affecting their peers on the personal side, generating a healthy environment for decision making and empathy with results orientation. The development of inter- and intrapersonal competencies allows subjects to understand the consequences of their activities, as well as innovative problem solving, and to address the implication of financial decisions on the future of the simulated companies.

The role of technology is very important according to the generation of the students. In the case of the research, most of the university students come from generation Z; they are digital natives and consider it as an ally in their teaching-learning processes, as indicated by Szymkowiak et al. [9]; precisely, simulators become STEM tools that favor learning because they promote the practical application of knowledge and allow the use of technological resources, capabilities, and skills of the university student to exercise their disciplinary competencies. In fact, university students improve their managerial competencies. A different vision is highlighted according to gender: women are collaborative and see the ability to analyze quantitative information as important, while men are competitive and work collaboratively; women indicate generating leadership in work teams and creating relationships for managerial decision making, while men consider seeking efficiency based on autonomous learning to create competitive teams.

Finally, simulators are important tools for Higher Education Institutions, there are ample possibilities to improve the skills of university students to validate all of their generic and professional knowledge, to facilitate insertion into the world of work, and improve the ability to analysis and their personal skills to perform in their jobs, where STEM methodologies are appropriate for the absorption of knowledge and its applicability in the simulated real world, therefore allowing the training of comprehensive professionals.

Acknowledgements

Special thanks to the Technological Units of Santander for providing the space for the use of the management simulators and to the Business Administration students who answered the questionnaire voluntarily.

Conflict of interest

The authors declare no conflict of interest.

Author details

Alba Guzmán-Duque* and Luisa Chalarca-Guzmán
Santander Technological Units, Santander, Colombia

*Address all correspondence to: aguzman@correo.uts.edu.co

References

[1] Heitzmann N, Fischer F, Hofmann R, Seidel T. Editorial for the special issue "Advances in simulation-based learning in higher education". Learning and Instruction [Internet]. 2023;**86**:101774. DOI: 10.1016/j.learninstruc.2023.101774

[2] Yao S, Li D, Yohannes A, Song H. Exploration for network distance teaching and resource sharing system for higher education in epidemic situation of COVID-19. Procedia Computer Science [Internet]. 2021;**183**:807-813. DOI: 10.1016/j.procs.2021.03.002

[3] Roman M, Plopeanu AP. The effectiveness of the emergency eLearning during COVID-19 pandemic. The case of higher education in economics in Romania. The International Review of Economics Education [Internet]. 2021;**37**:100218. DOI: 10.1016/j.iree.2021.100218

[4] García Otárola Á, Romero Valverde MG, Elizondo Soto M. Cambio paradigmático como plataforma para las oportunidades tecnológicas en la educación superior: un caso real a partir del uso de un simulador de gestión de negocios como técnica de enseñanza a nivel país. Revista RETOS XXI [Internet]. 2020;**4**(1):1-22. DOI: 10.33412/retosxxi.v4.1.2785

[5] Dinata YM, Tanjono C, Tanamal R, Tanuwijaya E. Business simulation training using monsoonsim. ABDIMAS [Internet]. 2022;**4**(2):1119-1123. DOI: 10.35568/abdimas.v4i2.1537

[6] Navarria L, González A, Zangara A. Laboratorios virtuales de electrónica básica para alumnos universitarios dentro del aula extendida. Trayectorias Universitarias [Internet]. 2021;7(13):075. DOI: 10.24215/24690090e075

[7] Guzmán Duque AP, del Moral Pérez ME. Percepción de los universitarios sobre la utilidad didáctica de los simuladores virtuales en su formación. Píxel-Bit. Revista de Medios y Educación [Internet]. 2018;**53**:41-60. DOI: 10.12795/pixelbit.2018.i53.03

[8] Craig CA, Petrun Sayers EL, Gilbertz S, Karabas I. The development and evaluation of interdisciplinary STEM, sustainability, and management curriculum. International Journal of Management Education [Internet]. 2022;**20**(2):100652. DOI: 10.1016/j.ijme.2022.100652

[9] Szymkowiak A, Melović B, Dabić M, Jeganathan K, Kundi GS. Information technology and Gen Z: the role of teachers, the internet, and technology in the education of young people. Technology in Society [Internet]. 2021;**65**:101565. DOI: 10.1016/j.techsoc.2021.101565

[10] Southworth J, Migliaccio K, Glover J, Glover J, Reed D, McCarty C, et al. Developing a model for AI Across the curriculum: transforming the higher education landscape via innovation in AI literacy. Computers & Education [Internet]. 2023;**4**:100127

[11] Goralski MA, Tan TK. Artificial intelligence and poverty alleviation: emerging innovations and their implications for management education and sustainable development. International Journal of Management Education [Internet]. 2022;**20**(3):100662. DOI: 10.1016/j.ijme.2022.100662

[12] Jackson NC. Managing for competency with innovation change in higher education: examining the pitfalls and pivots of digital transformation.

Business Horizons [Internet]. 2019;**62**(6):761-772. DOI: 10.1016/j.bushor.2019.08.002

[13] Membrillo-Hernández J, Cuervo Bejarano WJ, Mejía Manzano LA, Caratozzolo P, Vázquez Villegas P. Global shared learning classroom model: a pedagogical strategy for sustainable competencies development in higher education. International Journal of Engineering Pedagogy (iJEP) [Internet]. 2023;**13**(1):20-33. DOI: 10.3991/ijep.v13i1.36181

[14] Petrun Sayers EL, Craig CA, Gilbertz S, Feng S, Karam RT, Bohman A. Advancing STEM-based business sustainability: mending the curricular gap. Management Teaching Review [Internet]. 2019;**5**(1):82-93. DOI: 10.1177/2379298119852313

[15] Tang YM, Chen PC, Law KM, Wu CH, Lau YY, Guan J, et al. Comparative analysis of Student's live online learning readiness during the coronavirus (COVID-19) pandemic in the higher education sector. Computers and Education [Internet]. 2021;**168**:104211. DOI: 10.1016/j.compedu.2021.104211

[16] Chugh R, Turnbull D, Cowling MA, Vanderburg R, Vanderburg MA. Implementing educational technology in Higher Education Institutions: a review of technologies, stakeholder perceptions, frameworks and metrics. Education and Information Technologies [Internet]. 2023:1-27. DOI: 10.1007/s10639-023-11846-x

[17] Mirata V, Awinia C, Godson E, Bergamin P. The future of technology-based learning at the Open University of Tanzania. International Journal of Emerging Technologies in Learning (iJET) [Internet]. 2022;**17**(15):28-42. DOI: 10.3991/ijet.v17i15.33273

[18] Yang TC, Chen JH. Pre-service teachers' perceptions and intentions regarding the use of chatbots through statistical and lag sequential analysis. Computers and Education [Internet]. 2023;**4**:100119. DOI: 10.1016/j.caeai.2022.100119

[19] Diwan C, Srinivasa S, Suri G, Agarwal S, Ram P. AI-based learning content generation and learning pathway augmentation to increase learner engagement. Computers and Education [Internet]. 2023;**4**:100110. DOI: 10.1016/j.caeai.2022.100110

[20] Aljawarneh SA. Reviewing and exploring innovative ubiquitous learning tools in higher education. Journal of Computing in Higher Education [Internet]. 2019;**32**(1):57-73. DOI: 10.1007/s12528-019-09207-0

[21] Jaiswal P. Integrating educational technologies to augment learners' academic achievements. International Journal of Emerging Technologies in Learning (iJET) [Internet]. 2020;**15**(02):145. DOI: 10.3991/ijet.v15i02.11809

[22] Meissner D, Zhou Y, Fischer B, Vonortas N. A multilayered perspective on entrepreneurial universities: looking into the dynamics of joint university-industry labs. Technological Forecasting and Social Change [Internet]. 2022;**178**:121573. DOI: 10.1016/j.techfore.2022.121573

[23] Priyanto P, Suhandi NP. Indonesian higher education institutions competitiveness and digital transformation initiatives. Jurnal Pendidik Indonesia (JPIn) [Internet]. 2022;**11**(1):86-95. DOI: 10.23887/jpi-undiksha.v11i1.34263

[24] Schmeller R, Stoll R, Lifer JD. Strategy simulations: the relationship

of technical scores and socio scores. Journal of Education for Business [Internet]. 2021;**96**(8):485-497. DOI: 10.1080/08832323.2020.1858017

[25] Flegr S, Kuhn J, Scheiter K. How to foster STEM learning during Covid-19 remote schooling: combining virtual and video experiments. Learning and Instruction [Internet]. 2023;**86**:101778. DOI: 10.1016/j.learninstruc.2023.101778

[26] Strohmaier AR, Ehmke T, Härtig H, Leiss D. On the role of linguistic features for comprehension and learning from STEM texts. A meta-analysis. Educational Research Review [Internet]. 2023:100533. DOI: 10.1016/j.edurev.2023.100533

[27] Minichiello A, Lawanto O, Goodridge W, Iqbal A, Asghar M. Flipping the digital switch: affective responses of STEM undergraduates to emergency remote teaching during the COVID-19 pandemic. Proj Leadersh Soc [Internet]. 2022;**3**:100043. DOI: 10.1016/j.plas.2022.100043

[28] Faisal N, Chadhar M, Goriss-Hunter A, Stranieri A. Business simulation games in higher education: A systematic review of empirical research. Human Behavior and Emerging Technologies [Internet]. 2022;(2022):1-28. DOI: 10.1155/2022/1578791

[29] Mubaraz S, Heikkilä J. Business simulations as effective virtual and experiential learning environment. In: 16th International Technology, Education and Development Conference [Internet]; 7-8 de marzo de 2022; Valencia. España: Online Conference. IATED; 2022. DOI: 10.21125/inted.2022.1705

[30] Cristofaro M, Giardino PL, Leoni L. Reflective and intuitive thinking: how do they influence learning and performance in simulation gaming. International Journal of Information and Operations Management Education [Internet]. 2021;**7**(1):45. DOI: 10.1504/ijiome.2021.114737

[31] Loon M, Evans J, Kerridge C. Learning with a strategic management simulation game: a case study. International Journal of Management Education [Internet]. 2015;**13**(3):227-236. DOI: 10.1016/j.ijme.2015.06.002

[32] Rozhkov M, Alyamovskaya N, Levina T. Modeling perishability in MIT beer game business simulator. IFAC-PapersOnLine [Internet]. 2022;**55**(10):1882-1886. DOI: 10.1016/j.ifacol.2022.09.673

[33] Cosenz F, Noto G. Fostering entrepreneurial learning processes through dynamic start-up business model simulators. International Journal of Management Education [Internet]. 2018;**16**(3):468-482. DOI: 10.1016/j.ijme.2018.08.003

[34] Babin B, Tatham RL, Anderson RE, Black B, Hair JF. Multivariate Data Analysis. Madrid, España: Pearson Education, Limited; 2005. p. 928

Chapter 20

STEM Education and the Role of the HEAC in Uganda

Charles Tushabomwe-Kazooba, Imelda Kemeza and Calvin Okello

Abstract

The Higher Education Access Certificate programme, initiated in Uganda in 2016, is a bridging pathway for students who do not meet the conventional entry requirements for tertiary education. This study investigates the impact and implementation of the programme, particularly its effectiveness in increasing enrollment in STEM fields, and identifies best practices for implementation. The research involved a review of existing studies on STEM education alongside the collection of qualitative data through focus groups, case studies, and surveys, aimed at understanding the experiences and perspectives of programme participants. The findings indicate that HEAC, with a focus on science and innovation, has provided a valuable alternative route to higher education. Enrollment in HEAC certificate programmes has notably increased due to the programme's accreditation and adoption. Nevertheless, challenges have emerged, including concerns about graduation quality, bureaucratic approval processes, and tuition-related issues. The HEAC programme is recognized as a viable strategy for addressing educational inequalities and promoting Sustainable Development Goal 4 by offering underprivileged students an opportunity for further education and supporting socioeconomic development through STEM fields. The chapter underscores the importance of sustained funding and expansion of HEAC to ensure equitable access to high-quality education and the growth of a diverse STEM relevant workforce.

Keywords: bridging-pathway, education, HEAC, role, STEM, Uganda

1. Introduction

In 2015, the 191 United Nations states including Uganda, adopted the 17 Sustainable Development Goals (SDGs), popularly known as the Global goals of action which are designed to end poverty, protect the planet, and end discrimination against the Most-At-Risk-Populations (MARPs), more so the women and girls [1, 2]. The implementation and achievement of the 169 targets of the SDGs will ensure that by 2030 all people enjoy peace and prosperity. Because of the crucial role of education in socio-economic transformation, the SDGs has it as a stand-alone goal: number 4-ensures inclusive and equitable quality education and promotes lifelong learning opportunities for all. Emphasis, however, is laid on the significant importance of Science, Technology, Engineering and Mathematics (STEM) considered as the driving forces of economic growth [3]. Studies and trends indicate gender gaps, gender inequality and disempowerment of girls and women in STEM subjects [4–6]. This results in the dominance of boys and men in science related employment fields [7, 8]. In Uganda, this is more pronounced in the poorly facilitated districts because of issues of insufficient materials and the faculty in the primary, lower secondary and upper secondary school levels. Consequently, the students get poor grades at advanced level.

On the other hand, Higher Education Access Certificate (HEAC) is a sub-component of Higher Education Access Programme (HEAP). HEAP, a 10-year programme, was started in 2016 and is implemented by Forum for African Women Educationalists Uganda (FAWEU) with support from Master Card Foundation (MCF). Both STEM and HEAC are interventions meant to increase the profile in the fields of Science, Technology and Innovation (STI) in the country, at different education levels. Whereas STEM targets girls at the lower secondary school level to embrace and advance in STEM subjects in the known male dominated fields, the target population of the HEAC are both male and female students who failed to attain the minimum requirements for admission to undergraduate degree or diploma programmes. The minimum is Uganda Advanced Certificate of Education (UACE) or its equivalent with at least one principle pass and two subsidiaries. Therefore, STEM and HEAC respond to the Comprehensive Agenda 2030 [9] for sustainable development hinged on its transformative principle-leave nobody behind. The resultant effect of this is that students are acclimatized in science subjects to develop and possess the relevant profiles. Faced with the odds from multiple subjects, STEM could be challenged at building girls' confidence and giving them the opportunities to see themselves through science subjects. Additionally, HEAC contextualizes learning of science subjects which contributes both student motivation [10] and making of meaning. The implication is that students are not spread over multiple disciplines to thin out concentration or downsizing the field on interaction. What is more, it widens the scope of shared views for deeper understanding, internalizing and transferring learned knowledge, skills and values to propel success in real life.

2. Policy initiatives to advance and promote STEM education

Over the years, there have been a number of policy initiatives in the country focusing on advancing and promoting STEM education. These initiatives aim at improving the curriculum, providing necessary resources, supporting teacher training and development, collaborating with international partners, and emphasizing research and development. For example, Uganda is shifting its curriculum away from reinforcing gender stereotypes and portraying females as inferior to males. Efforts are being made to create a more gender-inclusive curriculum. The Uganda Bureau of Statistics (UBOS) and the National Curriculum Development Centre (NCDC) are working to remove gender biases from mathematics and science textbooks. Practical learning through experiential methods, experiments, and projects is being introduced to make STEM subjects more engaging and useful.

Similarly, various policies and strategies are being implemented to provide STEM labs and educational institutions with the necessary tools and supplies for conducting experiments. Access to digital materials and technologies is being expanded to ensure that instructors and students have equal access to computers and the internet. Thirdly, initiatives have been put in place to provide STEM instructors with opportunities for professional development. These programs address social factors such as negative gender stereotypes and biased teacher attitudes that favor boys over girls in STEM disciplines. Specialized training is also offered to enhance subject matter expertise. Additionally, the government increased the monthly pay for science teachers by 300 percent effective July 2021 to the equivalent of US $1066 for a graduate and US $800 for Grade V teacher. The aim is to attract and retain talented STEM educators in the related subjects.

Fourthly, the government is keen on promoting science, technology and innovation to create a knowledge-based economy. To achieve this, it collaborates with international organizations and foreign governments to access funding, expertise, resources, and best practices in advancing and promoting STEM education. For example, the Uganda Millennium Science Initiative, funded by the World Bank, is a notable example of this collaboration and partnership. In addition to the above, by recognizing the importance of research and development in STEM education, the government emphasizes a reliable method of gathering information on the outcomes of STEM education. The data collected aids in decision-making and improvements. In this agenda, monitoring and evaluation systems are vital for collecting data on STEM education outcomes and using it to continuously improve and adjust policies.

In summary, Uganda is committed to advancing STEM education through a number of all-inclusive policy initiatives. Some of these are tailored curriculum reviews, targeted resource budgeting and provision, teacher training, collaborations with international partners, and an emphasis on research and development. The aim is to foster a more inclusive, engaging, and effective STEM education system in the country.

3. Background information

3.1 The genesis of HEAC

The HEAC programme is the first of its kind in Uganda, initially championed by FAWEU with funding from MasterCard Foundation. It is in partnership with three Universities; Busitema University (BU), Gulu University (GU) and Mbarara University of Science and Technology (MUST). The other consortium institutions are Jinja School of Nursing and Midwifery, and Medical Laboratory Training School, Jinja. FAWEU is a non-governmental organization launched in February 1997 with a mission to promote gender equity in Uganda by influencing policies and nurturing positive gender practices and attitudes toward girls' education. FAWEU intends to have the HEAC program adopted and rolled out in 20 universities by 2025 to all 54 universities in Uganda with information on HEAC.

HEAC is a pre-entry bridging programme accredited by the National Council for Higher Education (NCHE) in Uganda. Its duration is a structured one-academic year comprising two semesters lasting 34-weeks including the examination period. The HEAC is an intensive full-time programme, designed to bridge the knowledge gap of students who might not have assumed knowledge required for a professional course or who might not have been confident with their preparation for university and tertiary level education study. It is designed with the assumption that the applicant has some basic knowledge of the subjects for which they are being bridged. The programme is a preparation stage to enable access to university and tertiary education programmes. It runs for two semesters (1 year), and focuses on preparation of candidates in specific subjects that provide generic academic skills to succeed in higher education levels of the desired programmes. Upon successful completion of the programme, and depending on their performance, the candidates will be given the opportunity to access specific degree/diploma programmes of their choice in university and other tertiary education.

The overall aim of the HEAC Programme is to provide entry opportunity for candidates who performed dismally at UACE. Also, they do not have appropriate qualifications based on current National admission criteria for higher education. Furthermore, those students with foreign qualifications but who do not merit admission to programmes at higher institutions of learning in the Ugandan Education system can also benefit. The targeted learners include UACE holders with at least two subsidiary passes or its equivalent. At least one of the subsidiary pass shall be from a principal subject. Students whose foreign secondary school qualifications are not considered equivalent to UACE but permit admission into Higher Education in the foreign country. Such students shall have passed at least five subjects with a minimum overall grade of C (equivalently a Ugandan grade 5). In addition, foreign students whose studies were not conducted in English show proof of proficiency in the English Language from English proficiency centers accredited by NCHE.

3.2 Differences and synergies of STEM and HEAC

Both STEM and HEAC are core practical approaches to learning with a focus on students retaining more information and understanding concepts in science subjects. Whereas STEM draws its population from students who enroll for the first 4 years of lower secondary school education leading to UCE, HEAC on the other hand targets students who narrowly missed on the university entry mark for various science programmes; and it is an intensive programme. As NCHE [11] points out, the programme also targets students from other countries whose secondary school qualifications are not considered equivalent to UACE but permit admission into Higher Education

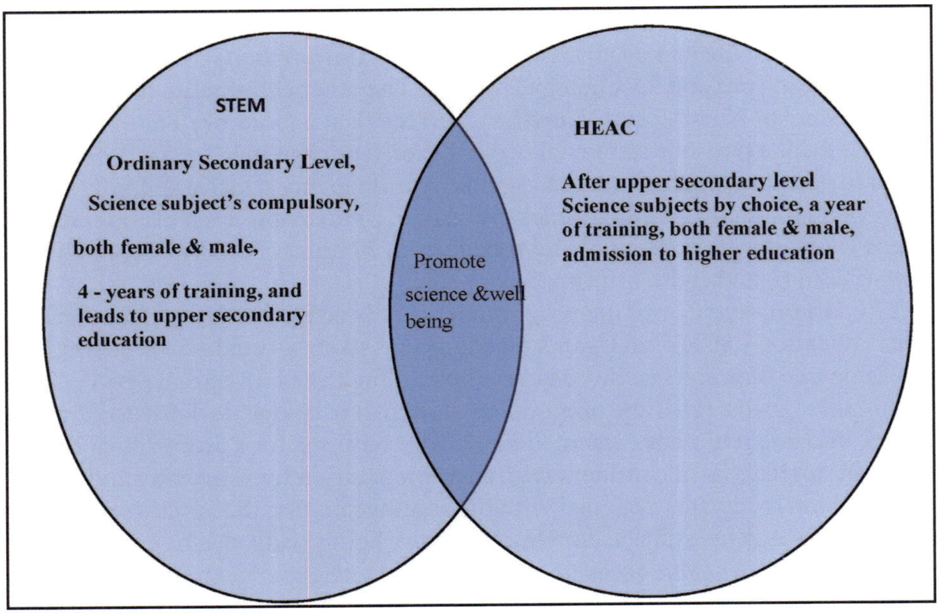

Figure 1.
Linkages and differences between STEM and HEAC.

in those countries. Additionally, it prepares students to equally compete with the direct entry students once admitted on higher education study courses. The differences notwithstanding, in our study, STEM and HEAC promote science, technology, innovation and well-being as presented in **Figure 1**.

3.3 How FAWE Uganda supports the process of HEAC adoption and implementation

HEAC is now institutionalized and can be sustained by the accredited institutions. FAWE offers technical support to the Universities interested in adoption of the HEAC bridging program. First, FAWEU together with subject experts from the partner institutions has developed standard curriculum for different academic tracks that can be adopted by institutions of higher learning. Secondly, there is popularization of HEAC in all Universities in Uganda to increase adoption and acceptance. Also, it supports accreditation of the courses by the NCHE. Fourthly, it advocates for Ministry of Education and Sport, curriculum development bodies, higher education standards bodies, professional bodies, employers, and employment bodies to recognize students who graduate via the HEAC route. Other support areas are; institutional support for Universities to adopt HEAC including; adoption of the curriculum to meet the Universities requirements and standards; training of HEAC tutors/lecturers in the implementing Universities; embedding gender responsive pedagogy in the training of lecturers, rolling out the extended orientation program for students enrolled in HEAC programs; provision of learning materials including textbooks and equipping laboratories of partner institutions; identifying focal point persons in the institutions to support the roll out and implementation of HEAC in the universities; mentorship training for the staff managing the HEAC students; and convening HEAC

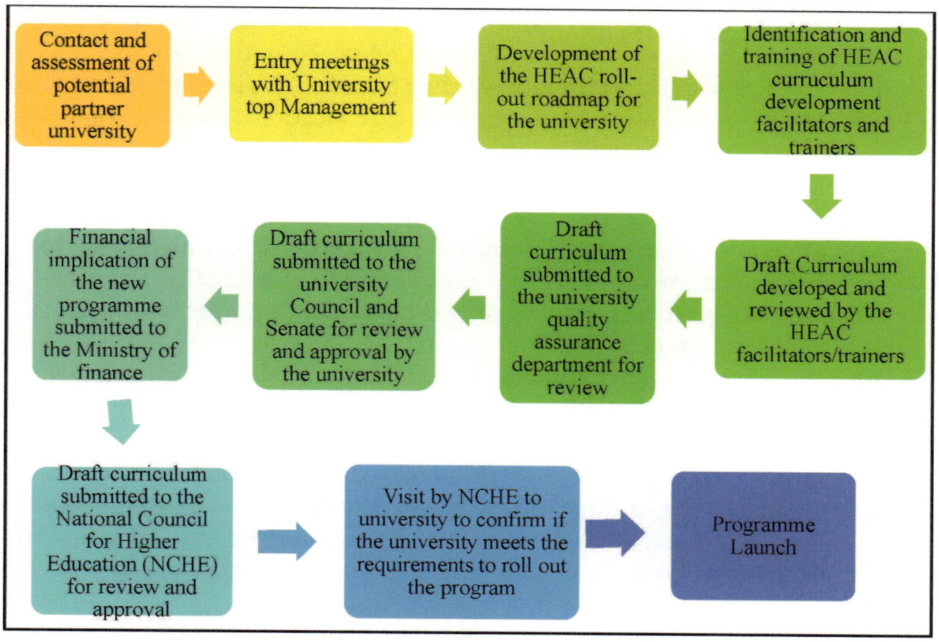

Figure 2.
Roll-out process in a university setting (adopted from FAWE(U) working documents).

stakeholders review meeting to discuss progress. Furthermore, FAWE addresses stigma and negative attitudes toward students enrolling through bridging by holding awareness sessions with universities, employer groups, communities and on national radios and televisions plus other mass media outlets.

3.4 The HEAC roll-out process in a university setting

From the review of the documents, we analyzed the process followed while implementing the HEAC. In **Figure 2** below, we present the roll-out process in a University setting.

4. Methodology

This section outlines the steps for investigating the efficacy and impact of the Higher Education Access Certificate (HEAC) in promoting STEM education and increasing access to higher education. The research objectives include: examining the potential benefits and challenges of implementing the HEAC, evaluating the effectiveness of the HEAC in increasing enrollment of underrepresented groups in STEM disciplines, and proposing strategies for optimizing the implementation and impact of the HEAC.

An extensive review of existing literature on STEM education, higher education access policies, and initiatives related to increasing diversity in STEM fields was conducted. The researchers utilized academic databases, journals, and reputable sources to ensure the reliability and credibility of the literature review. Best practices and challenges related to the HEAC as an alternative access pathway to promoting STEM were identified.

The research employed a qualitative approach to gather comprehensive data [12]. Qualitative insights were collected through focus group discussions and case studies [13]. A comprehensive survey instrument was developed to gather data on students' perceptions, attitudes, and aspirations regarding STEM education and the HEAC. The survey included questions related to demographics, educational background, motivations, and barriers faced by students in accessing higher education.

Focus group discussions were conducted with students, educators, policymakers, and representatives from relevant organizations to gain qualitative insights into their perceptions and experiences with the HEAC. Institutions or universities that pioneered implementation of the HEAC were selected for case studies. Data were collected on enrollment trends, student success rates, and the impact of the HEAC on underrepresented groups in STEM.

Quantitative data from surveys and case studies were analyzed using statistical tools to derive meaningful insights. Qualitative data from focus groups were subjected to content analysis techniques to identify themes and patterns [14]. The findings from different data sources were compared to triangulate results and validate conclusions. The study acknowledges certain limitations, such as potential biases in survey responses, limited generalizability of findings from specific case studies, and constraints in accessing comprehensive data from all institutions implementing the HEAC.

Ethical guidelines were followed throughout the research process to ensure the confidentiality, anonymity, and informed consent of the participants [15]. Institutional review and approval were sought where necessary.

5. Key results

5.1 Eligibility for a HEAC programme

Eligible students for admission to HEAC program are those who did not meet the national minimum statutory requirements for direct entry into university for degree or diploma courses. Additionally, foreign students whose qualifications are not considered equivalent to the Uganda's national qualification, but which allow university entry into their country-of-origin national exam. The HEAC programme is also designed to enroll holders of vocational qualifications and expose them to rigorous theoretical foundations that will enable them to pursue desired diploma or degree programs.

5.2 HEAC value proposition

Previous bridging programs were narrow in focus in that they allowed students to make up for gaps in their A-level results specific courses that they had failed. For example- a learner who did not pass mathematics could do a bridging course in mathematics to allow them attain a grade that would enable them qualify for university. The HEAC course is more generic and provides open access to a diverse range of courses dependent on the subjects selected in the bridging course. The students are eligible to join not just the university where they are taking the bridging course but any other that recognizes HEAC's accreditation by NCHE.

Following accreditation by NCHE, the HEAC program is more easily accepted and recognized by other institutions as harmonization of curriculum and training of lecturers ensure a standard quality across the different universities. HEAC too helps address the disparities and inequalities existing in education by providing "second chances" to students who are disadvantaged by external circumstances beyond their control. Thus, the bridging program enables students' transition into university better prepared than those who join directly from secondary schools. The certificate course prepares them to cope with life at university as well as they rediscover their potential in science subjects. The skills and knowledge acquired by the students are not just in academics but the training also includes twenty-first century skills and other basic skills that prepare students for higher education and the world of work. Hence, the HEAC programme is synonymous with increasing the student's population at the partner universities and access to the Higher Education system.

5.3 HEAC as model on the input and output aspects

The programme is clear on who it enrolls, the type of training offered, the level of knowledge gained across universities and the ultimate process output aspects as is shared in **Figure 3**. In light of this, the HEAC implementation process is very involving.

5.4 HEAC as an alternative entry route, sustainability and progress

In this study, we analyzed three HEAC programmes accredited by NCHE. These are the programmes being implemented by the universities. Our interest was to document the category of the certificate, eligibility criteria, the programme contents and the linkage to STEM in terms of the subjects. We present our results in **Table 1** below.

The results in the above table reveal that HEAC plays an important role in the continuity of STEM.

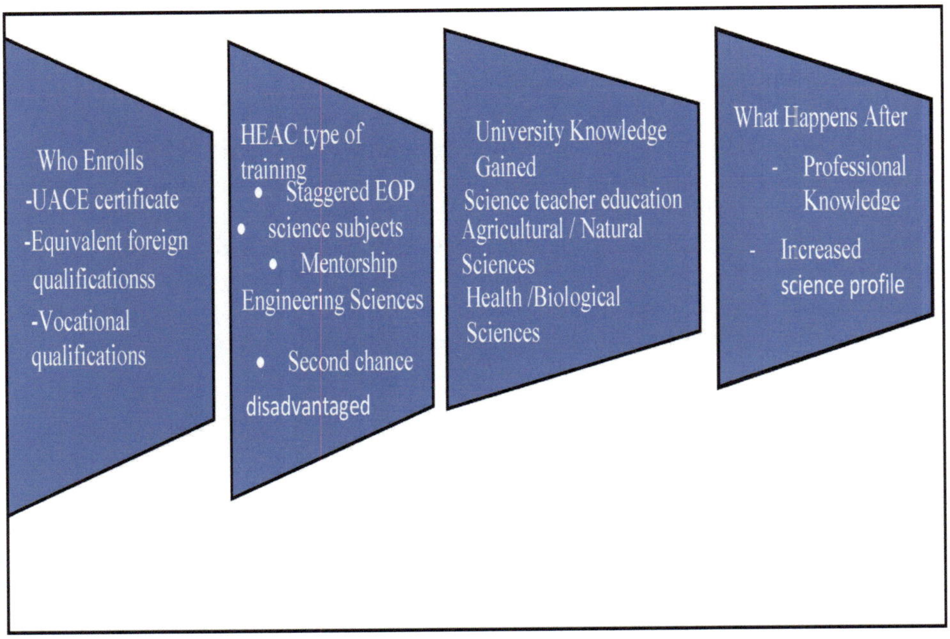

Figure 3.
HEAC model on the input and output aspects.

Results of this study indicate that the HEAC programme has been accredited by the NCHE, Uganda as a fifth entry route to enroll for Diploma or degree programmes, the other routes being "Direct" or A-Level, Diploma, degree and Mature Age Entry. The programme has become institutionalized and can be sustained by institutions through admission of private students and has already attracted privately sponsored students and those sponsored through other scholarships. HEAC has attracted privately sponsored by parents and other well-wishers. Gradually, the professional bodies are recognizing the HEAC programme.

HEAC bridging programme has been developed and institutionalized at a total of nine universities, namely: BU, MUST, GU, Kabale University, Muni University, Lira University, Bishop Stuart University, Kyambogo University and Mountains of the Moon University. To date, 697 students have enrolled on the HEAC program in five universities, namely BU, MUST, GU, Kabale University and Muni University. FAWEU is supporting Mountains of the Moon University and Lira University to redefine and customize the new curriculum in preparation for its rollout.

Four new universities, namely: Ndejje University, Soroti University, Kyambogo University and Uganda Christian University have also adopted the HEAC program. This brings the total number of partner universities implementing HEAC in Uganda to 13. For example, all the 53 HEAC students of the first cohort from BU were admitted to university-level study programs-2 on Medicine and Surgery, Education (13), Agricultural mechanization (3), Agro processing (2), Agricultural science (16), Fisheries (1), Nursing (12) and Anesthesia (4). This information has been consolidated in **Table 2** below.

The above are impressive figures for a new programme and concept within higher education. Upon completion of the certificate, the students have the requirements for admission to the various degree and diploma programmes in higher institutions of learning.

Programme	Eligible undergraduate programmes	Contents, linkage to STEM and duration	
		HEAC generic courses	**HEAC specific courses (duration:-34 weeks including the examination period)**
HEAC in chemistry & mathematics	Chemical engineering, teacher education—chemistry & mathematics, applied sciences—chemistry & mathematics, computer sciences, information technology	Information literacy, communication skills, basic mathematics, twenty-first century skills, development studies, computer skills	**Mathematics**-mathematical courses, introductory calculus, introductory linear algebra; **Chemistry**-inorganic chemistry, physical chemistry, organic chemistry; **Linkage to STEM**-Science, Technology, Engineering & mathematics
HEAC in biology & chemistry	Health sciences, teaching education—biology, chemistry, physical education, agricultural sciences, environmental sciences, Forestry, veterinary medicine, laboratory sciences, biological sciences, applied health programme, sports sciences & applied sciences	Information literacy, communication skills, basic mathematics, foundation physics, twenty-first century skills & computer skills	**Biology**-biology sciences, fundamentals of biology; **Chemistry**-functional biology, ecology, organic chemistry-inorganic chemistry, physical chemistry; **Linkage to STEM**-Science, Technology, Engineering & biology
HEAC in physics & mathematics	Engineering sciences, teaching education—physics & mathematics, statistics, Computer science, information technology	Information literacy, communication skills, computer skills, twenty-first century skills, development studies	**Mathematics**-basic mathematics II, Introductory to calculus mechanics, foundation chemistry; **Physics**-optics, waves, heat and thermos dynamics, introductory to linear algebra, electro magnetism and applied electronics, atomic, particles, modern physics; **Linkage to STEM**-Science, Technology, Engineering & mathematics

Table 1.
Current accredited HEAC programmes.

Location	Accreditation status	Year programme started	First enrolment student numbers		
			Female n(%)	**Male n(%)**	**Total**
BU	Accredited	2018	25(34.7)	47(65.3)	72
MUST		2020	30(41.7)	42(58.3)	72
GU		2019	8(20)	32(80)	40
KAB		2020	25(36.8)	43(63.2)	68
Grand totals			88 (35.0)	(65)164	252

Table 2.
HEAC first enrolment numbers in universities (Source: Universities & FAWEU records).

5.5 How students apply to benefit from the HEAC program

Students apply to study courses offered by the HEAC program at the university of their choice following an application call published in the institution's recommended

No	Institution	Academic year	Female n(%)	Male n(%)	Total enrolment	Accreditation status by NCHE
1	Mbarara University of Science and Technology	2020/2021	30(41.7)	42(58.3)	72	Accredited
		2021/2022	39(48.8)	41(51.2)	80	
		2022/2023	COVID-19 affected the enrolment. There was no Advanced level in 2021			
2	Gulu University	2020/2021	6(20.7)	23(79.3)	29	
		2021/2022	14(29.2)	34(70.8)	48	
		2022/2023	8(20)	32(80)	40	
3	Busitema University	2019/202	25(34.7)	47(65.3)	72	
		2020/2021	28(42.4)	38(57.6)	66	
		2021/2022	49(35.5)	89(64.5)	138	
		2022/2023	26(46.4)	30(53.6)	56	
4	Kabale University	2020/2021	25(36.8)	43(63.2)	68	
		2021/2022	39(46.4)	45(53.6)	84	
		2022/2023	7(21.2)	26(78.8)	33	
	Muni University	2022/2023	9(42.9)	12(57.1)	21	
5	Uganda Martyrs University, Nkozi	2021/2022	9(52.9)	8(47.1)	17	
Overall totals			**275(37.1)**	**469(63.2)**	**742**	
HEAC curriculum development trainings (staff trained)			**Female**	**Male**	**Total**	
1	Mountains of the Moon	2022	12(37.5)	20(62.5)	32	
2	Soroti University	2022	12(38.7)	19(61.3)	31	
3	Ndejje University	2022	21(40.4)	31(59.6)	52	
4	Lira	2022	9(34.6)	17(65.4)	26	
Overall totals			**54(38.3)**	**87(61.7)**	**141**	

Table 3.
HEAC student registered over academic years 2019/2020—2022/2023.

media or its website. Submitted applications are screened and successful candidates are enrolled as indicated in **Table 3**.

Records in universities indicate that 700 candidates responded to the MUST call on HEAC, 1000 applied at Busitema University and 800 applied at Gulu University. We note from the initial enrolment in **Table 3**, the HEAC programme started off on a good note. Moreover, all those that were admitted passed with good grades. Students on the HEAC programme are assessed academically through assignments, tests, essays, projects and written examination.

5.6 Balancing theory and practice

Through HEAC, university education is extended to students without the requisite educational qualification for direct entry. As HEAC students transit into a degree programme, it is envisaged that respective course organizers furnish the University with monitoring, academic performance and attrition rates information in comparison with students who enrolled through other entry routes. Facilitators/lecturers align what they have to deliver to fit the programme level standards and expectations as they bring students on board.

Support is extended to HEAC students through the Extended Orientation Programme (EOP), authentic role modeling, mentoring and constantly addressing their challenges. The relevance of the access programme to the country as one of the higher education entry routes worth embracing is showcased in order to increase admission chances into the higher education system. Rekindling HEAC students' hope to pursue university education improves their self-image and boosts their confidence levels in their academic endeavors. Implementing the HEAC programme in a university setting exposes students to understand and relieve a university natural setting and study with ease. Career guidance sessions conducted as part of psycho social support stimulate and sustain HEAC students' interest to pursue university degrees in preferred science fields.

5.7 HEAC programme opportunities

This study has also consolidated the programme opportunities. Overall, the HEAC is an exceptional bridging programme in Uganda that is intended to give students who do not match the minimal standards for direct admission into university degree or diploma programmes an opportunity to enroll. Similarly, BU, GU and MUST, are the three universities in which the programme is implemented in partnership. Thus, the HEAC programme can now reach a wider audience because of such partnerships. Additionally, it seeks to close the knowledge gap for students who are uncertain about their university readiness or who do not possess the presumptive knowledge needed for professional courses. Candidates can enroll in particular degree and diploma programmes in the university and post secondary education system upon successful completion of the HEAC. Moreover, the programme is designed to cater for a number of different groups of learners, such as holders of the UACE who have completed and obtained at least two subsidiary passes, vocational students, international students who possess qualifications comparable to the UACE, and international students who possess qualifications that allow them to enroll in post secondary education in their home countries.

Another opportunity of the programme is the increasing access to STEM Programmes. HEAC encourages research and innovation by providing opportunities for students who narrowly missed the cutoff points for university admission to a variety of science programs. Additionally, it facilitates underrepresented groups' entry to STEM fields. More so, Universities interested in implementing the HEAC program can receive technical assistance from FAWEU, which includes course accreditation and curriculum creation. Additionally, Professional associations, the Ministry of Education and Sports, and other institutions now recognize and acknowledge HEAC. This programme is being adopted by more Ugandan universities, which will improve access to higher education. Assignments, tests, essays, projects, and written exams are used to evaluate the academic progress of HEAC students. Studies are being carried out to assess the effectiveness and influence of the HEAC initiative in advancing STEM education and broadening the pool of candidates for higher education. Both a qualitative and quantitative method is used in this study. The chapter outlines problems with the HEAC programme identifies challenges associated with it and suggests fixes, including improving graduates' perceived quality, increasing awareness, and offering sufficient course materials.

Finally, Goals 4 and 17 of the Sustainable Development Agenda (SDGs) are aligned with the implementation of the HEAC. These goals center on lifelong learning, inclusive and equitable quality education, and tackling inequities and inequalities in education. HEAC students receive mentoring, more clarifications, and practical experiences from universities and personnel. Universities can expand their sources of incoming students by putting the HEAC programme into practice. The overall goals of the HEAC programme are to close the achievement gap, advance STEM education, and give students access to post secondary education, with a focus on science and innovation. It also shows a dedication to raising Uganda's educational standards.

5.8 HEAC programme challenges

This study also documented the programme challenges. Some Universities have expressed concerns over the perceived quality of graduates attaining degrees through the HEAC route and were afraid that the professional bodies like the Uganda Engineers Association, Uganda Dental Practitioners and Uganda Nurses Council would not recognize qualifications of students who did not attain degrees via direct entry to the universities. This is addressed through awareness creation on HEAC. The acceptance of the HEAC programme by the Ministry of Education and Sports has also improved the programme's acceptance by professional bodies. The slow pace of the internal approval processes of the developed HEAC curriculum by the Universities' Council and Senate is addressed by soliciting buy-in from the universities' top leadership before rolling out the programme. Transition to the University degree programmes challenges met by the HEAC students are linked to lack of tuition and other related academic requirements for enrollment and completion. Lack of adequate HEAC reference materials (textbooks and lab equipment) to support the programme is ameliorated by the partner Universities being provided with basic learning materials to start the programme. There are also plans to engage the HEAC technical team, National Curriculum Development Centre and the NCHE to develop more learning materials for the HEAC program. Stigmatization of the HEAC students by some lecturers at the partner Universities, is managed through provision of pyscho-social support and mentor-ship as well a continuous engagement and sensitization of the academic staff about the programme. Arising from this study, we suggest:

i. NCHE and the HEAC technical working group in collaboration with NCDC strengthen their activities to rectify inadequate learning and provide reference materials to support the programme.

ii. Limited knowledge about the existence of the programme in some communities in the country should be minimized through continuous sensitization and trainings.

iii. Publicity of and by significant organizations and institutions paves way for gradual acceptance of the programme by some professional bodies.

iv. Tuition challenges to transit to university degree programmes after completing the program could be addressed through extending government sponsorship slots to the HEAC graduates.

v. Adherence to a standardized HEAC Curriculum across the Universities be adopted as a precaution against some universities running the programme without a standard HEAC

vi. Strengthening staggered orientation extension programme and mentor-ship is suggested to help HEAC students overcome stigmatization issues.

5.9 Lessons learned from HEAC implementation

Implementation of HEAC does not only anchor in SDG 17 of ending poverty, protecting the planet, and addressing discrimination against MARPs but also SDG 4 of ensuring inclusive and equitable quality education and promoting lifelong learning opportunities for all. Science, Technology, Engineering and Mathematics are highlighted as crucial driving forces for achieving these goals. Moreover, they are a strong-bond for advancing the industrialization sector. On the other hand, HEAC is a bridging programme in Uganda, accredited by the NCHE and designed to provide entry opportunities for students who did not meet the minimum requirements for direct entry into the University. The HEAC programme aims at bridging the knowledge gap for students who might not have sufficient cognizance required for university professional courses, increasing science profiles among scholars, and addressing disparities and inequalities in education by providing second chances to disadvantaged students. While HEAC was championed by the FAWE-U with funding from the MCF, it has been institutionalized in five Ugandan universities so far.

The programme targets students with qualifications such as Uganda Advanced Certificate of Education (UACE) holders with at least two subsidiary passes, vocational qualifications and equivalent foreign qualifications. HEAC provides open access to a diverse range of courses in partner universities and prepares students for higher education and the workforce. The programme has been recognized and accepted by professional bodies and the Ministry of Education and Sports. All those involved, in both teaching and laboratory staffs are encouraged to spare a few minutes to encourage students during some lecture/practical hours. As implementation of HEAC takes off, teachers/facilitators provision of additional explanations and individuals/groups discussions enhances students' comprehension and unlocks focused interpretation of subject content. Organized workshops directed to orient laboratory and other teachers/facilitators motivate staff and students to hold mixed and separate regular meetings, conduct effective and constructive HEAC teaching and hands- on activities.

Studies are yet to be conducted to access the HEAC products and their performance. Sketchy results though indicate that the HEAC students transitioning into university are better prepared than those who join directly from secondary schools. Interestingly, the graduates of HEAC display their zeal to maintain their worth of admission to regular degree programmes. Moreover, those universities implementing the HEAC programmes are widening their source of enrolment.

5.10 Strategies to scale up for the HEAC

FAWE-U has declared her stand to support the adoption of HEAC by all universities in Uganda in addition to expanding the programme to other countries in Africa, to develop standardized curriculum, and promoting inclusion of marginalized groups. The undersigned are some of the strategies FAWE-U envisages to scale up HEAC. First, to continue supporting adoption of HEAC by all 54 universities in the country. The immediate target has been all 12 public universities in the country. FAWE-U is also working with the private universities to support the adoption of HEAC. Currently FAWE-U is rolling out HEAC in Rwanda, an East African Community Country with a pilot at INES-Ruhengeri University. Furthermore, FAWE-U intends to support 10 MasterCard Foundation partner universities in Africa to adopt HEAC in their programmes. Another strategy is to roll out the HEAC program in eight other countries in Africa by 2030 namely: the Democratic Republic Congo-DRC, Mali, Ethiopia, Senegal, Burkina Faso, Ghana, Kenya and Nigeria. A standard curriculum is being developed that meets international standards is being developed and is to be tailored to national curriculum requirements of different countries. Lastly, strategies are being worked upon to strengthen the inclusion component of the HEAC program to encourage the enrolment of marginalized groups including more women, refugees/IDPs and youth with disabilities.

5.11 Emerging issues

First, this study indicates that the HEAC curriculum was reviewed by the National Council for Higher Education together with the FAWE-U technical working group in Uganda. Relatedly, a new draft HEAC Curriculum which bridges both Arts and Science programs was developed. Thirdly, the new draft Programme will be presented to the key stakeholders for validation. Another emerging issue is the adherence to a standardized HEAC Curriculum across the Universities adopting the programme.

6. Conclusion

Criteria for gaining access to university education has been broadened from the traditional UACE results to include HEAC for students initially without the requisite educational qualifications for direct entry to universities. The current admission requirements to higher education programmes is at least two principal passes for degree programmes and one principal pass and two subsidiaries for a diploma programme. Initially STEM subjects were being emphasized. The implementation of the HEAC coupled with support to the candidates such as EOP, role modeling, mentoring and constantly addressing candidates' challenges have provided the opportunity to prove the relevance of the access programme to the country. As one of the higher education entry routes which integrates STEM subjects, it needs to be embraced to increase opportunity for scholars into the higher education system especially those from disadvantaged backgrounds.

Authors' Contributions

Charles Tushabomwe-Kazooba conceived the study idea, participated in the study design, data collection, analysis and interpretation, and manuscript drafting and revision. Imelda Kemeza participated in drafting data collection tools, data acquisition, analysis, and interpretation, manuscript drafting and revision. Calvin Okello supported conception of the idea, manuscript drafting and critically reviewed the manuscript. All the authors made further edits. The final manuscript was read by all authors who then gave their approval.

Disclosure

The authors have no conflict to disclose in this study.

Author details

Charles Tushabomwe-Kazooba[1]*, Imelda Kemeza[1] and Calvin Okello[2]

1 Mbarara University of Science and Technology, Mbarara City, Uganda

2 Forum for African Women Educationalists (FAWE) Uganda Chapter, Uganda

*Address all correspondence to: tkazooba@must.ac.ug

References

[1] United Nations. 17 Sustainable Development Goals. 2015. Available from: https://sdgs.un.org/goals [Accessed: October 20, 2023]

[2] Emas R. Brief for GSDR 2015 the Concept of Sustainable Development: Definition and Defining Principles. Florida International University; 2015. Available from: https://sustainabledevelopment.un.org/content/documents/5839GSDR%20 2015_SD_concept_definiton_rev.pdf. [Accessed: August 13, 2023]

[3] The Republic of Uganda. The Uganda Vision 2040. (n.d). Available from: http://www.npa.go.ug/wp-content/uploads/2021/02/VISION-2040.pdf. [Accessed: August 04, 2023]

[4] Hill C, Corbett C, St. Rose A. Why So Few? Women in Science, Technology, Engineering, and Mathematics. Washington, DC: American Association of University Women; 2010

[5] Hill C, Corbett C. Graduating to a Pay Gap: The Earnings of Women and Men One Year after College Graduation. Washington DC: American Association of University Women; 2012

[6] Diekman AB, Steinberg M. Navigating social roles in pursuit of important goals: A communal goal congruity account of STEM pursuits. Social and Personality Psychology Compass. 2013;**7**(7):487-501. DOI: 10.1111/spc3.12042

[7] Alam A, Tapia IS. Mapping Gender Equality in STEM from School to Work. 2020. Available from: https://www.unicef.org/globalinsight/media/1361/file [Accessed: July 27, 2023]

[8] Makarova E, Aeschlimann B, Herzog W. The gender gap in STEM fields: The impact of the gender stereotype of math and science on secondary students' career aspirations. Frontiers in Education. 2019;**4**:60. DOI: 10.3389/feduc.2019.00060

[9] United Nations. Transforming Our World: The 2023 Agenda for Sustainable Development A/RES/70/1. (n.d). Available from: https://sustainabledevelopment.un.org/content/documents/21252030%20Agenda%20for%20Sustainable%20Development%20web.pdf. [Accessed: October 19, 2023]

[10] Allchin D. Problem-and case-based learning in science: An introduction to distinctions, values, and outcomes. CBE—Life Sciences Education. 2013;**12**(3):364-372

[11] NCHE. Minimum Standards for Higher Education Certificate Programme in Uganda. Kampala City, Uganda; 2020. Available from: http://www.unche.or.ug

[12] Berg BL. Qualitative Research Methods for the Social Sciences. 4th ed. Long Beach, CA: Allyn and Bacon, California State University; 1989

[13] Ochieng NT, Wilson K, Derrick CJ, Mukherjee N. The use of focus group discussion methodology: Insights from two decades of application in conservation. Methods in Ecology and Evolution. 2018;**9**:20-32. DOI: 10.1111/2041-210X.12860

[14] Tümen-Akyıldız S, Ahmed KH. An overview of qualitative research and focus group discussion. Journal of Academic Research in Education. 2021;7(1):1-15. DOI: 10.17985/ijare.866762

[15] Eungoo K, Hee-Joong H. The importance of anonymity and confidentiality for conducting survey research. Journal of Research and Publication Ethics. 2023;**4**(1):1-7